COMPUTATIONAL INTELLIGENCE IN DESIGN AND MANUFACTURING

COMPUTATIONAL INTELLIGENCE IN DESIGN AND MANUFACTURING

ANDREW KUSIAK
Department of Industrial Engineering
The University of Iowa

A WILEY-INTERSCIENCE PUBLICATION

JOHN WILEY & SONS, INC.

New York / Chichester / Weinheim / Brisbane / Singapore / Toronto

Library of Congress Cataloging-in-Publication Data:

Kusiak, Andrew.
    Computational intelligence in design and manufacturing / by Andrew Kusiak
        p.   cm.
    Includes bibliographical references and index.
    ISBN 0-471-34879-1 (cloth: acid-free paper)
        1. Production management. 2. Computational intelligence. 3. Computer integrated manufacturing systems. 4. Engineering design. I. Title.

TS155.K895 2000
658.5—dc21

                                                        99-054630

10 9 8 7 6 5 4 3

CONTENTS

PREFACE

The industry is undergoing profound changes, with knowledge being on the forefront of business success. An enterprise of the future will be highly computerized, and its competitiveness will be expressed with knowledge-related measures. Much tighter integration will be seen across diverse functional areas such as product development, manufacturing, supply chain, and customer satisfaction as well as external liaisons. An enterprise of the future is likely to be agile, extended, virtual, model and knowledge based, and integrated in time and space.

The goal of this book is to present recent advances in modeling and applying computational methods to enterprises. The emphasis is on model integration and using computational intelligence approaches to solve problems across many areas of an enterprise. No single formalism, technique, or tool can generate useful decisions in a modern enterprise; rather, a magnitude of carefully crafted computational intelligence approaches are needed. It is important that the models and solution approaches appeal to a human user who irrespective of the degree of automation and computerization becomes a focus.

The material included in the 17 chapters of this book falls into three categories: (1) background on principles of basic functional areas ranging from the design of parts and process planing to manufacturing systems design and production management; (2) models and computational intelligence tools and techniques applicable to all functional areas; and (3) examples and case studies based on actual industrial projects.

Chapter 1 introduces the reader to the basic functional areas and technologies of a modern enterprise. It discuses issues ranging from manufacturing technology to computational aspects, design concerns, standards, and organizational issues.

Chapter 2 falls into the category of computational tools and discusses knowledge-based systems with all the details important in an industrial environment. Based on the content presented in this chapter, a reader should be able to design comprehensive intelligent application. The knowledge-based content of this chapter is utilized to a various degree in the subsequent chapters.

Chapter 3 is concerned with features that are a medium between a product and a manufacturing process. Features are fundamental to communication between design, manufacturing, and other functional systems and human experts working in these areas. As parts and products might be designed in one enterprise, manufactured in another, and yet distributed by a third enterprise, proper attention has to be given to the forms and standards related to the product and part information. Although the chapter emphasizes mechanical parts and metal-cutting processes, the presented principles apply to other components and processes as well.

Chapter 4 discusses the application of reason maintenance in conceptual design of products. Early use of the features defined in the previous chapter in the design of a product is a good indicator of its success. The design of parts making up a product is often expressed with the futures, being transformed by process plans (Chapter 5) into manufacturing features.

Chapter 5 presents a comprehensive set of models and techniques useful in process planning. Knowledge-based and optimization approaches are seamlessly integrated into a comprehensive computational framework. The content of this chapter can be easily applied to almost any other process, for example, electronics or health care.

Waist is an enemy of any business, including manufacturing. A dominant way of waist manifestation in manufacturing is through excessive setup costs. Formal ways for reducing the setup costs by design and management strategies are discussed in Chapter 6.

Chapter 7 contains a comprehensive treatment of key operational areas, production planning, capacity balancing, and manufacturing scheduling. The functionality of widely used production planning systems is discussed. Capacity balancing could be a function of an MRP or an ERP system or a stand-alone application. Models reflecting various objectives and constraints are discussed for discrete manufacturing systems. Scheduling models and algorithms ranging from the simple ones to the most comprehensive and applicable to almost any system are broadly discussed.

The production planning concepts included in Chapter 7 fall into the class of "push" systems, while Chapter 8 deals with various forms of kanban systems representing the "pull" production concept. This chapter provides information on issues pertaining to design and operations of kanban systems for various processes.

The first eight chapters deal largely with the definition of the interface between design parts and manufacturing and management and operations issues in manufacturing systems. The next seven chapters emphasize manufacturing system design. This group of chapters begins with Chapter 9, discussing formal approaches to the selection of manufacturing equipment and reduction of unnecessary manufacturing resources. The latter falls into the "waist" elimination category of problems discussed in Chapter 6 in the context of setups.

Chapter 10 introduces the reader to one of the most widely recognized issues in manufacturing—group technology. Over years group technology has been fundamental in different business initiatives ranging from classification and coding to just-in-time, focused manufacturing, and lean concepts. Despite wide coverage in the literature, group technology appears to be not well understood. This chapter presents a comprehensive set of models that can be applied in any manufacturing setting. This wealth of information can be utilized far beyond the manufacturing sector, as it breaks complex problems into manageable pieces based on well-justified principles.

Chapter 11 introduces two most widely used types of neural networks, backpropagation and self-learning networks. The concepts and algorithms are illustrated with numerical examples, mostly in the group technology context. The analogy of neural networks to expert systems and fuzzy systems is provided.

Chapter 12 discuses models and algorithms for determining layout of machines and facilities. It is shown that differences exist between models for machine and facility layout as well as the layout pattern depends on the type of material handling system. High computational complexity of the layout problems implies that only heuristic algorithms are practical. Exploiting special properties of the layout models has led to the development of computationally efficient heuristic algorithms.

Chapter 13 is an industrial case study demonstrating a model for allocation of inventory. The model presented in the chapter provides for a proper balance of the space requirement for in-process storage and final stage inventory. Issues that are related to reengineering the storage space layout and material handling optimization are considered.

The two previous chapters dealt with layout of manufacturing facilities, emphasizing machines, material handling systems, and inventory space. In Chapter 14 an industrial case study involving the design of a warehouse is described. A computational procedure for determining layout of a class-based warehouse is developed.

In a typical industrial practice, the relationship between design and manufacturing is not well articulated. The design for a manufacturing paradigm is normally limited to manufacturing processes. Chapter 15 discusses foundations of design in a much broader context—design for agility. The concept of improving manufacturing operations through design of products and systems is discussed.

In modern manufacturing activities outside of a design and manufacturing floor are as important as the design and manufacturing activities itself. The case study discussed in Chapter 15 illustrates application of a systems engineering approach to the development of a tool for supplier evaluation.

Chapter 17 presents fundamentals of data mining methodology, which is growing in popularity as a viable tool for extracting meaningful content from large volumes of data and information. Data mining is compared with other computational methodologies. Basic models and algorithms for data mining and data farming are presented.

The book is written to meet the needs of senior undergraduate and graduate engineering students, designers, systems analysts, managers, and other practitioners. As the book emphasizes modeling, analysis, and computational methods, it will be of interest to numerous disciplines, including industrial, mechanical, electrical, and systems engineering. Significant portions of the material apply to the service sector.

The main motivation of writing this book comes from multiyear collaboration with various industries. The book content reflects the needs of mechanical, electronics, and software companies as well as numerous service organizations, including health care. Many examples and case studies presented in the book are coming from industry. The models and algorithms presented have been widely used in practical applications.

Although most of the material presented in this book is based on the author's personal research conducted over many years, some material presented was a result of collaborative projects. The author expresses appreciation to many colleagues, visiting researches, and graduate students with whom he has collaborated in recent years. The ideas on setup reduction and features are due to my collaboration with J. Feng and P. G. Li. The design for agility rules were developed jointly with D. He. Y. K. Chung's research is reflected in the neural network content of this book. Collaborative research with C. C. Huang was fundamental to the presentation of the concept of kanban systems. My thanks to K. Park and for his collaboration on topics related to group technology and machine layout algorithms. R. Vujosevic and E. Szczerbicki have contributed ideas on product synthesis. Finally, the research of my former graduate student T. N. Larson was instrumental in the preparation of industrial case studies.

Many thanks go to my undergraduate and graduate students and many colleagues from industrial corporations and participants of professional seminars and workshops for discussing various ideas incorporated in the book.

Andrew Kusiak
The University of Iowa
Iowa City, Iowa

COMPUTATIONAL INTELLIGENCE IN DESIGN AND MANUFACTURING

CHAPTER 1

MODERN MANUFACTURING

1.1 INTRODUCTION

Integration is frequently considered an important competitive factor in the manufacturing industry. Some of the benefits realized from integration include (He and Kusiak, 1994):

- Improved quality
- Greater flexibility and responsiveness
- Improved competitiveness
- Reduced lead time
- Increased productivity
- Decreased work-in-process

In an integrated enterprise, engineering design, manufacturing, planning, purchasing, sales, and support functions are integrated in a single system. The functional areas of an integrated manufacturing system are shown in Figure 1.1 (Kusiak, 1990). In a modern enterprise, the functional areas indicated in Figure 1.1 are likely to be computerized. Each of the five areas of a computerized enterprise is briefly characterized next.

1. *Design.* A computer-aided design (CAD) system, one of many tools in an integrated enterprise, supports the function of design of components, products, tools, and fixtures.

2. *Process Planning.* A computer-aided process planning (CAPP) system determines sequences of operations and resources required for manufacturing a

1

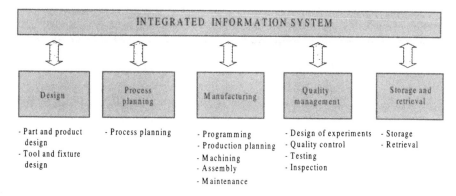

Figure 1.1. Basic functional areas of an integrated enterprise.

part. The process plan for a part should minimize production cost, manufacturing time, and machine idle time and maximize production rate and quality of parts.

3. *Manufacturing.* Computer-aided manufacturing (CAM) involves, for example, programming numerically controlled (NC) machines and material handling carriers, material requirement planning, and production planning and scheduling. Based on the component designs created by the CAD system and process plans, the programs for NC machines and material handling carriers are generated. Machining, assembly, and other manufacturing functions are performed to the best efficiency.

4. *Quality Management.* The total quality management (TQM) concept involves design of experiments, quality control, and testing and inspection of the manufactured parts and products. It ensures that the overall effort beginning in the early product (part) design phase be continued through manufacturing, process planning, production, purchasing, and sales to improve the total quality of the manufactured parts and products.

5. *Storage and Retrieval.* Storing items is not well received in manufacturing; however, the level of inventory stored must match the level of service expected by internal or external customers. Using an automated storage and retrieval (ASR) system, raw materials, finished products, and in-process inventory are stored and retrieved automatically.

Eliminating integration barriers among the functional areas in Figure 1.1 results in significant improvements in the quality, efficiency, cost, and competitiveness of a modern enterprise. Examples of initiatives increasing the level of integration in an enterprise are concurrent engineering, integrated product and process development, and TQM.

Computers play an important role in the automation and integration of manufacturing hardware components, that is, machines, robots, automated guided vehicles (AGVs), and the software components of the manufacturing business system.

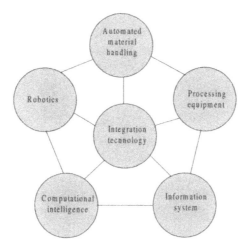

Figure 1.2. Relationship among components of a modern manufacturing system.

The major trends influencing manufacturing technology stem primarily from advances in computers. The important role of computer technology and the relationship among manufacturing system components is shown in Figure 1.2.

In this chapter, the following trends and advances in manufacturing technologies are examined: integration, robotic technology, automated material handling and storage technology, information system technology, and computational intelligence. The impact of integrated technology on areas such as product life cycle, management, and labor are discussed. The future trends in integrated manufacturing are discussed next.

1.2 INTEGRATION

Integration (sometimes referred to as fusion) brings together different functional areas of an enterprise into a unified system. This combination of technologies bridges the gap between design and manufacturing, allowing a path from an initial concept to a finished product within one system.

Computational intelligence is an important component of the integration technology. An intelligent system should be able to prevent certain errors, for example, design of an object that could not be manufactured or a faulty electrical circuit board. It should generate alternative process plans to accommodate for machine breakdowns. Furthermore, the intelligent system would facilitate the designer's work by performing such functions as comparing a current design to the designs stored in a database.

One of the major difficulties with integration in manufacturing is due to incompatibility of equipment and software. The compatibility issue arises when systems developed by different vendors cannot communicate with each other.

Manufacturing equipment allows one to store and execute sophisticated programs in less time and at a lower cost. In the future, the content of manufacturing data bases will expand to include more information on process plans, fixtures and tools, NC programs, quality control and inspection programs, robot programs, and so on.

1.3 ROBOTICS

Robots in a manufacturing environment are commonly applied in the following areas:

- Part and tool handling
- Loading/unloading of AGVs and machines with parts and tools
- On-line and off-line inspection
- Hazardous environments

Robots integrate with the manufacturing environment through the material and information flow (see Figure 1.3).

Productivity and flexibility of robots increases with the application of:

- Intelligent vision and tactile sensing systems
- Voice recognition
- Natural language processing
- Off-line programming

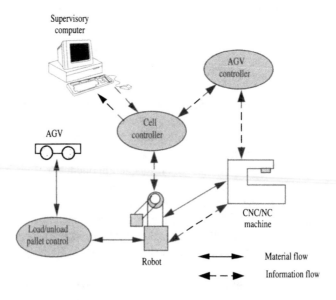

Figure 1.3. Robot in a manufacturing cell.

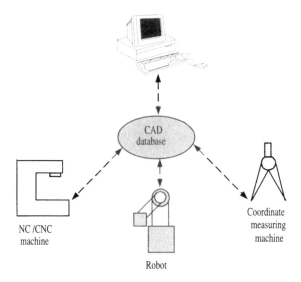

Figure 1.4. Concept of off-line programming.

The basic concept behind off-line programming is that an NC or a robot program is generated and simulated prior to its actual use, without interruption of the equipment. The concept of off-line programming is shown in Figure 1.4.

Robots are becoming more versatile and have greater intelligence through increased sensory capability. Not only has vision become more frequently used, but also "intelligent" tactile sensors are being located on the gripper that are able to distinguish the shape of the part picked.

Advanced robot systems are equipped with intelligent sensors that inform the robot of the state of its environment, controllers that interface with the sensors, and robot software that adapts to reflect the changing state of the environment. Intelligent robots are more adaptable to product changes as fewer fixtures are involved. Also, a gripper to handle a drill, laser, or any other machine tool can be easily changed.

1.4 MATERIAL HANDLING AND STORAGE TECHNOLOGY

There has been impressive progress in automated material handling and storage systems. Material handling systems are used to increase the speed of material movement, weight carried, distance traveled, and ability to deal with a harsh environment. This section discusses the major trends in material handling and storage technology.

1.4.1 Material Handling Technology

The future developments in material handling aim at better control of inventory, improved production flexibility, and fully integrated manufacturing systems. Small loads and low inventory levels are becoming a common occurrence. Material handling equipment is to transport totes and individual items rather than the traditional pallet loads. The integration of a storage system with transporter systems such as chain- or belt-driven bidirectional transport conveyors and microprocessor-controlled air-actuated transfers can lead to significant improvement in production rate, inventory reduction, and space utilization.

Automated guided vehicles are flexible and often used in manufacturing. The trend in AGV development is toward asynchronous movement and includes their uses as assembly platforms and for performing general transport functions; for example:

- Intelligent monorails transporting parts between workstations
- Transporter conveyors to control and dispatch work to individual workstations
- Robots to perform machine loading, case packing, palletizing, assembly, and other material handling tasks
- Microload automated storage and retrieval machines for material transport, storage, and control functions
- Cart-on-track equipment to transport material between workstations and manual carts for low-volume material handling activities

Autonomous AGVs able to communicate with other vehicles and with the factory control system are emerging. They will have the capability to detect malfunctions and to order and perhaps replace the assemblies that are defective. Also, they will be able to sense and warn operators of danger, sabotage, and unauthorized intrusion.

1.4.2   Automated Storage Systems

The developments in material storage technology are strongly affected by the need to reduce the amount of material to be stored. The application of just-in-time delivery has lessened demand for large-scale ASR systems. The equipment used in storage applications includes miniloads, microloads, tote stacks, and vertical and horizontal carousels.

A number of microload ASR systems have been introduced in recent years (Hunt, 1989). The microload ASR system is used to store, move, and control individuals tote boxes of materials. Rather than performing pick-up and deposit operations at the end of the aisle, a microload machine supplies and retrieves material along each side of the storage aisles.

A miniload is a fully automated computer-controlled storage/retrieval machine designed to deliver small parts in bins to an end-of-aisle order picker. Real-time inventory control in a miniload system can significantly improve inventory accuracy, material accessibility, and integration with internal data processing systems. A miniload system automatically updates inventory records, prints management reports, and provides inventory status on demand.

A storage technology alternative that deserves attention is a carousel conveyer. Combining carousels with transporters, robotic loaders and unloaders, and other devices offers not only faster, more efficient material flow to production but also the tighter stock control that allows reduction in inventory.

1.4.3   Control Systems

The progress in material handling control systems includes refinement of sensors, motor starters, communication links, and other methods for direct machine control as well as advances in hardware and software.

Automatic identification is perhaps the fastest growing technology in material handling. Among the various sensors to support automatic identification are a wide range of bar code technologies, optical character recognition, magnetic code readers, radio frequency and surface acoustical wave transponders, machine vision, fiber optics, voice recognition, tactile sensors, and chemical sensors.

Another development in control of material handling equipment is wireless guidance of AGVs. Equipped with enhanced sensors, the AGV can work as a mobile robot and function in a path-independent mode. With the use of computational intelligence techniques, AGVs are able to perform more than routine transport tasks without human intervention. Sophisticated hardware and software technology allows performing tasks such as automatic loading and unloading of delivery trucks.

1.5   INFORMATION SYSTEMS

An information system interconnects and integrates manufacturing process equipment with other systems. The information system technology plays a crucial role in the integration of modern manufacturing. In the next section computer networks, interface, and data standards are discussed.

1.5.1   Network Compatibility

The ultimate goal of an integrated manufacturing system is to deliver produced-to-order products. With the integration of engineering design, manufacturing, and supply chain management, orders can be filled quickly. Parts are designed with a CAD system, automatically analyzed for performance, and checked for manufacturability and other constraints. Process plans are generated using CAPP systems. Information on items such as the bill of materials, part routings, numerical control programs, and tools and material required is readily available whenever required on the shop floor. To enable information exchange between various subsystems, communication networks must be developed. Networks provide a protocol (an agreed-upon standard) for communication between heterogeneous systems. The manufacturing automation protocol (MAP) is a communication standard designed specifically for manufacturing systems. Another network standard developed is the technical office protocol (TOP), a specification for nonproprietary multiple-vendor data communications for technical and office environments. The MAP and TOP standards are particular implementations of the open interconnection model (OSI) and the International Organization for Standardization (ISO) seven-layer

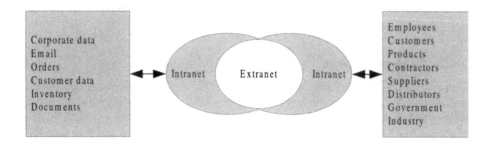

Figure 1.5. Intranet vs. Extranet.

system architecture standard. The ISO/OSI standard is the basis of the open-system architecture. It is a component of a family of open-system environment standards under the acronym MUSIC (Hannam, 1997):

M—Management: Largely under development

U—User interface, e.g., AEP (Application Environment Profile) and ISP (International Standard Profile)

S—Service interfaces and programs, e.g., POSIX (Portable Operating System Interfaces), an Institute of Electrical and Electronics Engineers (IEEE) standardization effort

I—Information and data formats, e.g., EDIFACT (Electronic Data Interchange Administration, Commerce, and Transport) specification and STEP (Standard for the Exchange of Product Model Data)

C—Communication interfaces, e.g., ISO/OSI and TCP/IP (Transfer Communication Protocol/Internet Protocol)

In recent years we have witnessed rapid development of exchanging information among industrial and service corporations through Intranets and Extranets (see Figure 1.5). These two electronic commerce technologies have had profound impact on business practices. The timeliness and accuracy of information exchanged has been significantly improved. An Intranet links business partners over the Internet using the TCP/IP standard. General Motors, Ford, and Chrysler and their suppliers have developed some of the largest Intranets. An Extranet is an extended Intranet and typically operates business to business; however, it may also be used for customer–enterprise exchange of information. An example of an individual customer–business Intranet is Dell's Computer Corporation network.

1.5.2 Interface Standards

Networks provide the physical language and format but do not address the semantics or effective use of the information communicated. Interface standards are needed to facilitate the effective communication of meaningful data. One example of such an interface standard is the STEP for design and manufacturing data exchange (see Figure 1.6). This standard has enabled previously incompatible design systems, with data stored in radically different formats, to communicate the data while preserving most of the meaning. This standard continues to evolve, pointing the way to wider data integration.

The lack of interface standards is a major impediment to achieving complete integration of various subsystems of an enterprise. A modern manufacturing system involves not only the integration of different kinds of machines and systems but also the integration of humans with machines. In addition to the interface standards for machines, interface standards for integration of humans and programmable systems are needed.

The progress made in object-oriented technology has resulted in the development of CORBA (Common Object Request Broker Architecture). This architecture has gained a considerable interest in various corporations.

The Internet is bringing meaningful changes to interface standards with a profound impact on integration among production and other systems. One of the most promising standards is XML (eXtensible Markup Language) from the XML/EDI (EDI = Electronic Data Interchange) group. This new standard jointly with JAVA and other Internet standards offers a flexible medium for exchanging information.

Different vendors provide most of the currently available programmable systems, and these systems communicate through a proprietary programming language. Therefore, it is difficult for a programmer to master a variety of languages and input devices found in a modern factory.

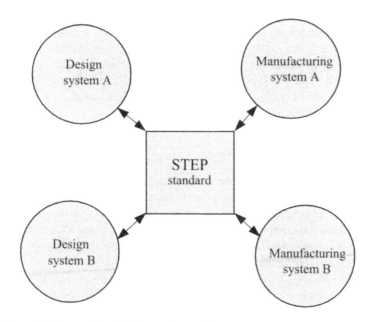

Figure 1.6. Integration of design and manufacturing systems by STEP standard.

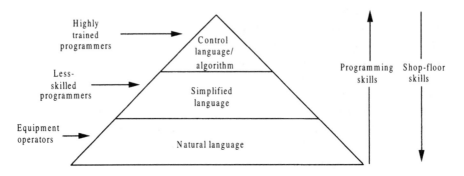

Figure 1.7. Hierarchy of programming languages.

The interface problem between nonprogrammers, such as shop-floor operators, and the increasingly complex programmable automation systems can be remedied with a hierarchical system (see Figure 1.7). At the top level of the hierarchy, trained programmers are required to deal with the control language and control algorithms. At the intermediate level of the hierarchy, less programming skills are needed. At the bottom level, virtually no programming is required.

Another trend aimed at improving human–machine interfaces is the development of task-level control languages using computational intelligence. Task programming systems significantly reduce the programming effort.

The use of a graphical user interface featuring high-resolution color graphics, pointing devices, and touch screens on the shop floor will continue to increase in the future. In the control room, voice output systems will replace bells and blinking lights. On the shop floor, voice recognition systems will help to speed inventory control and quality reporting by eliminating the need for manual data entry. The above-discussed advances will provide new generations of specialized, user-friendly computer-integrated manufacturing subsystems that will make the most of factory personnel (Hunt, 1989).

1.5.3 Intelligent Data Systems

The storage and retrieval of data in an integrated system are more complex than in traditional manufacturing. Integration involves areas such as market analysis, product design, product manufacturing, quality control, factory management, and so on. The activities in these areas are highly interrelated and large volume of data flows among them. Also, the information in an integrated system may be stochastic, fuzzy, and incomplete; therefore, intelligent database systems are required.

One of the main difficulties in integration is the information incompatibility between subsystems. Each subsystem, such as an NC/CNC (CNC = Computer Numerical Control) machine, robot programming, manufacturing scheduling, process planning, production control, has its own representation of information and knowledge. Sometimes, the data required are available in an existing database but in a different form. For example, CAD data are frequently incompatible with the information needed in the process planning. Computer-aided design has become effective in capturing the geometry of parts, including the description of dimensions, shapes, and surfaces. Process planning provides detailed information regarding the way the parts are manufactured, assembled, and so on. The CAD database does not include knowledge or specification of materials, but process planning needs material in the creation of NC programs, and material handling needs this information for selection of the material from inventory.

An intelligent database system can manipulate complex objects, maintain complex relationships among data, take corrective actions under certain conditions, and make deductions.

One of the keys to the data integration problem lies in the development of flexible data schemas. Future databases will have more flexible schemas so that, for example, material information can be added to the CAD database by an engineer at a manufacturing floor station or automatically by a knowledge-based system that contains knowledge of materials.

Electronic warehouses to store and retrieve data, intelligent applications, and intelligent user interfaces have entered modern enterprises.

1.6 COMPUTATIONAL INTELLIGENCE

One of most difficult problems to overcome in modern manufacturing is the inability of production systems to mimic such basic human capabilities as adjusting appropriately to differences in the size, shape, and/or orientation of parts. A technology that should help resolve this dilemma is computational intelligence (CI).

Computational intelligence has been applied in a number of manufacturing settings. It allows automated systems, e.g., robots, to duplicate such human capabilities as vision and language processing. It improves operations of robots and other types of equipment. Finally, computational intelligence can improve the capability of information management systems used in modern manufacturing.

Computational intelligence technology emerges as an integral part of nearly all areas of manufacturing automation and decision making. The CI products that have a significant impact on manufacturing include:

1. Knowledge-based systems in which the decision rules of human experts are captured and used for automated decision making (for details see Chapter 2)

2. Planning, testing, and diagnostic systems (see Chapters 11 and 17)

3. Ambiguity resolves attempting to interpret complex, incomplete, or conflicting data (see Chapter 2)

As human experts with years of experience become scarce, the knowledge-based system allows capturing and "cloning" the human expertise. Knowledge-based systems and other forms of CI are often embedded in manufacturing systems. These new intelligent systems perform decision-making tasks previously performed by humans. People without knowledge of the data schema will routinely query the system for information needed to make decisions. On the factory floor, speech recognition systems are helping to speed inventory control and quality inspection by eliminating the need for manual data entry. The data retrieval system will have to determine what is important to the user and then retrieve the appropriate data. The person may even want the system's "opinion," or the system may ask for the person's opinion. The mechatronic factory of the future will regard personnel and intelligent systems as partners in a dialogue that should encourage sound decision making.

One area of CI applications in modern manufacturing is robotics and vision systems. The current capabilities of robots and machine vision systems are limited compared to those of humans. Therefore, future development in this area will focus on linking vision, perception, and consequent actions of the robotic systems. The future robotic and vision systems will increase a robot's flexibility, improve error recovery, enhance its interface with people, and enable it to learn as it operates.

Another area of CI applications in manufacturing is simulation. Simulation tools, such as ARENA, SIMPLE++, PRO-MODEL, and TAYLOR II, have been used for years. With the addition of CI, a modeling system can become an optimization tool, guiding human managers to the most productive, most cost-effective, or highest quality utilization of resources. With such optimization capability, managers will be able to analyze in real time the various possibilities to determine optimal solutions. The availability of accurate information on cost, time, and quality will eliminate much of the guesswork in manufacturing decision making.

It is clear that the power of computers will continue to increase at a reduced cost. The answer to the question "What comes next?" in technical terms may well be that the next development in the application of computing technology in manufacturing industry will be the widespread use of CI technologies.

1.7   IMPACT OF MANUFACTURING TECHNOLOGY

The growth of manufacturing technology will have a significant influence on the way people work and live. The impact of new technologies on product life cycle, management, and labor is discussed next.

1.7.1   Product Life Cycle

With the development of new technologies, the manufacturing environment and the markets have become highly competitive. Many new products and new models are being introduced in the market every year. The most modern product today will become obsolete tomorrow. As new products with better quality and performance will replace the old ones, and also because of the pressure of competition, every company must introduce new models with improved technical performance. With this trend, product life times tend to be shorter, particularly those at the high-technology end.

1.7.2   Management

A traditional manager uses his or her own judgment and experience to make decisions. Modern managers will be using technology to manage technology. The wide use of computer technology in manufacturing and services, office automation, telecommunications technology, and production technologies and systems tends to increase the flow of information. Managers receive more information and are to make decisions more effectively. As a result, technology has increased the centralization of power. However, power is simultaneously decentralized down to lower levels of employees, who make a limited number of decisions on the basis of their expertise and judgment by using the technology (e.g., personal computers).

There is likely to be less emphasis on production itself. More time will be committed to the research and development (R&D) and postproduction areas of the business, such as marketing. Due to the pressure of competition, companies have to improve their products by investing much more time and money into R&D. Also, more products have to be sold to recover the costs involved; therefore the postproduction area will get more attention.

1.7.3   Human Dimension

The main objective of automation is to produce better quality goods at a lower cost. To some people automation means substitution of machines for people in the factory. It is well known that the trend in robotics is to make machines that act like human beings. As a result, automation will reduce the need for the direct involvement of humans in the production process. However, some argue that automation in manufacturing industries itself will create more new job opportunities in areas other than production.

It is obvious that employment in direct manufacturing will decrease. The production workforce will shift from jobs in direct labor to those in indirect labor. Since many manual activities (e.g., loading, unloading) and tasks in assembly systems will be performed by industrial robots and computerized systems, fewer workers will be needed. However, more professionals and technicians will be required in such areas as planning and maintenance. Also, automation will create more job opportunities in the high-technology industry. Some examples include computer electronics, software development, machine vision, and semiconductor technology. Employment in the information sector will be increasing while employment in more traditional industries might be decreasing.

The changing technology has significantly influenced the working environment in the factory. There are fewer workers per workstation and they are less in contact with each other and have a larger span of responsibilities.

1.8   DESIGN OF MODERN MANUFACTURING SYSTEMS

The basic characteristics of machining systems as well as the differences between modern machining systems and the equivalent classical machining systems are presented next in the form of eight characteristics. These characteristics are based on an analysis of numerous existing modern machining systems; for each characteristic a brief justification is provided. It should be noted that two machining systems are called *equivalent* if they provide the same output, such as production rate or part mix.

1.8.1 Machining Systems

Characteristic 1. The degree of automation of machines and material handling systems in a modern machining system is much higher than in an equivalent classical machining system. The characteristic follows from the definition of a modern machining system.

A modern machining system can be defined as a set of machines linked by a flexible material handling system (e.g., robot, automated guided vehicle), all controlled by a computer system.

Characteristic 2. A modern machining system consists of fewer machines than an equivalent classical machining system.

The classical machining system in Figure 1.8 includes a drilling machine (M_1), two horizontal milling machines (M_2, M_3) and two vertical milling machines (M_4, M_5). The five machines are integrated into a modern machining cell that consists of two machining centers (MC_1, MC_2) linked by an AGV. The modern machining cell in Figure 1.8 might be a component of a larger manufacturing system, shown in Figure 1.9, that consists of the following:

- Three machining cells
- Two assembly cells
- A functional machining facility
- An assembly facility
- An ASR system
- Three AGVs

Characteristic 3. The type of material handling equipment used determines the layout pattern of machines in a modern machining system.

Those most commonly used are articulated arm robots, gantry robots, AGVs, and stacker cranes. To support characteristic 3, consider the four basic machine layouts for flexible machine cells (Figure 1.10). Machines served by an AGV tend to be arranged along a straight line (Figures 1.10*a* and *b*), whereas the reach constraints of a robot forces machines to be arranged in a circle or in a cluster (Figures 1.10*c* and 1.10*d*, respectively).

An interesting variant of a double-row machine layout is shown in Figure 1.11, where machines are located along racks of the ASR system. The advantage of the system in Figure 1.11 is that the stacker crane fulfills the following functions:

- Loading machines with material or semifinished parts and tools

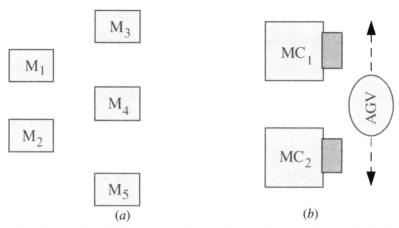

Figure 1.8. Classical machining system and equivalent modern machining cell: (*a*) classical system; (*b*) modern cell.

- Picking machined parts and tools
- Storing and retrieving material, semifinished parts, machined parts, and tools

Characteristic 4. The number of setups in a process plan designed for a modern machining system is significantly smaller than in an equivalent classical process plan.

In classical process planning, a small number of operations (very often only one) were assigned to one setup. This was due to the limited versatility of traditional machines and material handling systems as well as the specialization concept prevalent in classical production management. However, the much greater versatility and

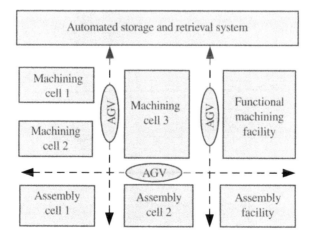

Figure 1.9. Manufacturing system layout.

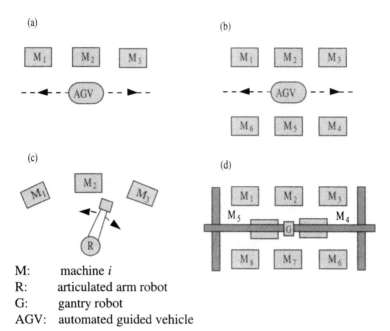

M: machine i
R: articulated arm robot
G: gantry robot
AGV: automated guided vehicle

Figure 1.10. Basic types of machine layouts: (*a*) linear, single-row; (*b*) linear, double-row; (*c*) circular (articulated arm robot); (*d*) cluster (gantry robot)

efficiency of automated machines and material handling systems forced production managers to change their approach to process planning. In process planning for a modern manufacturing system, one attempts to assign as many operations to one setup as possible. This approach can be referred to as an aggregation approach, as opposed to the specialization approach used in classical manufacturing systems.

To illustrate the difference between classical process planning and process planning for a modern manufacturing system, consider the part in Figure 1.12, for

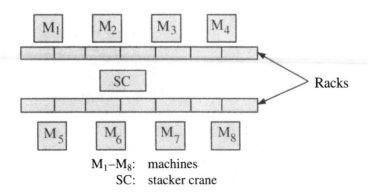

M_1–M_8: machines
SC: stacker crane

Figure 1.11. Double-row machine layout.

which two process plans have been designed (Tables 1.1 and 1.2). For a part to be machined in a modern manufacturing system, one often designs the basic process plan and alternative process plans. However, the number of alternative process plans is usually limited because of the additional costs involved for NC programs, material handling programs (i.e., control programs for robots, AGVs), tools, pallets, fixtures, and process design and maintenance.

The process plan in Table 1.2 for the modern machining system requires three times fewer setups than the classical process plan in Table 1.1. The number of fixtures is also reduced. However, the fixtures used in modern manufacturing are typically more complex and more expensive than the fixtures in classical machining systems. Therefore, to reduce their cost, modular fixtures have been introduced (Kusiak, 1985).

> **Characteristic 5.** In a modern machining system, the processing time per machine load is much longer than it would be in an equivalent classical machining system.

This is due to the differences in process planning that were just discussed and an attempt to load machines in a modern machining system with a number of identical parts clamped onto a fixture. The latter measure reduces the loading–unloading time per part, and since a number of identical parts is loaded at a time, the machining time per machine load increases.

> **Characteristic 6.** The volume and flow of information in a modern machining system are much higher than in an equivalent classical machining system.

This is due to:

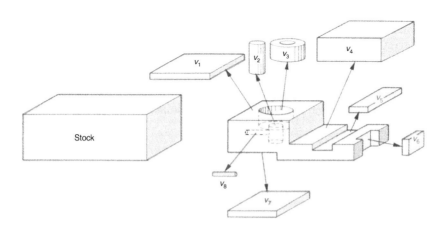

Figure 1.12. Part with material volumes $v_1, \ldots, v_8$ that have been removed.

TABLE 1.1. Classical Process Plan

Setup Number	Machine Number	Operation	Volumes to Be Removed	Tools	Fixtures
1	M_1	Milling	v_1	v_4	T_1, T_2
2	M_2	Drilling	v_2, v_3	T_3, T_4, T_5	F_1
3	M_3	Milling	v_5	T_6	F_1
4	M_4	Milling	v_6	T_7	F_2
5	M_5	Milling	v_7	T_8	F_1
6	M_6	Drilling	v_8	T_9	F_3

- Introduction of new elements into modern machining systems, such as automated material handling systems, fixtures, and pallets
- Increased part mix and the installation of sensors to detect tool breakage and to check availability of machines, tools, and so on

Characteristic 7. In a modern machining system, batch sizes result from order sizes, the capacity of fixtures, and the limited life of tools, rather than being determined by optimization procedures similar to those used in classical manufacturing systems.

To justify this characteristic, one can use the following arguments:

- The basic concept of modern machining systems is to support small-batch production. In addition, the average batch size has been reduced due to progress in hardware development (e.g., machines and material handling systems) and integration of CAD, CAPP, and CAM systems as illustrated in Figure 1.1. This has made the implementation of design changes much more effective. The design of products and parts can be modified frequently according to the changing market requirements and progress in their innovation. Naturally, these frequent changes tend to reduce the batch sizes of products and parts.

TABLE 1.2. Process Plan for a Modern Machining System

Setup Number	Machine Number	Operation	Volumes to Be Removed	Tools	Pallet and Fixture
1	MC_1	Milling and drilling	$v_1, v_2, v_3, v_4, v_5,$ v_6	$T_1, T_2, T_3, T_4, T_5,$ T_6, T_7	PF_1
2	MC_2	Milling and drilling	v_7, v_8	T_8, T_9	PF_1

- A typical fixture has a capacity of more than one part. This reduces the unit (per-part) cost of fixtures and the unit fixture loading (setup) time. In a modern machining system, it is efficient to run parts in batch sizes equal to exact multiples of fixture capacity.

- Some machines are equipped with changeable tool magazines with limited tool capacity (e.g., 25 tools). Each tool can manufacture a limited number of parts. When one of them is worn out or broken, the whole tool magazine has to be changed, and this involves changeover costs. To reduce these costs, it is desirable to manufacture parts in batch sizes that require a tool availability time shorter than the expected lifetime of tools loaded into a tool magazine.

Characteristic 8. The design of a modern machining system has an impact on its operation.

To justify this, consider the modern machining cell shown in Figure 1.13. Assume that this cell has to machine a batch of 60 identical parts in an unmanned mode. There are two parts clamped onto a fixture, and it takes 2 minutes to machine a pallet with two parts (1 minute per part). Simple calculation shows that in order to provide 1 hour of unmanned operation, 30 pallets and 30 fixtures are required. Considering the significant cost of pallets and fixtures, this is obviously an inefficient design of a modern machining module.

In the revised design (Figure 1.14) a robot loads parts onto the machining center. To perform 60 minutes of unattended machining, only one pallet and one fixture are required. In this case an extension of unattended operation to, for example, 8 hours requires a very small investment: eight pallets, a conveyor to feed the pallets, and a robot gripper to handle the parts and pallets. The modern machining cell in Figure 1.14 can be easily linked with the rest of the manufacturing facility by an AGV.

1.8.2 Assembly Systems

Most of the characteristics formulated in the previous section for modern machining systems also apply to assembly systems (ASs). Any differences that may arise are in

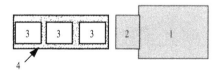

1: machining center
2. pallet shuttle
3. pallet and fixture
4. conveyor belt

Figure 1.13. Modern machining cell.

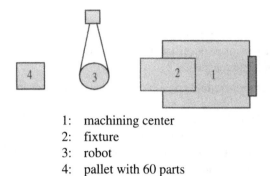

1: machining center
2: fixture
3: robot
4: pallet with 60 parts

Figure 1.14. Revised design of modern machining cell in Figure 1.13.

the terminology rather than conceptual. Instead of rephrasing the characteristics formulated for modern machining systems, this section discusses process planning and design of products and parts for modern assembly systems.

1.8.2.1 Process Planning for Assembly Systems. Consider product C illustrated in Figure 1.15, which consists of four parts P_1, P_2, P_3, and P_4. This product has two assembly plans, as presented in Tables 1.3 and 1.4. In the classical assembly line, product C is assembled according to the process plan in Table 1.3 on three serial assembly stations (Figure 1.16). In a modern assembly system, on the other hand, the versatility of the assembly stations allows many operations to be performed on a single assembly station. Therefore, in a typical modern assembly system the number of serial assembly stations is smaller than in the equivalent classical assembly system. An example modern assembly system that consists of two parallel assembly stations is presented in Figure 1.16.

Each of the two assembly stations in Figure 1.17 is based on robot R.

Product C from Figure 1.15 can be assembled on one of the two assembly stations. Since the changeover cost for each assembly station is small, all products can be

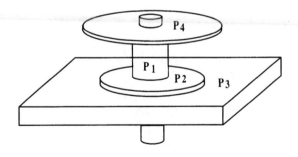

Figure 1.15. Examples product C.

TABLE 1.3. Classical Assembly Plan

Assembly Station	Parts and Subassemblies
AS_1	$\{P_1, P_2\}$
AS_2	$\{\{P_1, P_2\}, P_3\}$
AS_3	$\{\{P_1, P_2, P_3\}, P_4\}$

TABLE 1.4. Assembly Plan for a Modern Assembly System

Modern Assembly Station	Parts to Be Assembled
FAS_1 or FAS_2	$\{P_1, P_2, P_3, P_4\}$

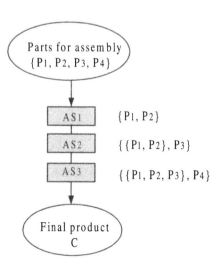

Figure 1.16. Classical assembly system.

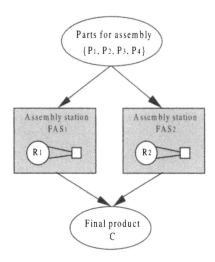

Figure 1.17. Modern assembly system.

assembled in small batches. The cost of fixtures in modern assembly systems tends to be higher than in classical assembly lines.

The differences between modern manufacturing systems and classical manufacturing systems are useful in understanding the design and management issues in modern manufacturing.

1.9 ORGANIZATION AND PRODUCT EVALUATION STANDARDS

The ISO completed the quality management system standard ISO 9000 in 1987 (Besterfield et al., 1999). In 1992 the ISO formed a technical committee (TC 207) charged with the development of standards for an environmental management system. Similar to the ISO 9000 standards that do not address the performance of the product or service, an attempt has been made to address the process rather than the end goal. The ISO 14000 series of standards has been developed consisting of the following categories:

- ISO 14001, "Environmental Management Systems—Specifications with Guidance for Use," specifying the elements that organizations seeking registration are to conform
- ISO 14004, "Environmental Management Systems—Guidelines on Principles, Systems, and Supporting Techniques," providing supplementary material
- ISO 14010, "Guidelines for Environmental Auditing—General Principles on Environmental Auditing," providing information for internal and external auditing

- ISO 14011, "Guidelines for Environmental Auditing—Audit Procedures—Auditing of Environmental Management Systems," covering information on planning and conducting audits
- ISO 14012, "Qualification Criteria for Environmental Auditors Performing Environmental Management Systems Audits," providing information on auditor qualifications, training, personal attributes, and skills
- ISO 14032, "Guidelines for Environmental Performance Evaluation," presenting information on recording information to track performance

In conjunction with the above-mentioned organization evaluation standards, a set of product evaluation standards has been developed (Besterfield et al., 1999):

- ISO 14020, "Environmental Labeling—Basic Principles for all Environmental Labeling"
- ISO 14021, "Environmental Labeling—Self Declaration of Environmental Claims: Terms and Definitions"
- ISO 14022, "Environmental Labeling—Symbols"
- ISO 14023, "Environmental Labeling—Testing and Verification Methodologies"
- ISO 14024, "Environmental Labeling—Practitioner Programs: Guiding Principles, Practices, and Certification Procedures for Multiple Criteria Programs"

1.10 THE FUTURE OF MANUFACTURING ENTERPRISES

1.10.1 Enterprise Attributes

Future enterprises can be described with the following attributes (NGMS, 1999):

- Adaptability
- Agility
- Modularity
- Standardization
- Collaborative
- Distributed
- Simplicity
- Knowledge orientation
- Human orientation
- Environmentally aware

A future enterprise needs to be capable of reconfiguring and adapting to the changing business environment. Many future enterprises are likely to be small in size, modular in concept, and able to interface with other units. The latter demands robust business interfaces, protocols, and standards. Collaboration and teamwork will be essential. The concept of *virtual enterprise* is likely to serve as a viable model of a future company. A virtual enterprise is likely to be a temporary alliance of enterprises that come together to share skills and resources in order to better respond to business opportunities. Networking, cooperation, logistics, transportation, and supply chain are the key components of any virtual enterprise. Future enterprises will strive for simplicity. Life-cycle costs can be greatly affected by a design that minimizes on-going maintenance and operational costs. The ability to utilize new information and knowledge and learning will be of paramount importance as the volume of information and knowledge is increasing at a rapid rate. A future enterprise may be constrained by the laws of physics for certain hardware applications; however, no boundaries or constraints are on the horizon in the area of utilizing information and knowledge. Enterprises will become highly human oriented, supporting life-long learning, integration, and connectivity appropriate to the domain. They also must be environmentally aware by using less consumable resources and generating less waste.

1.10.2 Manufacturing Technology

There is no doubt that manufacturing technology will undergo significant changes in the future. The markets will become more competitive. Products will have to be of better quality and produced at a lower cost. A computer-integrated manufacturing system created by the integration of manufacturing processes and systems has the following elements:

1. Information communication utility that accesses data from the constituent parts of the system and serves as an information communication and retrieval system.

2. Information-sharing utility that integrates data across system elements into a unified database.

3. Analysis utility that provides a mathematical model of a real or hypothetical manufacturing system. Employing simulation and, when possible, optimization, this utility is used to characterize the behavior of the modeled system in various configurations.

4. Resource-sharing utility that employs optimal or heuristic algorithms to plan and control the allocation of resources to meet a demand profile.

5. Higher order entity that integrates information and processing functions into a more capable, effective processing system.

Several key trends in numerical control directly affect the growth of modern manufacturing. Distributed numerical control (DNC) and flexible manufacturing systems controlled by computers will become a part of the plantwide management information system.

Flexible and cellular systems will continue to change manufacturing in several ways. Machining operations will be grouped less by the type of operation but rather by the type of part. Improvements will also be made in speeding operations outside the actual processing, for example, machining. The elements of programming, maintaining, tooling, and fixturing both machines and workpieces will require improved communications. There is a growing trend toward the availability of machine diagnostics and improved control diagnostics. Two other trends are:

1. Systems that recognize voice commands will be widely used.
2. Systems that model the geometry of objects as solids rather than surfaces will allow designers to be more productive. In the near future designers are also expected to use holography for solids modeling. The trend toward completely synthesized design systems will allow the designer to concentrate on the task of specifying general characteristics.

One of the most important issues is increasing productivity of small batch manufacturing through automation. The productivity is closely tied to the issue of quality, which in turn has been identified as a key factor in the renaissance of industry.

Both horizontal and vertical growth of integration can be expected in the future. Enterprisewide integrated solutions, for example, Enterprise Resource Planning (ERP) and Internet systems, will become commonly used. Integration will expand beyond singular enterprise boundaries.

It is clear that the future development in computer technology and its resulting impact on the existing manufacturing technologies will emphasize automation and integration. Computational intelligence technology will make each manufacturing component more intelligent, which implies less human involvement, high speed and accuracy, high quality of product, and high productivity. It will offer manufacturing companies a competing edge with more flexibility and fast responsiveness to the market.

The increasing global competition will serve as the driving force for manufacturing implementation. To be competitive, companies need to produce manufactured goods at a lower cost and adequate quality. In automated manufacturing, the ratio of direct labor (labor applied to the actual manufacturing cost of producing the product) to the total cost of the product will be reduced. The highest single cost is the material cost. Hence, areas other than direct labor need to be found for cost improvement. The most obvious is reduction in the level of work-in-process.

The future development in manufacturing technologies will require highly trained personnel. Management not only has to acquire enough technical background but also must be familiar with the postproduction areas. Computational intelligence techniques will be employed for efficient training of human resources.

The integrated manufacturing concept may be extended to service systems. There are applications other than the manufacturing industry, where stock levels, resources, and conflicting demands have to be carefully scheduled, for example, hospitals and government organizations. As facilities become more automated and computerized, planning and control will be increasingly important.

REFERENCES

Besterfield, D. H., C. Besterfield-Michna, G. H. Besterfield, and M. Besterfield-Sacre (1999), *Total Quality Management*, Prentice-Hall, Upper Saddle River, NJ.

Hannam, R. (1997), *Computer Integrated Manufacturing: From Concepts to Realization*, Addison Wesley Longman, Harlow, Essex, England.

He, W., and A. Kusiak (1994), Future trends in computer-integrated manufacturing, in G. Salvendy and W. Karwowski (Eds.), *Design of Work and Development of Personnel in Advanced Manufacturing*, John Wiley, New York, pp. 553–569.

Hunt, D. V. (1989), *Computer-Integrated Manufacturing Handbook*. Chapman and Hall, London.

Kusiak, A. (1985), Flexible manufacturing systems: A structural approach, *International Journal of Production Research*, Vol. 23, No. 6, pp. 1057–1073.

Kusiak, A. (1990), *Intelligent Manufacturing Systems*, Prentice-Hall, Englewood Cliffs, NJ.

NGMS (1999), *Promotional Materials*, Next Generation Manufacturing System Program, Consortium for Advanced Manufacturing—International, Arlington, TX.

QUESTIONS

1.1. What are the major functional areas of an enterprise?

1.2. Do the functions of an enterprise change or remain the same over time?

1.3. What is the role of integration in an enterprise?

1.4. What are the basic manufacturing technologies of a modern enterprise?

1.5. What is the Intranet?

1.6. What is the Extranet?

1.7. Name a few manufacturing standards.

1.8. What is XML?

1.9. What are the basic characteristics of a modern enterprise?

1.10. What are the basic attributes of a modern enterprise?

PROBLEMS

1.1. The bold lines in Figure 1.18 indicate material handling routes. The machine loading/unloading points are A, C, and E for machines M_1, M_2, and M_3. Assign a suitable material handling equipment to each of the four manufacturing cells.

1.2. Corporation AA considers purchasing the following machines for producing printed circuit boards (PCBs):

(a) Three type 1 insertion machines

(b) Three soldering machines

(c) Three type 2 insertion machines

(d) Three inspection stations

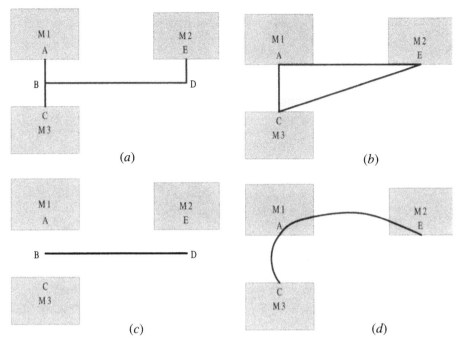

Figure 1.18. Material handling routes.

The sequence of processing the PCBs is insertion–soldering for type 1 and insertion–inspection for type 2. It has been estimated that the utilization rates are 84% for each insertion machine of type 1, 94% for each insertion machine of type 2, 88% for the soldering machine, and 54% for the inspection station.

Draw two different layouts of the manufacturing system and discuss the advantages and disadvantages of each.

1.3. Six parts are to be machined on six machines as indicated in Table 1.5, e.g., part 1 is machined in the sequence machine 2–machine 3; part 3 is processed in the sequence machine 3–4–5–6.

TABLE 1.5. Part Routings

Part \ Machine	1	2	3	4	5	6
1		*	*			
2	*	*	*			
3			*	*	*	*
4				*	*	*
5			*	*	*	
6		*	*	*	*	

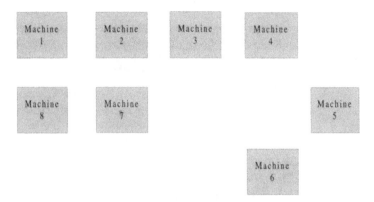

Figure 1.19. Manufacturing cell.

Parts 1 and 2 can be handled by an articulated arm robot, while parts 3, 4, and 5 can be handled by an AGV. Part 6 can be handled by the robot and the AGV. Design a layout of the manufacturing system with six machines.

1.4. Corporation BB has decided to automate material handling in the manufacturing cell shown in Figure 1.19. Every part manufactured in the cell visits each of the eight machines at least once. Due to a large volume of part flow, all machines are to be tended by an automated material handling. It is also known that relocating machines is expensive.

The corporate database includes the following equipment: belt conveyors (Figure 1.20), roller conveyors, arc roller conveyors, AGVs, overhead gantry robots, articulated arm robots, slide conveyors, storage racks (Figure 1.20), fixtures, pallets, and pallet stands. Select material handling system(s) and other necessary equipment for the machining cell in Figure 1.19.

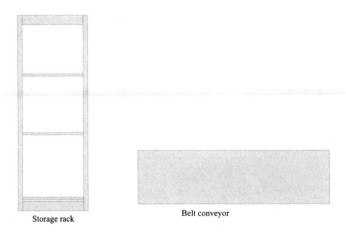

Storage rack   Belt conveyor

Figure 1.20. Examples of equipment available in corporate database.

Figure 1.21. Specialized monitor.

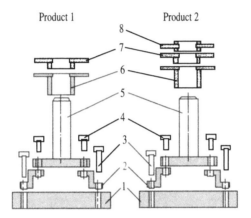

Figure 1.22. Two products.

TABLE 1.6. Description of Assembly Operations

Operation Number	Operation Name
1	Place base 1
2	Position part 2 on both sides of base 1
3	Fasten the left screw 3
4	Fasten the right screw 3
5	Position part 5
6	Fasten the left screw 4
7	Fasten the right screw 4
8	Position and secure part 6
9	Position and secure part 7
10	Position and secure part 8

    (a)   Draw two different layouts of the material handling system(s).

    (b)   Discuss strengths and weaknesses of each of the two design alternatives in (a).

1.5. Corporation CC assembles 280 specialized touch-screen monitors (Figure 1.21) per year. The flow of products during the assembly process is to be unidirectional. The corporate database includes the following equipment: belt conveyors (Figure 1.20), roller conveyors, arc roller conveyors, AGVs, overhead gantry robots, articulated arm robots, slide conveyors, storage racks (Figure 1.20), fixtures, pallets, and pallet stands. Justify and draw the layout of the assembly system.

1.6. The two products P1 (with eight parts) and P2 (eight parts) in Figure 1.22 are to be produced in the same assembly system. Draw two alternative layouts of the assembly system and discuss the strengths and weaknesses of each system design. The assembly operations are listed in Table 1.6.

CHAPTER 2

KNOWLEDGE-BASED SYSTEMS

2.1 INTRODUCTION

A knowledge-based system is a computer program that contains the expertise required for solving a problem. Knowledge-based systems derive their power from the knowledge stored in the knowledge base.

The architecture of a typical knowledge-based system includes the following components (Figure 2.1):

- Knowledge base
- Working memory
- Inference engine
- Knowledge acquisition module
- User interface module

2.2 KNOWLEDGE REPRESENTATION

Effective representation of knowledge is one of the key issues in knowledge-based systems. The knowledge may have many forms, including descriptive definitions of domain-specific terms, descriptions of individual objects, classes of objects and their interrelationships, and criteria for making decisions.

The knowledge acquired should be represented in a computer implementable form. The next sections review the most frequently used knowledge representation schemes.

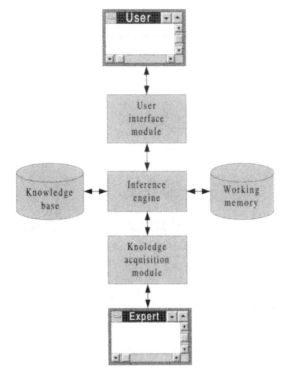

Figure 2.1. Basic components of a knowledge-based system.

2.2.1   First-Order Logic

Much of the early research in computational intelligence in general and knowledge representation in particular was concerned with first-order logic. First-order logic deals with the relationship between assumptions and conclusions. It is not concerned with the truth, falsity, or acceptability of individual phrases but with the relationship among them.

First-order logic sentences can be expressed as a collection of clauses. Formally speaking, clauses can be defined and interpreted as follows. A clause is an expression in the form

$$B_1, \ldots, B_m \leftarrow A_1, \ldots, A_m \qquad m, n \geq 0$$

where $A_1, \ldots, A_m$ are conditions of the clause and $B_1, \ldots, B_m$ are alternative conclusions of the clause.

Both conclusions and conditions are expressions in the form $P(t_1, \ldots, t_k)$ called an atom, where P is a k-argument predicate symbol and $t_1, \ldots, t_k$ are terms. A term is a variable, a constant symbol, or a function in the form $f(t_1, \ldots, t_q)$, where f is q-argument function symbol and $t_1, \ldots, t_q$ are terms. If the clause contains the variables $x_1, \ldots, x_k$, then it can be interpreted as stating that, for all $x_1, \ldots, x_k$,

$$B_1 \text{ or } \ldots \text{ or } B_m \text{ if } A_1 \text{ and } \ldots \text{ and } A_m$$

A variety of knowledge can be expressed using the above defined clausal form of logic. For example:

```
REDUCE (INCREASE_SAFETY_STOCK, SYSTEM NERVOUSNESS) ←
  Increased safety stock reduces the system nervousness)
BETTER (FREEZE_MPS, PLANNED ORDER) ←
  Freezing MPS is better than a planned order)
```

The logic representation is useful in providing formal proofs as it offers clarity, is well defined, and is easily understood.

2.2.2 Production Rules

Production rules are essentially a subset of predicate calculus with an added prescriptive component indicating how the information in the rules is to be used during reasoning. A production rule has the following basic form:

IF (conditions)

THEN (conclusions)

When the IF portion of a production rule is satisfied by the conditions, the action specified by the THEN portion is performed. When this happens, the rule is said to fire. A rule interpreter compares the IF portion of each rule with the facts and executes the rules whose IF portions match the facts, as shown in Figure 2.2 (Waterman, 1986).

Consider, for example, the following production rule:

IF Part Pi is to be dispatched to machine Ma that is occupied by another part Pj

THEN Check availability of an alternative machine Mb

There are several advantages of expressing knowledge declaratively as production rules (Kusiak, 1990):

- A rule's use in specific circumstances can be automatically explained and defended to the system user.
- Developers and users can realistically expect to modify a few rules without breaking the entire system.
- New knowledge can be added to the system simply by adding new rules without worrying about how they will fit in. This is essential to a system that is designed to benefit (or learn) from the new knowledge or past experience.

A system in which knowledge is represented with rules is called a rule-based system.

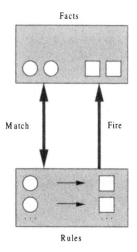

Figure 2.2. Execution of rules through a match–fire sequence.

2.2.3 Frames

Frames were suggested by Minsky (1975) as an alternative to predicate calculus. A frame provides a concise structural representation of useful relations. It can be viewed as the generalization of a property list that provides a structured representation of an object or a class of objects. A frame may represent each individual object or class of objects. The descriptions in a frame can obtain an abstract template providing a skeleton that describes any instance and a set of defaults for typical members of the class.

The most important features of a frame are as follows (Kusiak, 1990):

- It provides a structured representation of objects and classes of objects in an application domain.
- It provides a mechanism (called inheritance) that guides description movement from class descriptions to individual descriptions.
- It allows the specification of procedures (called demons) for computing descriptions, a feature known as procedural attainment.
- It allows one to determine descriptions in the absence of specific knowledge, a feature referred to as the default feature. Frames can represent not only objects being reasoned about but rules as well. Rules represented as frames can be grouped into classes, and the description of a rule can include arbitrary attributes of the rule.

A frame is viewed here as a nested association list with a number of levels of embedding:

```
(Frame
 (slot (facet (datum (label message . . . ) )
              (datum . . . ) . . . )
```

```
        facet . . . ) . . . )
           . . . )
(slot . . . )
```

A slot is essentially an attribute of a frame. It can have multiple values and a set of properties called facets. The value of a slot is a datum under a $VALUE facet. The data in slots may be function calls that may be evaluated automatically, known as procedural attachment. Nested substructures, such as the list (label message) after a datum, are optional. For example, consider the following frame:

```
(Freeze_MPS
 (AKO ($VALUE (Dampening_mechanism)))
 (Method_of_lot_size_MPS ($VALUE Wagner_Whitin_method)))
 (Planning_horizon_length ($VALUE (N (unit: k multiple of
  natural cycle length))))
(Replanning_frequency ($VALUE (R (unit: the number of
  periods between replanning cycle))))
 (Freeze_interval ($VALUE (P (unit: the proportion of
  the overall planning horizon n that
  remains fixed in each planning cycle))))
(Planning_information ($VALUE (Order_based))))
```

This frame describes a dampening mechanism, called Freeze_MPS, in a master planning system. The first slot AKO (short for "a kind of") has one facet $VALUE, followed by the datum Dampening_mechanism. This represents the item of information that "Freeze_MPS is a kind of dampening mechanism." The second slot Method_of_lot_size_MPS represents the available scheduling models that can be used to lot-size future orders in the MPS. Here, the algorithm used is the Wagner–Whitin method. Note that the optional structure (label message) has been used in specifying the slots Planning_horizon_length, Replanning_frequency, and Freeze_interval.

Each slot may have any number of procedures attached to it. It could be, for example, a mechanism for triggering procedures that support the process of creating an instance of a frame and filling in values for its slots. There are three useful types of procedures that might be attached to slots, as illustrated in Figure 2.3 (Waterman, 1986):

- If-added procedure: Executes when new information is placed in the slot.
- If-removed procedure: Executes when information is deleted from the slot.
- If-needed procedure: Executes when information is needed from the slot.

In some cases, instead of storing the data associated with the $VALUE facets, it may be possible to inherit it from a generic frame. The process of acquiring information from a more generic frame is called "inheritance." For example, Freeze_method1 could inherit information from its generalization, the Freeze_MPS frame:

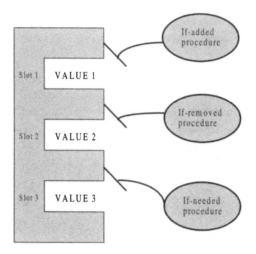

Figure 2.3. A frame.

```
(Freeze_method1
(AIO ($VALUE (Freeze_MPS)))
 .  .  .
 .  .  .
(Planning_information ($VALUE (Period_based)))))
```

Here, AIO means "an instance of" and describes the relationship between an individual and its category, whereas AKO is a relationship between generic objects or between categories of objects.

This discussion illustrates two central ideas that separate frames from simple property lists. First, a frame can inherit knowledge from a more general frame; therefore, it can be related hierarchically for more efficient storage of information. Second, one could associate various procedures (also referred in the literature as "demons") with properties instead of values. The mechanism for attaching demons to properties can also be used to attach other kinds of auxiliary information about the value in a slot.

Frames provide various ways of attaching procedural information expressed in some other language (e.g., LISP, PROLOG). This procedural capability enables one to model the behavior of objects in an application domain. The procedures are attached to the slots of a frame rather than to assertions in an arbitrary form. This is one way a frame-based language provides a more structured environment than languages such as PROLOG.

One of the most common forms of inferencing in a frame-based system is inheritance. Any frame can have a member link or a subclass link to one or more classes of frames. The assertion and retrieval mechanisms for frame-based languages use the member and subclass links to augment the descriptive information in a frame via inheritance.

2.2.4 Semantic Networks

Semantic networks provide a good representation scheme to assert relationships between objects or a class of objects in the form of planar graphs. A graph G is said to be *planar* if there exists a geometric representation of G that can be drawn on a plane so that no two of its edges intersect. Such formalism offers a wider scope for representing relationships than that offered by logic.

According to a semantic representation, the knowledge is a collection of objects and associations represented as a labeled directed graph. In its simplest form, a semantic network is a graph whose nodes represent individual objects and whose directed arcs represent binary relationships.

For example, the fact Freeze MPS is a Dampening_mechanism is illustrated in the semantic network in Figure 2.4a:

Freeze_MPS and Dampening_mechanism are represented as objects and is_a is the relationship describing that Freeze_MPS is a Dampening_mechanism.

The type of a network can be enhanced further with an object having more than one type of relationship (but with different objects). In Figure 2.4b, a component of Freeze_MPS is incorporated.

The semantic network representation of knowledge can be viewed as a graphical representation of the binary predicate version of predicate calculus, where an arc labeled R directed from node x to node y,

$$\underset{x}{O} \overset{R}{\longrightarrow} \underset{y}{O}$$

represents the assertion

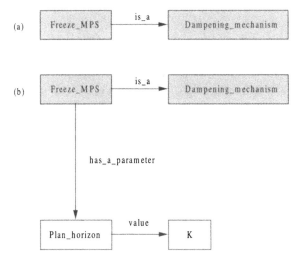

Figure 2.4. Semantic networks: (*a*) simple example; (*b*) expanded network.

$$R(x, y) \leftarrow$$

Hence the semantic network in Figure 2.4*a* is represented by the following assertion:

```
is_a (Freeze_MPS, Dampening_mechanism) ←
```

2.3   INFERENCE ENGINE

An inference engine is the part of a knowledge-based system that contains the general problem-solving knowledge. The inference engine applies the knowledge to the solution of an actual problem and acts as a control system.

Depending on the type of knowledge representation scheme adopted, different types of inference engines are possible. In a rule-based system, the inference engine examines facts and executes rules contained in the knowledge base according to the inference procedure selected. Some of the commonly encountered inference procedures are *modus ponendo ponens* (modus ponens), *modus tollendo tollens* (modus tollens), and *resolution* (Pham and Pham, 1988):

1. Modus ponens:
   P ⇒ Q (If P is TRUE Then Q is TRUE)
2. Modus tollens:
   P̄ ⇒ Q̄ (If Q is not TRUE Then P is not TRUE)
3. Resolution:
   (P ⇒ Q ; Q ⇒ R) ⇒ (P ⇒ R)
   ((If P is TRUE Then Q is TRUE) and (If Q is TRUE
   Then R is TRUE)) Then
   (If P is TRUE Then R is TRUE).

2.3.1   Basic Reasoning Strategies

Reasoning with production rules can proceed in different ways according to the inference procedure. One procedure is to start with a set of facts (or data) and to look for those rules in the knowledge base for which the IF portion matches the facts. When such rules are found, one of them is selected based upon an appropriate conflict resolution criterion and fired. This generates new facts in the knowledge base, which in turn causes other rules to fire. The reasoning stops when no more rules can be fired. This kind of reasoning is known as *forward reasoning* or *data-driven reasoning*.

An alternative approach is to begin with the goal to be proved and attempt to establish the facts needed to prove it by examining those rules that have the desired goal in the THEN portion. If such facts are not available in the knowledge base, they become subgoals. The process continues until all the required facts are found, in which case the original goal is proved, or the situation is reached when one of the subgoals cannot be satisfied, in which case the original goal is disproved. This method of reasoning is called *backward reasoning* or *goal-directed reasoning*.

Backward reasoning is also known as "top-down" search because it begins from the goal state and proceeds to the initial state, while "bottom-up" search, as implemented in forward reasoning, takes place in the opposite direction.

AND/OR trees are widely used in the computational intelligence literature as they enable understanding the inference process. The AND/OR tree is used to illustrate the inference process in the next three figures. Figure 2.5 shows three production rules represented with an AND/OR tree and demonstrates the relationship between forward and backward reasoning. Note that in the AND/OR tree an arc between edges denotes a logical AND relationship and no arc implies logical OR. There are no OR relationships in the tree in Figure 2.5.

Figures 2.6 and 2.7 illustrate forward and backward reasoning, respectively, with the AND/OR tree. Assume that three facts b, d, and e are given. The node of the AND/OR tree corresponding to a given fact is marked with an arc (see Figure 2.6).

Given the fact e, rule R_3 in Figure 2.6 is fired. Now, knowing facts c and d (see the three facts given in Figure 2.6), rule R_2 is fired. Knowing the facts a and b, rule R_1 is fired and the goal is achieved. In other words, the three known facts b, d, and e allow to reach the goal (the hypothesis has been proven). The backward reasoning is a top-town approach, as opposed to the bottom-up approach in forward chaining. Given the goal and fact b, node a in Figure 2.7 becomes subgoal S_1. To prove the subgoal a with the fact d known, node c in turn becomes subgoal S_2. Subgoal S_2 is proven with the fact e.

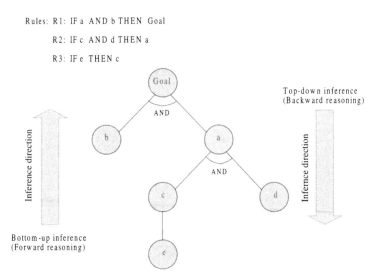

Figure 2.5. AND/OR tree representation of three production rules.

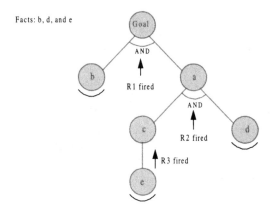

Figure 2.6. Forward reasoning.

Three factors influence the selection of the forward- or backward-reasoning strategy (Rich, 1983):

- The number of possible start states and goal states. One would like to move from a smaller set of states to the larger set of (thus easier to be determined) states.
- Value of the branching factor (i.e., the average number of nodes that can be reached directly from a single node). One would like to reduce the branching factor.

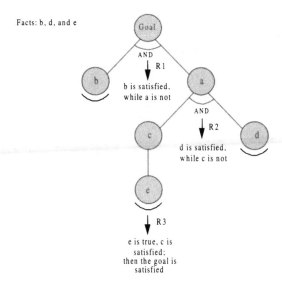

Figure 2.7. Backward reasoning.

- Justification of the knowledge-based system reasoning process by a user. When justified, it is important to proceed in the direction that matches closely the way the user is thinking.

Factors that impact the selection of a forward-reasoning approach over backward reasoning are provided (Durkin, 1994):

- All or most of the data are given in the initial problem statement.
- A large number of potential goals exist, but there are only a few ways to use the facts and the information given for a particular problem instance.
- It is difficult to form a goal (hypothesis).

The backward-reasoning approach is suggested if:

- A goal (hypothesis) is given as the problem statement or can be easily formulated.
- A large number of rules match the facts of the problem and thus produce an increasing number of conclusions or goals. Early selection of a goal can eliminate most of these branches, making a backward-reasoning approach more effective in pruning the search space.
- Problem data are not given but must be acquired by the problem solver. In this case, the backward-reasoning approach can be useful in guiding the data acquisition process.

In practice, forward and backward reasoning are sometimes integrated, and a process is used to join these opposite lines of reasoning at some intermediate point to yield a problem solution.

2.3.2 Uncertainty in Rule Bases

Often, in knowledge-based systems it necessary to consider the uncertainty aspect associated with the knowledge elicited. The uncertainty is handled by associating a value c belonging to the interval (0, 1), known as a *certainty factor*. The certainty factor can be assigned to each element in the condition phrase as well as to the production rule outcome. The formulas for calculating certainty factors are illustrated with four production rules.

Consider production rule R_1:

Rule R_1:   IF A_1 AND B_1 THEN D_1

and the corresponding values of certainty factors of the two elements in its condition clause:

$$CF(A_1) = C_{A1} \qquad CF(B_1) = C_{B1}$$

The value of the certainty factor of rule R_1 is computed as follows:

$$CF(D_1) = CF(R_1) = CF(A_1 \text{ AND } B_1)$$
$$= \min\{CF(A_1), CF(B_1)\}$$
$$= \min\{C_{A1}, C_{B2}\}$$

Analogously, given production rule R_2:

Rule R_2:   IF A_2 OR B_2 THEN D_2

the certainty factor is computed:

$$CF(D_2) = CF(R_2) = CF(A_2 \text{ OR } B_2)$$
$$= \max\{CF(A_2), CF(B_2)\}$$
$$= \max\{C_{A2}, C_{B2}\}$$

For production rule R_3:

Rule R_3:   IF A_1 AND B_1 THEN D_1 CF = c

In addition to the certainty factors for the elements A_1 and A_2, the certainty factor CF = c for the entire rule R_3 is specified.

The certainty factor for the rule outcome is computed as follows:

$$CF(D_1) = \min\{C_{A1}, C_{B1}\}\, c$$

In a similar way, for the outcome of production rule R_4:

Rule R_4:   IF A_2 OR B_2 THEN D_2 CF = c

$$CF(D_2) = \max\{C_{A2}, C_{B2}\}c$$

The calculation of certainty factors for a rule base is illustrated in Example 2.1.

Example 2.1.   Consider the following three production rules:

R_1: IF F AND G THEN D
R_2: IF D AND E THEN A
R_3: IF A AND B THEN C

Also consider the corresponding AND/OR tree in Figure 2.8. The certainty factors for the facts F, G, E, and B are given as

$CF(F) = .8$, $CF(G) = .9$, $CF(E) = .95$, and $CF(B) = .75$.

The certainty factors for rules R_1, R_2, and R_3 are specified as
$CF(R_1) = .85$, $CF(R_2) = .9$, and $CF(R_3) = .9$.

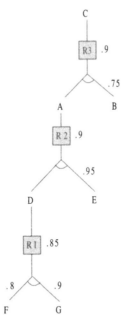

Figure 2.8. AND/OR tree for three rules.

The remaining certainty factors are computed using the above uncertainty formulas:

$$CF(D) = \min\{CF(F), CF(G)\} \cdot CF(R_1) = .68$$

$$CF(A) = \min\{CF(D), CF(E)\} \cdot CF(R_2) = .61$$

$$CF(C) = \min\{CF(A), CF(B)\} \cdot CF(R_3) = .55$$

In addition to evaluating certainty factors for singular production rules, it is also possible to evaluate the certainty of two or more rules, known in computational intelligence literature as combined evidence.

Consider the following two production rules with the corresponding certainty factors:

Rule R_1: IF A_1 AND B_1 THEN D CF $= c_1$
Rule R_2: IF A_2 OR B_2 THEN D CF $= c_2$

The combined evidence is computed as

$$CF(R_1, R_2) = c_1 + c_2 - c_1 c_2$$

$$= c_1 + c_2 (1 - c_1)$$

For three production rules the combined evidence is

$$CF(R_1, R_2, R_3) = CF(R_1, R_2) + CF(R_3) [1 - CF(R_1, R_2)]$$

The calculation of combined evidence is illustrated in Example 2.2.

Example 2.2.   Consider the following two production rules:

Rule R_1:   IF the inflation rate is less than 5%
              THEN stock market prices go up $CF = c_1 = 0.7$
Rule R_2:   IF unemployment rate is less than 7%
              THEN stock market prices go up $CF = c_2 = 0.6$

Now let us assume it is predicted that next year the inflation rate will be 4% and the unemployment rate will be 6.5% (i.e., we assume that the premise of the rule is true). The combined evidence is computed as

$$CF(R_1, R_2) = c_1 + c_2 - c_1 c_2 = 0.7 + 0.6 - 0.42 = 0.88$$

2.3.3   Other Search Strategies

Besides forward and backward chaining, numerous other search strategies have been used in computational intelligence. Some of the most commonly used approaches are discussed next.

2.3.3.1   Depth-First and Breadth-First Search Strategies.   In addition to specifying a search direction, a search algorithm determines the order in which states are examined in a tree. From any given goal state or initial state, there are usually several alternative paths leading to different solutions. The tree may be searched depth first or breadth first.

A depth-first search starts from the root node deep into the tree, considering a sequence of successors to a state until the path is exhausted. The search then proceeds to the next branch of the tree, exploring it in depth. Depth-first search is appropriate when a tree structure is not too deep. With breadth-first search, all possible alternatives at the root node are generated, then the alternatives at the next level are produced, and so on. Breadth-first search is suitable when the number of alternatives at the choice nodes is not too large. Depth-first and breadth-first searches are illustrated in Figure 2.9. In Figure 2.9a, the search order is (A, B, E, C, F, G, D, H), while in Figure 2.9b it is (A, B, C, D, E, F, G, H).

2.3.3.2   Optimization and Knowledge-Based Systems.   Optimization ap- proaches have been used for solving problems for a long time. The strength of optimization approaches lies in their formal treatment. However, the following difficulties make them not always easy to apply:

- The data may not be easily available.
- Their scope of applicability may be limited.
- Human expertise might be required.

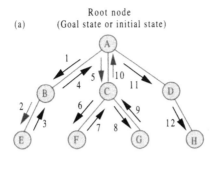

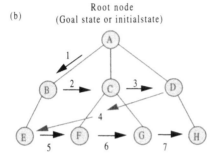

Figure 2.9. Search strategies: (*a*) depth-first; (*b*) breadth-first.

- The algorithms often do not provide optimal solutions because of the problem's complexity.

Knowledge-based approaches, on the other hand, have typically been used to solve problems that are either too complex for mathematical formulation or too difficult to solve using optimization approaches. They begin with the human expertise and capabilities; thus they can handle highly complex problems by using subjective and heuristic methods similar to those used by human experts. Knowledge-based approaches can be seen as alternatives to optimization approaches. It is beneficial to synthesize the two approaches.

Based on their operational mode, two classes of knowledge-based systems can be identified (Kusiak, 1990)—Stand-alone and tandem knowledge-based systems:

A *stand-alone knowledge-based system* uses data and constraints relevant to the problem and solves it using procedures similar to those used by human experts. An optimization approach, which involves modeling the given problem and solving the model using algorithms, is not used in this mode. Many existing knowledge-based systems belong to this class.

A *tandem knowledge-based system*, on the other hand, combines the optimization approach with the knowledge-based system approach (see Figure 2.10). The computing scenario in a tandem knowledge-based system is as follows: A

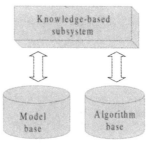

Figure 2.10. Tandem knowledge-based system.

suitable model is either selected or built for the given problem. To solve the model, an optimal or heuristic algorithm (available in the algorithm base) is selected or built.

2.4 KNOWLEDGE ACQUISITION

The first step in building a knowledge-based system is knowledge acquisition. Knowledge about the problem domain could be acquired from many sources, such as textbooks, reports, the study of published literature, and experts in the domain. In addition, humans generally posses some specific knowledge, for example, rule-of-thumbs and heuristics.

To build a knowledge base, an analyst (called also a knowledge engineer) has to acquire knowledge and incorporate it into the system. In fact, the knowledge acquisition is the transfer and transformation of problem-solving expertise from some knowledge sources to a program. A typical knowledge acquisition process for building a knowledge base is depicted in Figure 2.11.

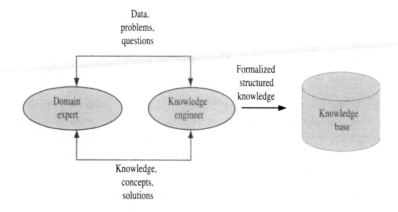

Figure 2.11. Typical knowledge acquisition process.

Different methods are used for knowledge acquisition. For example, questionnaires belong to the most widely used knowledge acquisition methods for rule-based systems. A traditional method of knowledge acquisition is *protocol analysis* (Hayes, 1988). Verbal protocol analysis has widely been used by cognitive psychologists. It can also be used for development of knowledge-based systems.

A protocol is simply a record of information, and protocol analysis is the process of taking a detailed record of an action and studying the behavior through analysis of that record. Typically the term *protocol* implies that the expert is solving a whole problem or problem segment using the approaches and tools that ordinarily would be used.

In other words, a protocol is a trace of the expert's thought processes behind the problem solving. The protocol shows the avenues that were explored and alternatives considered. Protocol analysis is the study of these mental footprints and the attempt to construct models of thinking from the paths that were taken during problem solving. This kind of protocol analysis is called a "think-aloud" protocol.

Autonomous knowledge acquisition is discussed in Chapters 11 and 17.

2.5 KNOWLEDGE CONSISTENCY

Production rules are a frequently used form of knowledge representation. Elicitation of production rules is a difficult step in building a knowledge-based system. Errors associated with knowledge elicitation result in various inconsistencies and redundancies among production rules. This is especially likely to happen while developing large knowledge bases, in particular when the knowledge is elicited from multiple sources. Some knowledge bases may include hundreds or thousands of production rules. As an example of a large-scale system, the XCON configuration system developed at Digital Equipment Corporation can be considered. It has taken almost 10 years to develop XCON and the related subsystems used for configuring hardware and software (Barker and O'Connor, 1989).

Consider a rule base that consists of a number of production rules R_i, each in the form

$$\text{IF } C(R_i) \text{ THEN } A(R_i)$$

where

$C(R_i)$ = condition clause of a production rule
$A(R_i)$ = action part of a production rule

Using example production rule R_1

IF machine M_1 is available

AND tool t_4 is loaded in tool magazine T_3

THEN release part p_6 for processing on machine M_1

where

$C(R_1)$ = machine M_1 is available AND tool t_4 is loaded in tool magazine T_3

$A(R_1)$ = release part p_6 for processing on machine M_1.

Researchers [see references in Liebowitz and DeSalvo (1989) and Gupta (1991)] have identified a number of anomalies among production rules that can be divided into two categories:

- Static anomalies that can be detected without inferencing rules
- Inference (dynamic) anomalies that are identified during the inference process

Some of the most commonly discussed types of static anomalies are [see Liebowitz and DeSalvo (1989) and Gupta (1991)]:

- *Type 1 (Potential Conflict).* Two rules with different conditions result in identical actions: for $C(R_i) \neq C(R_j)$, $A(R_i) = A(R_j)$.
- *Type 2 (Potential Conflict).* Two rules with identical conditions result in different actions: for $C(R_i) = C(R_j)$, $A(R_i) \neq A(R_j)$.
- *Type 3 (Redundancy).* Two rules with identical conditions result in identical actions: for $C(R_i) = C(R_j)$, $A(R_i) = A(R_j)$.
- *Type 4 (Subsumption).* Two rules have identical conditions, but one contains additional elements in the action clause: for $C(R_i) = C(R_j)$, $A(R_i) \subset A(R_j)$.

Four of the most frequently discussed inference (dynamic) anomalies are (Gupta, 1991):

- *Type 5 (Cycle).* Set of production rules forms a cycle: $A(R_1) \subseteq C(R_2) \subseteq A(R_3). \ldots A(R_n) \subseteq C(R_1)$
- *Type 6 (Unreachable Action).* If in the backward inferencing the action of a rule is neither a part of the possible query nor a part of the condition of another rule, then the rule is unreachable.
- *Type 7 (Dead-End-Query).* If in the backward inferencing the query does not match an action of one of the rules in the rule base, then the query is dead end.
- *Type 8 (Dead-End-Condition).* To satisfy the condition of a rule in the backward inferencing, either the condition must be askable or the condition must be matched by a conclusion of one of the rules in the rule base.

In the next two sections the first three types of anomalies are considered:

- Type 1 (potential conflict; different conditions but identical actions)
- Type 2 (potential conflict; identical conditions but different actions)
- Type 3 (redundancy; identical conditions and identical actions)

It should be emphasized that production rules falling into the first and second types are only suspected of inconsistency. For example, it is possible to have two production rules that have different conditions and the same actions without being in conflict.

To detect the three types of anomalies, two cases of production rules are considered:

- *Case 1.* Production rules with a simple (one-element) action clause
- *Case 2.* Production rules with a compound (more than one element) action clause.

Using the logic embedded in programming languages, a production rule with the compound action clause can be transformed into a production rule with a simple action clause, as illustrated in Examples 2.3–2.5.

Example 2.3. The production rule

IF C_1 OR (C_2 AND C_3) OR (C_4 AND C_5) THEN A_1 ELSE A_2

can be replaced by two production rules with simple action clauses:

IF C_1 OR (C_2 AND C_3) OR (C_4 AND C_5) THEN A_1

and

IF NOT C_1 AND (NOT C_2 OR NOT C_3) AND (NOT C_4 OR NOT C_5) THEN A_2

Example 2.4. Consider the following production rule with the nested IF structure:

IF C_1 THEN
    IF C_2 THEN
        IF C_3
            THEN A_1
        ELSE A_2
      ELSE A_3
   ELSE A_4

This can be rewritten as four production rules:

IF C_1 AND C_2 AND C_3 THEN A_1
IF C_1 AND C_2 AND NOT C_3 THEN A_2
IF C_1 AND NOT C_2 THEN A_3
IF NOT C_1 THEN A_4

Example 2.5. The following production rule has a conjunctive consequent:

IF C_1 AND C_2 AND C_3 THEN C_4 AND C_5

It can be rewritten as two production rules with simple action clauses:

IF C_1 AND C_2 AND C_3 THEN C_4

IF C_1 AND C_2 AND C_3 THEN C_5

For other examples illustrating the transformation of production rules refer to Pedersen (1989) and Richards (1989).

2.5.1 Detection of Anomaly Rules with Simple Action Clauses

In this section, production rules with simple action clauses (i.e., one element in the action clause) are discussed. Consider six production rules $R_1, \ldots, R_6$ with the condition and action clauses represented in Figure 2.12 as a bipartite graph (Gupta, 1991). Rather than using a bipartite graph to detect inconsistent or redundant production rules, a condition–action clause incidence matrix can be used (Gupta, 1991). The latter representation is more suitable for computer applications, especially for large-scale knowledge bases. The bipartite graph in Figure 2.12 is represented with the incidence matrix (2.1), where the incidence of a condition–action clause is indicated by an asterisk:

$$
\begin{array}{c}
\text{Action clause} \\[4pt]
\begin{array}{cccc}
 & A_1 & A_2 & A_3 \\
\begin{array}{l}
R_{1:}\ C_1 \text{ AND } D_1 \\
R_2:\ C_2 \text{ AND } D_2 \\
R_3:\ C_3 \text{ AND } D_3 \\
R_4:\ C_4 \text{ AND } D_1 \\
R_5:\ C_2 \text{ AND } D_2 \\
R_6:\ C_3 \text{ OR } D_3
\end{array}
&
\left[\begin{array}{ccc}
 & & * \\
* & & \\
 & * & \\
 & & * \\
* & & \\
 & * &
\end{array}\right]
\end{array}
\end{array}
\qquad \begin{array}{l}\text{Condition}\\\text{clause}\end{array} \qquad (2.1)
$$

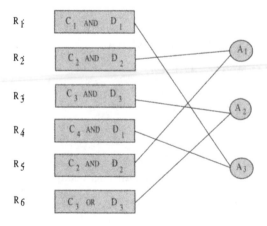

Figure 2.12. Bipartite graph representing six production rules.

Applying the cluster identification algorithm presented in the next section to matrix (2.1) results in matrix (2.2), which decomposes into three mutually separable submatrices. Each of the three submatrices (vectors) corresponds to two production rules. The anomalies discussed in Figure 2.12 are more clearly visible in matrix (2.2), which is partitioned into three mutually separable submatrices:

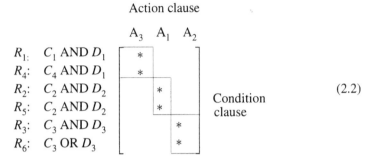

$$(2.2)$$

The bipartite graph corresponding to matrix (2.2) is shown in Figure 2.13. Note that each separate graph in Figure 2.13 corresponds to a cluster in matrix (2.2). One can conclude from Figure 2.13 that production rules R_2 and R_5 are identical because they share the same action clause A_1, and their condition clauses are identical. Production rules R_3 and R_6 have identical action clause A_2 but their condition clauses differ. The two rules are suspected of type 1 anomaly. The rules suspected of inconsistency have to be examined by the knowledge engineer. Rules R_1 and R_4 are also suspected of type 1 inconsistency.

The clustering problem for a general incidence matrix is NP complete (Lenstra, 1974). Matrix (2.1) has a special structure, namely each row includes only one asterisk. The latter is due to the fact that each production rule has only one condition

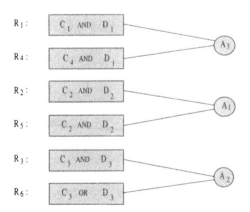

Figure 2.13. Bipartite graph representing matrix (2.2).

and one simple action that are incident. This property of the condition–action clause incidence matrix guarantees its decomposition into mutually separable submatrices.

The advantages the CI algorithm used to transform matrix (2.1) into (2.2) are its simplicity and low computational time complexity of order (mn), where m is the number of rows and n the number of columns in the incidence matrix. The CI algorithm is briefly presented below.

CI Algorithm

Step 0. Set iteration number $k = 1$.

Step 1. Select any row i of incidence matrix $[a_{ij}]^{(k)}$ and draw horizontal line h_i through it ($[a_{ij}]^{(k)}$ reads "matrix $[a_{ij}]$ at iteration k").

Step 2. For each asterisk crossed by a horizontal line h_i, draw a vertical line v_j.

Step 3. For each asterisk crossed once by a vertical line v_j, draw a horizontal line h_k.

Step 4. Repeat steps 2 and 3 until there are no more crossed-once asterisks in $[a_{ij}]^{(k)}$. All crossed-twice asterisks in $[a_{ij}]^{(k)}$ form condition clause group CC-k and action clause group AC-k.

Step 5. Transform the incidence matrix $[a_{ij}]^{(k)}$ into $[a_{ij}]^{(k+1)}$ by removing rows and columns corresponding to all the horizontal and vertical lines drawn in steps 1–4.

Step 6. If matrix $[a_{ij}]^{(k+1)} = \mathbf{0}$ (where $\mathbf{0}$ denotes a matrix with all elements equal to zero), stop; otherwise set $k = k + 1$ and go to step 1.

The CI algorithm is illustrated in Example 2.6.

Example 2.6. Consider the incidence matrix (2.1) with rows and columns numbered as follows:

$$
\begin{array}{c}
\text{Action clause} \\
\begin{array}{ccc} 1 & 2 & 3 \end{array} \\
\begin{array}{c} 1 \\ 2 \\ 3 \\ 4 \\ 5 \\ 6 \end{array}
\left[\begin{array}{ccc}
 & & * \\
* & & \\
 & * & \\
 & & * \\
* & & \\
 & * & \\
\end{array}\right]
\begin{array}{l} \\ \\ \text{Condition} \\ \text{clause} \\ \\ \end{array}
\end{array}
\qquad (2.3)
$$

In Step 1 of the CI algorithm row 1 is selected and horizontal line h_1 is drawn [see matrix (2.4)]. In Step 2 vertical line v_3 is drawn through the crossed-once asterisk as shown in the matrix

$$
\begin{array}{c}
\begin{array}{ccc} 1 & 2 & 3 \end{array} \\
\begin{array}{c} 1- \\ 2 \\ 3 \\ 4 \\ 5 \\ 6 \end{array}
\left[
\begin{array}{ccc}
\text{---} & \text{---} & * \\
* & & \\
& * & \\
& & * \\
* & & \\
& * &
\end{array}
\right]
\begin{array}{l} -\ h_1 \end{array} \\
\hspace{3.5em} v_3
\end{array}
\tag{2.4}
$$

In step 4 horizontal line h_4 is drawn as

$$
\begin{array}{c}
\begin{array}{ccc} 1 & 2 & 3 \end{array} \\
\begin{array}{c} 1- \\ 2 \\ 3 \\ 4- \\ 5 \\ 6 \end{array}
\left[
\begin{array}{ccc}
\text{---} & \text{---} & * \\
* & & \\
& * & \\
\text{---} & \text{---} & * \\
* & & \\
& * &
\end{array}
\right]
\begin{array}{l} -\ h_1 \\ \\ \\ -\ h_4 \end{array} \\
\hspace{3.5em} v_3
\end{array}
\tag{2.5}
$$

The crossed-twice entries in matrix (2.5) indicate:

- Condition clause group with conditions 1 and 4, CC-1 = $\{1, 4\}$
- Action clause group with action 1 only, AC-1 = $\{3\}$

Since there are no more crossed-twice asterisks left, in Step 5 matrix (2.5) is transformed into

$$
\begin{array}{c}
\begin{array}{cc} 1 & 2 \end{array} \\
\begin{array}{c} 2 \\ 3 \\ 5 \\ 6 \end{array}
\left[
\begin{array}{cc}
* & \\
& * \\
* & \\
& *
\end{array}
\right]
\end{array}
\tag{2.6}
$$

In the second and third iteration of the CI algorithm the five steps are repeated. After three iterations a matrix identical to matrix (2.2) with three condition clause groups and three action clause groups is generated.

The grouping performed for the condition–action clause incidence matrix can obviously be done for the action–condition clause incidence matrix. As previously discussed, production rules with compound condition clauses can be transformed into rules with simple condition clauses (see Example 2.7).

Example 2.7. The production rule with the disjunctive antecedent

IF C_1 OR C_2 THEN A_1 AND A_2 AND A_3

can be rewritten into the following two rules:

IF C_1 THEN A_1 AND A_2 AND A_3

IF C_2 THEN A_1 AND A_2 AND A_3

In fact, the rewriting of production rules can be automated. To transform the production rule in Example 2.7, the following rewrite rule is used:

IF   literals in the condition clause of a rule are connected by OR connectors

THEN rewrite the rule by separating it into simple rules that do not include OR connectors

2.5.2 Grouping Rules with Compound Condition and Action Clauses

The material presented in the previous section applies to the case when all production rules either are in the form of simple action (or condition) clauses or can be transformed into such a form. In this section, the anomalies are detected among production rules with compound action and condition clauses; that is, each action and condition contains more than one element linked by an AND and/or OR connector. To represent production rules, an incidence matrix is used with each entry defined as follows:

$$a_{ij} = \begin{cases} * & \text{for the first element in clause } j \text{ of rule } i \\ \text{AND}^2 & \text{for the second AND element in clause } j \text{ of rule } i \\ \text{OR}^2 & \text{for the second OR element in clause } j \text{ of rule } i \\ \text{AND}^3 & \text{for the third AND element in clause } j \text{ of rule } i \\ \text{OR}^3 & \text{for the third OR element in clause } j \text{ of rule } i \\ \bullet & \\ \bullet & \\ \bullet & \\ \text{AND}^k & \text{for the } k\text{th AND element in clause } j \text{ of rule } i \\ \text{OR}^k & \text{for the } k\text{th OR element in clause } j \text{ of rule } i \end{cases}$$

Using the above definition of a_{ij}, the six production rules

R_1: IF C_1 AND D_1 THEN A_5 AND A_1

R_2: IF C_2 AND D_2 THEN A_3 AND A_7 AND A_6

R_3: IF C_3 AND D_3 THEN A_2 AND A_4

R_4: IF C_4 AND D_1 THEN A_5 AND A_1

R_5: IF E_2 OR F_2 THEN A_7 AND A_6

R_6: IF C_3 AND D_3 THEN A_2 OR A_4

are represented in the condition–action clause incidence matrix

$$
\begin{array}{l}
\quad\quad\quad\quad\quad\; A_1 \quad A_2 \; A_3 \; A_4 \quad A_5 \quad A_6 \quad A_7 \\
\begin{array}{ll}
R_1: & C_1 \text{ AND } D_1 \\
R_2: & C_2 \text{ AND } D_2 \\
R_3: & C_3 \text{ AND } D_3 \\
R_4: & C_4 \text{ AND } D_1 \\
R_5: & E_2 \text{ OR } F_2 \\
R_6: & C_3 \text{ AND } D_3
\end{array}
\left[
\begin{array}{ccccccc}
\text{AND}^2 & & & & * & & \\
& & * & & & \text{AND}^3 & \text{AND}^2 \\
& & * & \text{AND}^2 & & & \\
\text{AND}^2 & & & & * & & \\
& & & & & \text{AND}^2 & * \\
& * & & \text{OR}^2 & & &
\end{array}
\right]
\end{array}
\quad (2.7)
$$

Applying the CI algorithm to matrix (2.7) results in the matrix

$$
\begin{array}{l}
\quad\quad\quad\quad\quad\; A_4 \quad A_2 \; A_5 \; A_1 \quad A_6 \quad A_3 \quad A_7 \\
\begin{array}{ll}
R_3: & C_3 \text{ AND } D_3 \\
R_6: & C_3 \text{ AND } D_3 \\
R_1: & C_1 \text{ AND } D_1 \\
R_4: & C_4 \text{ AND } D_1 \\
R_2: & C_2 \text{ AND } D_2 \\
R_5: & E_2 \text{ OR } F_2
\end{array}
\left[
\begin{array}{ccccccc}
\text{AND}^2 & * & & & & & \\
\text{OR}^2 & * & & & & & \\
& & * & \text{AND}^2 & & & \\
& & * & \text{AND}^2 & & & \\
& & & & \text{AND}^3 & * & \text{AND}^2 \\
& & & & \text{AND}^2 & & *
\end{array}
\right]
\end{array}
\quad (2.8)
$$

From the diagonally structured matrix (2.8) the following conclusions can be drawn:

- Production rules R_3 and R_6 are suspected of type 2 anomaly.
- Production rules R_1 and R_4 are suspected of type 1 anomaly.
- Production rules R_2 and R_5 show some similarity in the action clause, while their condition clauses differ.

Of course, rather than considering the condition–action clause incidence matrix, the grouping can be performed for the action–condition clause matrix.

The CI algorithm produces an ideal (with zero elements outside diagonal blocks) decomposition of an incidence matrix with the block diagonal structure embedded. To group the production rules when an ideal decomposition of an incidence matrix does not occur, the CI algorithm cannot be used. The extended CI algorithm presented in Kusiak (1990) could be used. Rather than the extension of the CI algorithm, an alternative approach using a similarity measure is presented. The disadvantage of the similarity-based approach is that it considers only two rules at a time, while the clustering approach incorporates the entire knowledge base at the same time. The similarity measure approach allows detecting different types of static anomalies.

Consider the two logical clauses

$$
e_i = P_{i1} \text{ AND } P_{i2} \text{ AND } \ldots \text{ AND } P_{if} \text{ AND } P_{i,f+1} \text{ AND } \ldots \text{ AND } P_{in'}
$$

$$
e_j = Q_{j1} \text{ AND } Q_{j2} \text{ AND } \ldots \text{ AND } Q_{jg} \text{ OR } Q_{j,g+1} \text{ OR } \ldots \text{ OR } Q_{jn''}
$$

presented in a disjunctive form. Note that in the disjunctive form a clause, for example, C_1 AND C_2 OR C_3 means $(C_1$ AND $C_2)$ OR C_3. While in the conjunctive form, the same clause means C_1 AND $(C_2$ OR $C_3)$. Arranging the elements in the second clause so that they correspond to the elements of the first clause gives

$$[e_j] = Q_{j[1]} \text{ AND } Q_{j[2]} \text{ AND } \ldots \text{ AND } Q_{j[g]} \text{ OR } Q_{j[g+1]} \text{ OR } \ldots \text{ OR } Q_{j[n'']}$$

For the two clauses a similarity measure is defined:

$$s(e_i, e_j) = \frac{\sum_{k=1}^{n} \delta(e_{ik}, e_{j[k]})}{n}$$

for

$$\delta(e_{ij}, e_{j[k]}) = \begin{cases} 1 & \text{for } P_{ik} = Q_{j[k]} \\ 1 & \text{for } op\_P_{ik} = op\_Q_{j[k]}, \ k = 2, \ldots, n \\ 0 & \text{otherwise} \end{cases}$$

where

n $= \max\{n', n''\}$

e_{ik} $= k$th element of clause e_i

$e_{j[k]}$ $= k$th element of clause $[e_j]$

op_P_{ik} = logical operator and element that follows that operator

The similarity measure is illustrated in Example 2.8.

Example 2.8.   Consider the following two logical clauses in a disjunctive form:

$$e_1 = A_1 \text{ AND } A_2 \text{ AND } A_3 \text{ OR } A_4 \text{ OR } A_5 \text{ OR } A_6 \text{ OR } A_7$$

$$e_2 = A_3 \text{ AND } A_1 \text{ AND } A_2 \text{ AND } A_5 \text{ OR } B_1 \text{ AND } A_8 \text{ AND } A_7$$

Ordering clause e_2 so that its elements correspond to the elements in e_1 results in

$$[e_2] = A_1 \text{ AND } A_2 \text{ AND } A_3 \text{ AND } A_5 \text{ OR } B_1 \text{ AND } A_7 \text{ AND } A_8$$

The similarity measure between the two clauses e_1 and e_2 is

$$s(e_1, e_2) = \frac{1+1+1+0+0+0+0}{7} = \frac{3}{7}$$

Each of the two clauses e_1 and e_2 may represent a condition or action clause of a production rule. In addition to AND and OR operators, negation operators could be considered.

The similarity measure defined above can be used to detect anomalies of types 1, 2, and 3 using the following three conditions:

TABLE 2.1. Solution of Example 2.9

	Similarity		
Pair of Rules	Conditions $s[C(R_i, R_j)]$	Actions $s[A(R_i, R_j)]$	Type of Anomaly
R_1, R_2	$s[C(R_1, R_2)] = 1$	$s[A(R_1, R_2)] = 2/3$	Type 2
R_1, R_3	$s[C(R_1, R_3)] = 1/2$	$s[A(R_1, R_3)] = 1$	Type 1
R_1, R_4	$s[C(R_1, R_4)] = 1$	$s[A(R_1, R_4)] = 2/3$	Type 2
R_2, R_3	$s[C(R_2, R_3)] = 1/2$	$s[A(R_2, R_3)] = 2/3$	None
R_2, R_4	$s[C(R_2, R_4)] = 1$	$s[A(R_2, R_4)] = 1$	Type 3
R_3, R_4	$s[C(R_3, R_4)] = 1/2$	$s[A(R_3, R_4)] = 2/3$	None

Condition 1: If $s[C(R_i, R_j)] < 1$ and $s[A(R_i, R_j)] = 1$, then type 1 anomaly exists for production rules i and j.

Condition 2: If $s[C(R_i, R_j)] = 1$ and $s[A(R_i, R_j)] < 1$, then type 2 anomaly exists for production rules i and j.

Condition 3: If $s[C(R_i, R_j)] = s[A(R_i, R_j)] = 1$, then type 3 anomaly exists and one of the two production rules i and j is redundant.

The three conditions are illustrated in Example 2.9.

Example 2.9. Determine anomalies in the following rule base:

R_1: IF C_1 AND C_2 THEN A_1 AND A_2 AND A_3
R_2: IF C_2 AND C_1 THEN A_2 AND A_1
R_3: IF C_1 OR C_2 THEN A_1 AND A_2 AND A_3
R_4: IF C_1 AND C_2 THEN A_1 AND A_2

The similarities between production rules and the types of anomalies are presented in Table 2.1. Based on the types of anomalies reported in Table 2.1, the knowledge engineer examines the production rules and determines the source of a possible error. The only rules in Table 2.1 that do not appear to be inconsistent are rules R_2, R_3 and R_3, R_4. Since the similarity between their actions is rather high, $s[A(R_i, R_j)] = 2/3$, the knowledge engineer should also examine the two pairs of production rules.

The similarity measure can also be used to detect other types of anomalies.

2.5.3 Inference Anomalies in Rule Bases

The analysis of inference anomalies is illustrated with the case of anomaly of type 5 (a cycle). The inference anomalies can be analyzed using a graph or matrix representation.

Example 2.10. The following production rules are represented with two AND/OR graphs in Figure 2.14.

R_1: IF c_1 AND c_2 THEN a_1
R_2: IF c_3 AND c_4 OR c_5 THEN a_2

The following rule base with four simple rules is represented with two graphs in Figure 2.15:

R_1: IF c_1 OR c_2 THEN a_1
R_2: IF c_7 AND a_1 THEN a_2
R_3: IF c_3 AND c_4 OR c_5 THEN a_3
R_4: IF a_3 AND c_6 THEN a_4

Assume that the rule base was expanded by incorporating the following two production rules:

R_5: IF a_2 THEN c_4
R_6: IF a_4 AND c_6 THEN a_3

The expanded rule base is illustrated in Figure 2.16. For computational purposes, it is convenient to represent the four rules in Figure 2.15 in the adjacency matrix $[b_{ij}]$ (Gupta, 1991), where an entry

$$b_{ij} = \begin{cases} + & \text{if } A(R_i) \subseteq C(R_j) \\ 0 & \text{no relationship between rules } R_i \text{ and } R_j \text{ exists} \end{cases}$$

The adjacency matrix for the rules $R_1, \ldots, R_4$ in Figure 2.15 is shown next:

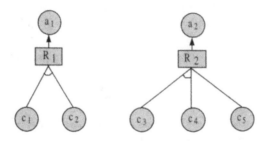

Figure 2.14. Representation of production rules with AND/OR graphs.

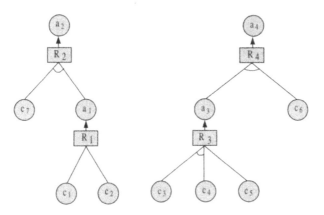

Figure 2.15. Graph representation of the knowledge base with four production rules.

$$
\begin{array}{c}
\quad\quad R_1 \quad R_2 \quad R_3 \quad R_4 \\
\begin{array}{c}
R_1 \\ R_2 \\ R_3 \\ R_4
\end{array}
\left[
\begin{array}{cccc}
+ & & & \\
+ & + & & \\
& & + & \\
& & + & +
\end{array}
\right]
\end{array}
\tag{2.9}
$$

Note that in matrix (2.9), the entries (R_2, R_1) and (R_4, R_3) are denoted with a +, because $A(R_1) \subseteq C(R_2)$ and $A(R_3) \subseteq C(R_4)$, respectively. For the convenience of interpretation of the adjacency matrix (2.9), each diagonal entry has been denoted with a +. The matrix (2.9) has a lower diagonal structure, or more precisely it includes two triangular submatrices, each corresponding to the graph in Figure 2.15.

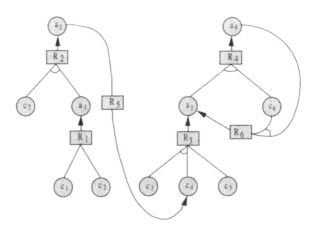

Figure 2.16. Graph representation of the knowledge base with six production rules.

Had the four production rules been converted into the matrix (2.9) in a different order, the triangular form might not have been visible. However, whenever a triangular form is embedded into a matrix, it can be retrieved by the triangularization algorithm presented in Kusiak (1999) and http://www.icaen.uiowa.edu/~ankusiak/process-model.html.

The graph in Figure 2.16 is represented with the following adjacency matrix:

$$
\begin{array}{c}
 \\
R_1 \\ R_2 \\ R_3 \\ R_4 \\ R_5 \\ R_6
\end{array}
\begin{array}{cccccc}
R_1 & R_2 & R_3 & R_4 & R_5 & R_6 \\
\left[\begin{array}{cccccc}
+ & & & & & \\
+ & + & & & & \\
 & & + & & + & \\
 & & + & + & & + \\
 & + & & & + & \\
 & & + & & & +
\end{array}\right]
\end{array}
\tag{2.10}
$$

Applying an existing triangularization algorithm to matrix (2.10) results in

$$
\begin{array}{c}
 \\
R_1 \\ R_2 \\ R_5 \\ R_3 \\ R_4 \\ R_6
\end{array}
\begin{array}{cccccc}
R_1 & R_2 & R_5 & R_3 & R_4 & R_6 \\
\left[\begin{array}{cccccc}
+ & & & & & \\
+ & + & & & & \\
 & + & + & & & \\
 & & + & + & & \\
 & & & + & + & + \\
 & & & + & & +
\end{array}\right]
\end{array}
\tag{2.11}
$$

Matrix (2.11) provides a number of interesting insights, namely:

1. The order in which the production rules can be fired is R_1, R_2, R_5, R_3, R_4.
2. The element (R_4, R_6) in the upper diagonal indicates that there exists a cycle between rules R_4 and R_6.

To cope with any anomaly, one has to carefully reexamine the rules involved in a conflict. In the cycle in Figure 2.16 [and matrix (2.11)], the interfaces of the two rules R_4 and R_6 involved in the cycle have to be considered first, that is, a_4 and a_3. It might be possible that a_4 is not an outcome of rule R_4 or that rule R_6 is not using a_4 as a fact; rather, it is based on another fact that has been confused with a_4. Similar reasoning could be applied to a_3. If the conflict could not be resolved by the analysis of local interfaces, one would have to check the validity of other production rules in the conflict neighborhood, for example, rule R_3 that might be generating an outcome different than a_3.

The two inference anomalies discussed above were concerned with production rules with simple action clauses, that is, each including only one element. All the preceding considerations apply to production rules with complex action and condition

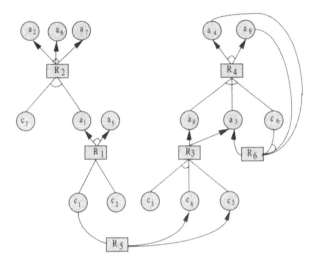

Figure 2.17. Graph representation of six production rules with complex clauses.

clauses, that is, each clause including more than one element. The case with complex action classes is illustrated in Example 2.11.

Example 2.11. Consider a knowledge base including the following six production rules:

R_1: IF c_1 OR c_2 THEN a_1 AND a_5
R_2: IF c_7 AND a_1 THEN a_2 AND a_6 AND a_7
R_3: IF $(c_3$ AND $c_4)$ OR c_5 THEN a_3 OR a_8
R_4: IF a_3 AND a_8 AND c_6 THEN a_4 AND a_9
R_5: IF c_1 THEN c_4 OR c_5
R_6: IF a_4 OR $(a_9$ AND $c_6)$ THEN a_3

In fact, the above rule base has been obtained from rule base illustrated in Figure 2.16 by changing in each rule a simple action clause with a complex action clause. The six production rules with complex action clauses are presented in Figure 2.17. The adjacency matrix representing the graph in Figure 2.17 is identical with matrices (2.10) and (2.11).

2.6 SUMMARY

In this chapter, the fundamentals of knowledge-based systems were presented. Different inference strategies were discussed. One of the most important issues in designing knowledge-based systems is to ensure that the production rules are consistent and nonredundant. Efficient methodologies for detecting static and inference (dynamic) anomalies were discussed. Two approaches for detecting static anomalies among production rules were developed. The first approach is concerned with production rules containing only one element in the action or condition clause. The second approach allows for detection of anomalies among production rules with many elements of the action and condition clauses. The inference anomalies were analyzed with AND/OR graphs and incidence matrices. The concepts presented were illustrated with examples.

REFERENCES

Barker, V. E., and T. O'Connor (1989), Expert systems for configuration at Digital: XCON and beyond, *Communications of the ACM*, Vol. 32, No. 3, pp. 298–317.

Durkin, J. (1994), *Expert Systems: Design and Development*, Prentice-Hall, Englewood Cliffs, NJ.

Gupta, U., Ed. (1991), *Validating and Verifying of Knowledge-Based Systems*, IEEE Computer Society Press, Los Alamitos, CA.

Hayes, C. (1988), Automated process planning system for prismatic parts, in A. Kusiak (Ed.), *Expert Systems: Strategies and Solutions in Manufacturing Design and Planning*, Society of Manufacturing Engineers, Dearborn, MI, pp. 151–183.

Kusiak, A. (1990), *Intelligent Manufacturing Systems*, Prentice-Hall, Englewood Cliffs, NJ.

Kusiak, A. (1999), *Engineering Design: Products, Processes, and Systems*, Academic Press, San Diego, CA.

Lenstra, J. K. (1974), Clustering a data array and the traveling salesman problem, *Operations Research*, Vol. 22, pp. 413–414.

Liebowitz, J., and D. A. DeSalvo, Eds. (1989), *Structuring Expert Systems: Domain, Design, and Development*, Prentice-Hall, Englewood Cliffs, NJ.

Minsky, M. (1975), A framework for representing knowledge, in P. H. Winston (Ed.), *The Psychology of Computer Vision*, McGraw-Hill, New York, pp. 211–227.

Pedersen, K. (1989), *Expert Systems Programming: Practical Techniques for Rule-Based Systems*, John Wiley, New York.

Pham, D. T., and P. T. N. Pham (1988), Expert systems: a review, in D. T. Pham (Ed.), *Expert Systems in Engineering*, IFS Publications, Kempston, United Kingdom.

Rich, E. (1983), *Artificial Intelligence*, McGraw-Hill, New York.

Richards, T. (1989), *Clausal Form Logic: An Introduction to the Logic of Computer Reasoning*, Addison-Wesley, Readings, MA.

Waterman, D. A. (1986), *A Guide to Expert Systems*, Addison-Wesley, Reading, MA.

QUESTIONS

2.1. What is a knowledge-based system?

2.2. What are the basic components of a knowledge based-system?

2.3. What are the basic schemes for representing knowledge? Give an example of each scheme.

2.4. What are the two most widely used reasoning (inferencing) strategies? When should they be used?

2.5. How is the uncertainty handled in knowledge-based systems?

2.6. Is a reasoning strategy a search strategy?

2.7. What types of knowledge-based systems do you know?

2.8. What is protocol analysis?

2.9. What are static and dynamic knowledge anomalies?

2.10. What methods are used to detect static and dynamic knowledge anomalies?

PROBLEMS

2.1. Act as an expert in machine tool products and a user of an expert system for selection of manufacturing resources (machines, tools, fixtures, and so on) that you want to develop. Elicit not less than four production rules used in the selection of machine resources:

(a) List the production rules.

(b) Draw an AND/OR tree.

(c) List the sequence of firing the production rules using the forward-chaining inference strategy.

2.2. Write a computer code (in any programming language) using backward-chaining reasoning and run it for the knowledge base that you have generated in Problem 2.1. Attach the source code.

2.3. Think of a scenario where an expert system could make decisions, for example, selecting a computer system for specific needs, predicting the length of a cycle time, and estimating the level of in-process inventory. Elicit production rules by talking to your partner or interview an expert in the area selected. The knowledge base should have at least six different production rules.

(a) Draw an AND/OR tree.

(b) Show two examples, one for forward reasoning and other for backward reasoning.

2.4. Write a computer code of a knowledge-based system with a forward-reasoning engine and run it for the knowledge base that you have generated in Problem 2.3. Attempt to design a user-friendly interface. Submit the source and execution codes and specify the programming environment.

2.5. Given the following five production rules:

R_1: IF A_2 OR C_1 THEN A_5

R_2: IF (C_8 AND C_4) OR C_5 THEN A_2

R_3: IF A_5 AND D_7 THEN A_4

R_4: IF C_9 THEN C_8

R_5: IF A_4 AND A_2 THEN A_6

(a) Draw an inference AND/OR tree

(b) Can the following two hypotheses be proven?

IF C_4 AND C_5 THEN A_6

IF D_7 AND C_9 AND C_4 THEN A_6

(c) Knowing certainty factors $CF(C_1) = .8$, $CF(A_2) = 1.0$, $CF(R_1) = .9$, and $CF(D_7) = .75$, compute $CF(A_4)$.

2.6. You will need to obtain a copy of an expert system shell (e.g., CLIPS, which is a public domain software). Use the expert system shell to develop an expert system for grading course work. The knowledge base should contain the following domain information:

- Grading scale:

  $90 \leq \%$ of total points ≤ 100 A

  $80 \leq \%$ of total points < 90 B

  $70 \leq \%$ of total points < 80 C

  $60 \leq \%$ of total points < 70 D

  $\%$ of total points < 60 F

- If the percentage of total points is greater than 100 or negative, an error has been made in grading.
- If the student's work is incomplete, a grade I is assigned.

The system must prompt the user to enter the percentage of total points (or incomplete) and assign the appropriate grade. If the user response is not consistent with the information in the knowledge base, the system must notify the user of the error and prompt him or her for another response. Other than these few requirements, you are free to design the expert system so that it is user friendly and accomplishes its objective.

Submit the following:

(a) A printout of the rules in the knowledge base

(b) A printout of sample runs that illustrate all possible recommendations and any safeguards built into the system

(c) A few sentences describing an interesting project that would involve an expert system

2.7. Given the following four production rules:

R_1: IF A_2 OR C_1 THEN A_5

R_2: IF $(C_8$ AND $C_4)$ OR C_5 THEN A_2

$R3$: IF A_5 AND D_7 THEN A_4

R_4: IF C_9 THEN C_8

R_5: IF A_3 OR C_6 THEN $A2$

(a) Draw an inference AND/OR tree for the above production rules.

(b) Can the following two hypotheses be proven?

IF C_6 AND C_1 THEN A_4

IF D_7 AND C_9 AND C_4 THEN A_4

2.8. For the AND/OR tree in Figure 2.18:

(a) List the minimum number of facts allowing one to derive subgoal H.

(b) Determine the minimum and maximum certainty factor of the outcome H, given the certainty factor of each condition branch of the AND/OR tree, $CF = .90$.

(c) Imagine that you are developing a computer system for forecasting the sales of new products. Would you recommend an expert system or a neural network approach? Justify the approach recommended.

2.9. Find the maximum certainty factor of obtaining the goal C for the following four production rules:

R_1: IF G AND H THEN D

R_2: IF A AND B THEN C

R_3: IF D AND E OR F THEN B

R_4: IF K AND L THEN H

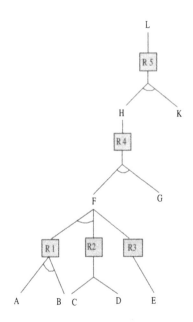

Figure 2.18. AND/OR inference tree.

Certainty factors for the elements in the conditions of the production rules are $c_A = .9$, $c_E = .8$, $c_F = .7$, $c_G = .9$, $c_K = .8$, and $c_L = .9$. The remaining certainty factors are 1.

2.10. You have been contracted by Gateway 2000 to design and develop an expert system to support customer service. The expert system will help customers select the system that matches their computing needs by asking several questions about the desired configuration. Table 2.2 lists the available models for the targeted user environments. The user should be asked to specify his or her user environment and price range. The expert system should then recommend an appropriate model and list the price. The system should contain safeguards for invalid input and conflicting requests (i.e., a multimedia machine that costs less than $2000). Other than these few requirements, you are free to design the expert system so that it is user friendly and accomplishes its objective. Submit the following:

(a)   A printout of the rules included in the knowledge base

(b)   A printout of at least four runs that illustrate possible recommendations and safeguards built into the system

(c)   A few sentences describing an interesting project suitable for an expert system application

(d)   A disk with the expert system file

2.11. Question yourself (or a partner) about diagnosing a broken car engine. Write down three to six production (IF–THEN) rules that could be incorporated into an expert system for engine failure diagnosis. Draw an AND/OR inference tree.

2.12. The expert system used to signal flow alarms for a two-column chemical process includes the following production rules:

Column 1:

R_1: IF temperature $T_1 = 1080$ AND pressure in the supply line $P_1 = 75$
     THEN column 1 output $V_1 = 174.0$

TABLE 2.2. Available Computer Models

User Environment	Model	Price
Home	P5-60 family PC	$2099
Home	P5-75 family PC	$2499
Home	P5-100 family PC	$2999
Business	P5-60	$1999
Business	P5-75	$2399
Business	P5-100	$2899
Multimedia	P5-100XL	$3699

Column 2:

R_2: IF temperature $T_2 = 940$ AND pressure in the supply line $P_2 = 84$
THEN column 2 output $V_2 = 135.2$
R_3: IF $V_1 + V_2 = 309.2 \pm 10$
THEN the process is normal

Express the above knowledge base in the form of fuzzy rules. Note that the following relationship holds:

$$\frac{PV}{T} = \text{const}$$

2.13. Given the following five production rules:

R_1: IF temperature is high
THEN pressure is high
R_2: IF pressure is high
AND fluid level is high
THEN status is dangerous
R_3: IF indicator is on
THEN temperature is high

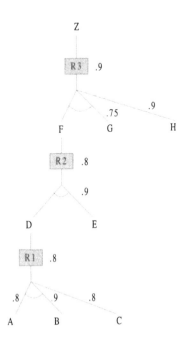

Figure 2.19. AND/OR tree.

R_4: IF status is dangerous

    THEN relay is on

R_5: IF relay is on

    AND standby unit is off

    THEN indicator is on

(a) Draw an AND/OR inference tree.

(b) List the anomalies, if any.

(c) Indicate a possible source of inconsistency and modify appropriate production rule(s) to eliminate the anomaly.

2.14. Given the certainty factors at the AND/OR tree in Figure 2.19:

(a) Determine the maximum certainty factor of attaining the goal Z.

(b) What would be the implication of adding another rule R_4—IF G AND H THEN E—to the inference tree.

(c) What changes to the inference tree would you recommend to properly reason with the four-rule knowledge base.

CHAPTER 3

FEATURES IN DESIGN AND MANUFACTURING

3.1 INTRODUCTION

The product life-cycle cost and performance are largely impacted by the decision made at its design stage (Feng et al., 1996). Efforts have been made to better understand the product development process. Design requirements and functions, a representation of the interaction between requirements and functions, and optimization in the functional space were presented in Kusiak and Szczerbicki (1992). Welch and Dixon (1992) discussed the representation of design information for a class of conceptual design problems. A conceptual design problem was represented as a set of functional parameters and dependencies among them. Ishii et al. (1988) used design compatibility analysis (DCA) to evaluate a candidate design from multiple viewpoints. Kannapan and Marshek (1990) applied algebraic and predicate logic for representation and reasoning in the design of machines.

In this chapter features in detail design of mechanical components are considered.

The activities at the detail design stage of a component may include (see Figure 3.1):

- Requirement–function transformation (transformation from component requirements into feature-related functions)
- Function–feature transformation
- Feature selection
- Feature compatibility analysis
- Tolerance design
- Feature–process transformation

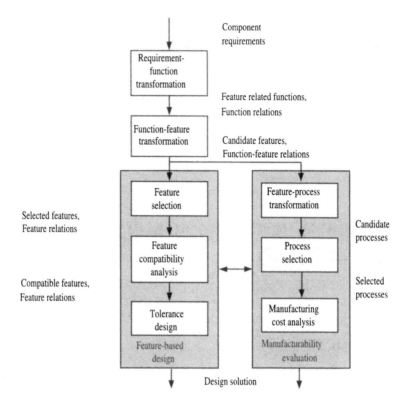

Figure 3.1. Typical activities in detail design of components.

- Process selection
- Manufacturing cost analysis

The component requirements are transformed into feature-related functions. A feature-related function is what a feature should perform. There are certain relationships among the feature-related functions of a component, which in this chapter are referred to as function relations. The function–feature transformation leads to the candidate features. This transformation should be based on the general function–feature relations. Other activities such as feature selection, feature compatibility analysis, and tolerance design are also required for "feature design." Manufacturability evaluation should be included in the framework of detail design.

Rosen (1993) proposed a common set concept to represent relationships among geometric entities; however, a detailed investigation has not been conducted. Henderson and Taylor (1993) developed a framework and prototype system for metamodeling of products throughout the design stages of planning, conceptual design, embodiment design, and detail design. This chapter focuses on the representation of relations among features and functions to facilitate detail design.

3.2 FUNDAMENTALS OF REQUIREMENTS, FEATURES, AND FUNCTIONS

3.2.1 Requirement Space

Design requirements are "demands" and "wishes" that define the design task (Pahl and Beitz, 1988). The design requirements are formed at various levels, such as product requirements, assembly requirements, and component requirements. The component requirements can be considered as the refined product and assembly requirements. In this chapter, it is assumed that the component requirements are known before detail design of a component begins.

For mechanical design, it is difficult to provide an exhaustive list of all requirements and functions. Without loss of generality, it is possible to consider a limited requirement space. Design of rotational parts can be further divided into, for example, design of a shaft or a gear. This decomposition imposes limits on the requirement and function space.

The component requirements can be represented as a tree (Kusiak and Szczerbicki, 1992). Figure 3.2 shows the AND/OR requirement tree for a machine tool spindle.

3.2.2 Fundamentals of Features

Features represent the engineering meaning of the geometry of a part or assembly. The engineering meaning may involve, for example, the formalization of the functions the feature serves, or how it can be produced, or how the feature behaves in various situations (Shah, 1992).

Features are classified into form features (nominal geometry), material features (material composition and condition), precision features (allowable deviations from nominal geometry), and technological features (information related to part performance and operation) (Shah and Rogers, 1988).

A mechanical design can be viewed from different perspectives: design, manufacturing, and geometry. Xue and Dong (1993) used three feature sets—*design*

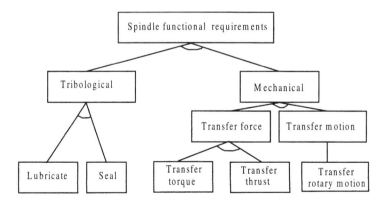

Figure 3.2. Requirement tree of a machine tool spindle.

features, manufacturing features, and *geometry features*—for modeling from three distinctive perspectives.

Features are application specific, and they support multiple views, for example, design, process planning, and assembly. Each application may have its own view of an object (Bronsvoort and Jansen, 1993). Transformation can be achieved between certain application-specific feature spaces (Shah, 1989).

Features can also be divided into primary and secondary. The primary features do not inherit any instance parameters from other features. The secondary features depend on their parents and are used to modify the shape of the part for the purpose of, for example, strengthening or smoothing the part (Chen et al., 1991). One secondary feature may have multiple parents.

The feature concept used in this chapter is in terms of geometry (form) and manufacturing.

3.2.3   Classification of Feature-Related Functions

A mechanical part includes a set of features. Although in most cases features are designed to satisfy requirements of a product or an assembly, some features are used for other purposes. This chapter defines three basic types of feature-related functions (see also Figure 3.3*a*):

- *Performance-Related Functions.* Functions corresponding to the product performance requirements.
- *Process-Related Functions.* Functions corresponding to the manufacturing process requirements.
- *Ergonomics-Related Functions.* Functions concerned with the ergonomic requirements.

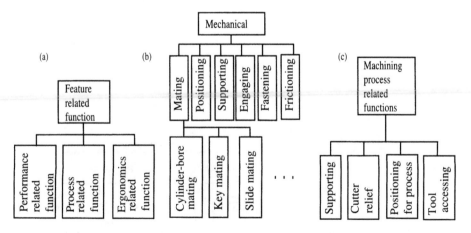

Figure 3.3. Classification of (*a*) feature-related functions, (*b*) mechanical functions, and (*c*) machining process-related functions.

The performance-related functions are obtained directly or indirectly from the functional requirements. The functional requirements of a mechanical product can be divided into mechanical, structural, thermal, fluid, and tribological requirements (Hodgson and Pitts, 1991). There are also corresponding *mechanical, structural, thermal, fluid*, and *tribological* functions. The feature-related functions are important in detail design of components. To study such functions, further classification of feature-related functions is required.

Consider the feature-related functions regarding the "mechanical" functions: *mating, positioning, supporting, engaging, fastening,* and *friction.* Further decomposition is possible; for example, as shown in Figure 3.3*b*, the "mating" functions can be split into "cylinder-bore mating," "key mating," "slide mating," and so on, while "positioning" can be divided into "centering," "linear positioning," and so on.

The functions of the second type are related to manufacturing processes. They are important in evaluating detail component designs due to requirements of the manufacturing process. Figure 3.3*c* shows functions corresponding to the machining process requirements. An example of the "supporting" function is two centering holes supporting a long shaft during machining. "Tooling relief" is a function that prevents a cutting tool from interference with a component. For example, a feature "groove" may be required for cutter relief while turning the thread. "Positioning for process" indicates a function to position or locate in the machining or inspecting process. Sometimes, a feature may have a function of "tool accessing" for machining another feature; otherwise the cutter cannot reach the desired machining section. For instance, in machining a hole in a certain section inside a box-type part, a "window" of the box may be required for the cutter to reach.

3.2.4 Mapping of Requirements and Functions

A set of features describes a component design. The functions implemented by the features of a component should meet the component requirements. At the beginning of detail design, a designer needs to transform component requirements into a set of elementary functions. Such a transformation is called a requirement–function mapping. Suitable features are then selected to implement the functions. Although automating the mapping for all possible mechanical designs seems to be impractical, it is possible to construct a mapping for a specific design domain (a specific group of components). The mapping can be achieved by using, for example, an expert system approach. Figure 3.4 shows a requirement–function tree for a machine tool spindle.

3.2.5 Relations

The flow of information between various types of relations (function relations, feature relations, and function–feature relations) and the modules of "feature design," "manufacturability evaluation," and "process planning" is of importance.

The feature design process may include "feature selection," "feature compatibility checking," "tolerance design," and some other modules (see Figure 3.1). "Feature selection" requires the information from "requirement–function relations" as a

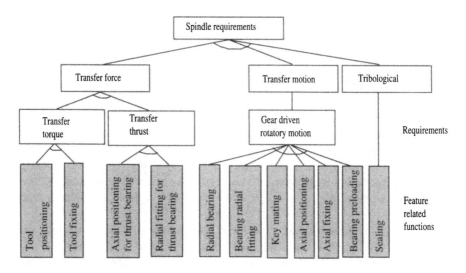

Figure 3.4. Requirement–function tree of a machine tool spindle.

designer selects features that meet the functional requirements. The information on the features selected is the input of "feature relations" and "function–feature relations," which are used in other applications such as "manufacturability evaluation" and "process planning." The "feature compatibility checking" module analyzes various combinations of features. If the features of a part are "compatible," the information is provided to "feature relations." In addition to "feature relations," both "function–feature relations" and "function relations" communicate with "tolerance

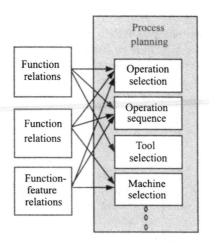

Figure 3.5. Information flow between relations and "process planning."

design," because the module "function relations" contains some geometric requirements.

Figures 3.5 demonstrate the application of function relations, feature relations, and function–feature relations in "process planning." Details about these relations and their representation schemes are discussed next.

3.3 FUNCTION RELATIONS

3.3.1 Problem Statement

There are three reasons to study function relations:

- To avoid omissions of certain functions and features in the design process
- Benefits to tolerance analysis
- Benefits to downstream applications (e.g., process planning)

It is not unusual that a feature with a certain function is not considered in a component design due to the lack of the designer's experience. Some functions, for example, "bearing mating" and "bearing preloading," may appear together. A proper representation of function relations guarantees that all functions relevant to a design are considered.

Often a strict geometric relation is required between two features in order to obtain a proper functionality. In other words, there may exist implicit geometric relations among the functions of features. In such cases, a proper functionality of features should be guaranteed by adding some geometric tolerances (especially position tolerances such as perpendicularity, parallelism, and concentricity) to the related features. Some functions corresponding to those features must meet certain precision requirements. Frequently, the precision of two functions must be compatible. For example, a high accuracy of "bearing mating" and a low accuracy of "axial positioning" may not be compatible. Furthermore, the precision of one function may be based on another function. For instance, the function "bearing mating" might be the datum (reference) of "axial positioning." The information of the function relations is useful for determining reference surfaces in tolerance design. Therefore, the description of function relations is beneficial in tolerance analysis.

The third advantage of function relations is their use in downstream applications (e.g., process planning). The data on functions and function relations are useful in process planning and evaluating manufacturability of the features. The information about the implicit geometry relations of functions facilitates the selection of processes and determines the sequence of operations so as to ensure proper geometry relations and functionality.

In the next two sections, the classification and representation of function relations are discussed.

3.3.2   Classification of Function Relations

The function relations are divided into explicit and implicit. The *explicit relation* depicts the dependency of functions or their common actions, while the *implicit relation* reveals the geometric relations and precision relations of the corresponding features. The implicit relation depends on the existence of the explicit relation. It is difficult to come up with the correct implicit function relationships just by the functions themselves. The classification tree of function relations is shown in Figure 3.6.

The explicit relations can be further divided into three types: *dependency, interdependency, and designated dependency.* The dependency relation indicates that a function exists only due to the presence of another function. For example, a cylinder and its keyway fulfill the functions "cylinder mating" and "key mating," respectively. The existence of the function "key mating" depends on the function "cylinder mating." In other words, if the function "cylinder mating" does not exist, the function "key mating" is meaningless. The interdependency relation between two functions illustrates that two functions depend on each other. For example, the functions "bearing mating" and "axial bearing positioning" depend on each other, as there is no proper functionality without either of the two functions. There are often cases in mechanical design that two functions may exist independently but function dependently. Such a function relation is referred to as a designated dependency relation. For example, the functions "centering" and "bearing mating" correspond to two features. Their existence does not depend on each other; however, "centering" cannot function well without proper "bearing mating." In other words, they function dependently.

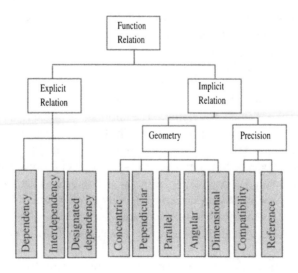

Figure 3.6. Classification tree of function relations.

Some relations embedded in functions are referred to as *implicit relations*. The implicit relations are categorized as geometry relations and precision relations. In most cases, geometry relations are the position geometry requirements, such as concentricity, perpendicularity, parallelism, and angularity. For example, the functions of "centering" and "axial positioning" imply that the relation between the corresponding features should be "perpendicular." There are also cases of implicit "dimensional" relations. One example is shown in Figure 3.7. The two functions "cylinder bore mating" and "key mating" impose a tolerance on dimension *a*.

The precision relation is categorized as *compatibility* and *reference relation*. The compatibility relation applies to dependency of precision data of two functions. Quite often, the implied precision data of two functions should be compatible. For example, for a machine tool spindle, the tolerance of "tool positioning" and "bearing mating" must be compatible. Another example of precision relation is a feature corresponding to "bearing mating" being a reference for "axial positioning."

3.3.3 Representation Scheme for Function Relations

Graph and matrix representations are used to describe the function relations. A *graph* $G = (V, E)$ includes a finite, nonempty set $V = (1, 2, \ldots, m)$ and set $E = \{e_1, e_2, \ldots, e_m\}$, $e_k = (i, j)$, where $i, j \in V$. The elements of V are called *nodes*, and the elements of E are called *edges*. In the graph representing function relations, the node and edge represent a function and a relation, respectively.

3.3.3.1 Representation of Explicit Function Relations. Colored graphs are used to represent explicit function relations. The nodes represent functions, the edges denote function relations, and the numbers along the edges represent the colors.

In a graph of explicit function relations, the color numbers 1, 2, and 3 represent the relation of "dependency," "interdependency," and "designated dependency," respectively. Consider some of the functions of the rotational part in Figure 3.8 and

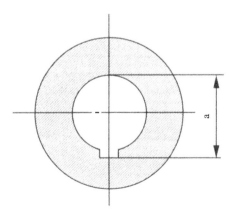

Figure 3.7. Example of implicit "dimensional" relation.

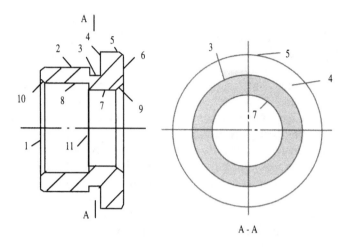

Figure 3.8. Rotational part with features.

the corresponding features shown in Table 3.1 (only the functions that are relevant are listed). It can be seen from the graph in Figure 3.9 that the function "tool relief" ③ is in a dependency relation with function "centering" ②; the functions of "axial positioning" ④ and "bearing preloading" ⑤ are in the interdependency relation; the functions of "tight mating" ① and "bearing mating" ⑥ are in the designated dependency relation; and so on.

The graph for the explicit relations can be represented by the matrix $R^E = [r_{ij}^E]$ with $k \times k$ entries (k is the number of functions), where

$$r_{ij}^G = \begin{cases} 1 & \text{dependency relation between functions } i \text{ and } j \\ 2 & \text{interdependency relation between functions } i \text{ and } j \\ 3 & \text{designated dependency relation between functions } i \text{ and } j \\ 0 & \text{otherwise} \end{cases}$$

TABLE 3.1. Performance and Machining Process-Related Functions and Corresponding Features of the Part in Figure 3.8

Function	Corresponding Features
1. Tight mating	2
2. Centering	2
3. Tool relief	3
4. Axial positioning	4, 6, 11
5. Bearing preloading	6, 11
6. Bearing mating	8

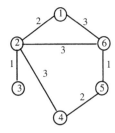

Figure 3.9. Graph of explicit function relations.

3.3.3.2 *Representation of Implicit Function Relations*

Geometry Relations. Similarly to the representation of explicit function relations, a graph of implicit geometry relations can be constructed. The only difference between the two graphs is in color assignment. Here, five colors are assigned to the implicit relations:

    Color 1: concentric
    Color 2: perpendicular
    Color 3: parallel
    Color 4: angular
    Color 5: dimensional

Figure 3.10 is the implicit relation graph of the part in Figure 3.8 and the functions in Table 3.1.

The functions ① and ② (corresponding to feature 2 of the part in Figure 3.8) have relations with other functions. If two functions connected by an edge of color 1 or 3 (relation "concentric" or "parallel") have the same relations with all the other functions, it is necessary to check whether these two functions could be implemented by one feature.

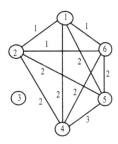

Figure 3.10. Graph of implicit function relations.

The matrix of implicit function relations $R^G = [r_{ij}^G]$ can be constructed similar to R^E, where

$$r_{ij}^G = \begin{cases} 1 & \text{if edge color is 1 (concentric relation) between functions } i \text{ and } j \\ 2 & \text{if edge color is 2 (perpendicular relation) between functions } i \text{ and } j \\ 3 & \text{if edge color is 3 (parallel relation) between functions } i \text{ and } j \\ 4 & \text{if edge color is 4 (angular relation) between functions } i \text{ and } j \\ 5 & \text{if edge color is 5 (dimensional relation) between functions } i \text{ and } j \\ 0 & \text{otherwise} \end{cases}$$

Precision Relations. To represent the implicit precision relations between two functions, the following three cases are specified:

- There exist both reference relations and compatibility relations.
- There exist compatibility relations but no reference relations.
- Neither reference nor compatibility relations exist.

Furthermore, it is necessary to specify the degree of compatibility.

A reference relation of two functions can be represented as a directed edge. A directed graph and an undirected graph can represent the reference relations and compatibility relations, respectively.

For simplicity, a hybrid graph consisting of both directed edges and undirected edges is used to represent the precision relations of a component.

The graph of precision relations has the following properties:

- For a reference relation, an edge is directed to a reference node. Usually, there also exist a compatibility relation between two functions with a reference relation.
- An undirected edge indicates that there is only a compatibility relation.
- A weight w ($0 \le w \le 1$) is assigned to an edge. The more compatible the two functions, the closer to 1 is the weight w.

For the part in Figure 3.8, the function "centering" is the reference of "bearing mating," "axial positioning," and "bearing preloading." All the functions in Table 3.1, except "tool relief," are in the compatibility relation.

The adjacency matrix $R^{PB} = [r_{ij}^{PB}]$ of the precision relation graph is defined as follows:

$$r_{ij}^{PB} = \begin{cases} w_{ij} & \text{if there is a directed edge from node } i \text{ to node } j \\ -w_{ij} & \text{if there is an undirected edge from node } i \text{ and node } j \\ 0 & \text{otherwise} \end{cases}$$

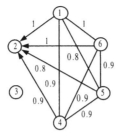

Figure 3.11. Graph of precision relations.

The graph of precision relations for the example part in Figure 3.8 is shown in Figure 3.11. The matrix of precision relations has the following properties:

- A zero entry ($r_{ij}^{PB} = 0$) does not necessarily imply that there is no precision relation between functions i and j. No precision relation exists between two functions i and j only if $r_{ij}^{PB} = r_{ji}^{PB} = 0$.
- If all the elements in the i-th row and the ith column are zeros, then the ith row and i-th column can be removed from the relation matrix.
- If $r_{ij}^{PB} = -w_{ij}$, then = $r_{ji}^{PB} = -w_{ij}$. It means that all negative elements in R^{PB} are symmetric.

3.4 FEATURE RELATIONS

3.4.1 Basic Concepts

Chen et al. (1991) defined three types of relations between features, is-in, is-on, and adjacent-to. The is-in relation indicates a spatial relationship between a positive and a negative feature, that is, the negative feature is within the positive one. The is-on relation may occur in the case of two positive or two negative features. It also implies a geometric or function dependency between two features. The adjacent-to relation simply indicates a geometric adjacency between two features.

Some overlap may exist between is-on and adjacent-to relations. Although a claim has been made that the two relations do not overlap (Chen et al., 1991), an ambiguity exists. On the other hand, there are cases that are difficult to represent by any of the three relations. For instance, the relation between an inner fillet and a hole is difficult to represent using the above definitions.

Three feature types, *positive, negative,* and *face,* are defined in this chapter. The positive and negative features have the same meaning as those used in Chen et al. (1991). In this chapter, the positive and negative features are called *entity features* since they represent certain geometric entities. *Faces* indicate the boundary faces or the adjacent faces of two entity features no matter whether they are positive or negative. There are three reasons for introducing a face feature. First, one face may require a specific machining operation to form it. Second, a face is often used as a reference plane. Finally, it is easier to establish the geometry relation of a part using face features, as one entity feature may be adjoined to another entity feature through a face.

The representation schemes of geometry relations and precision relations of features are discussed in the next section. Three types of geometry relations are defined: *union, subtraction,* and *side-to,* denoted by ∪, −, and ∣, respectively. Consider the relations for two features f_a and f_b. The union feature f_u contains the elements of f_a and f_b, while the subtraction feature f_s contains only the elements remaining after taking f_b away from f_a. The relation side-to indicates the relation between a face feature and an entity feature.

The precision relations of features are defined similarly to that of functions mentioned in Section 3.3.2, simply by replacing features with functions.

3.4.2 Graph Representation of Feature Relations

3.4.2.1 Geometry Relations. The symbols ⊕, ⊖, and ○ are used to denote the positive, negative, and face features, respectively. They are the nodes of a feature relation graph. The three-feature relations union, subtraction, and side-to are represented by a double arrowhead, single arrowhead, and line segment of a graph, respectively. For subtraction, the arrowhead points to the feature that should be taken away, for example, $⊕_i → ⊕_j$, meaning $⊕_i − ⊕_j$.

The following properties hold for geometry relations:

(a) $⊖ ∪ ⊖$

(b) $⊖ ∪ ⊕$ or $⊕ ∪ ⊖$

(c) $⊕ ∪ ⊕$

(d) $⊕ − ⊕$

(e) $⊖ ∣ ○$ or $○ ∣ ⊖$

(f) $⊕ ∣ ○$ or $○ ∣ ⊕$

Each operation type should be carefully designated when constructing a graph of geometry relations for a part. For a cylinder with a through hole, if the hole is considered a negative feature, the operation between the cylinder and the hole is ∪ (union). If the hole is considered a virtual cylinder, that is, a positive feature, then the operation is − (subtraction).

To reduce the complexity of a graph, only the features that are of interest to a designer should appear on the graph.

Figure 3.12 shows an example geometry relation graph for the part in Figure 3.8. Some chamfers are not depicted in this graph.

Constructing a relation graph, one must consider that a negative feature should be in relation not only with a face but also with a positive feature. Figure 3.12 shows that feature 8 is in relation with the positive feature 2 in addition to the relation with face feature 1. However, the path between two adjacent positive features (e.g., features 2 and 5 in Figure 3.12) should cross their common face feature (here feature 4 in Figure

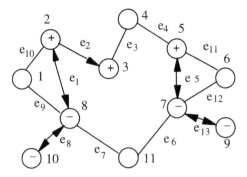

Figure 3.12. Geometry relation graph for the part in Figure 3.8.

3.12). In other words, the *union* relation should not be applied to the two adjacent positive features.

The definition of primary and secondary feature is adopted from Chen et al. (1991). Based on the definitions of features and relations, the following properties are derived:

Property 1. A side-to relation can only exist between one face and one entity feature (either positive or negative).

Property 2. There exists at least one path connecting all the primary features of a part if no primary feature is deliberately omitted.

Based on the part geometry, a feature cannot be independent of other features. Correspondingly, a feature is reachable for any other features in a geometry relation graph of a part. The edges $e_1, e_2, \ldots, e_7$ form a path connecting the primary features 2, 5, 7, and 8 as shown in Figure 3.12.

Property 3. If two primary features have side-to relations to the same face, then:

(a) Any two negative or two positive features (not virtual) are not directly connected.

(b) Any positive and negative features are in a union relation.

Two positive or two negative features with a common face are actually adjacent, and their relation is shown only as a side-to relation. However, for one positive and one negative feature with a common face, the negative one is actually contained in the positive one. The union relation does apply in this case.

Property 4. For a part with a through hole or hollow cavity, there is at least one closed-loop path connecting all the primary features.

Consider two end faces in a side-to relation with a through hole or a hollow cavity. According to Property 2, there is a path connecting the two end faces without considering the hollow cavity. In addition, one can find another path through the hollow cavity. Therefore, these two paths form a closed loop.

Property 5. A face feature is always dependent on at least one entity feature, that is, at least one edge of a side-to relation connects the face feature to an entity feature.

Face is not an entity feature and cannot exist independently.

Property 6. A secondary feature remains in the union or subtraction relation with at least one primary feature.

This property is quite obvious, as a secondary feature always depends on its parents based on the definition of the secondary feature.

The above properties are useful in checking the validity of the graph of geometry relations.

3.4.2.2 *Precision Relations.* In a graph representing feature precision relations, the nodes represent features while the edges represent precision relations, similar to the precision relations of functions. The graph of feature precision relations for the part in Figure 3.8 is presented in Figure 3.13.

3.4.3 Matrix Representation of Feature Relations

Although a graph representation is explicit enough to describe the feature relations, it is not convenient for computer processing. The matrix representation is more appropriate for computer applications.

3.4.3.1 *Geometry Relations.* Assume that a part has m features and n relations. To represent the geometry relation graph of features, the incidence matrix $C = [c_{ij}]$ with $m \times n$ entries is used, where

$$c_{ij} = \begin{cases} 1 & \text{if edge } j \text{ connects node } i \text{ (relation } j \text{ applies to feature } i) \\ 0 & \text{otherwise} \end{cases}$$

The incidence matrix of the graph of geometry relations in Figure 3.12 is presented below.

Each column of C has exactly two 1's. The incidence matrix is not sufficient to completely represent all feature relations since the matrix does not reflect the relation type. Therefore, a *weight vector* $W = [w_i]$ is introduced, where

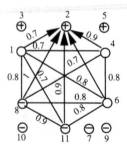

Figure 3.13. Graph of feature precision relations.

$$w_i = \begin{cases} 1 & \text{if edge } i \text{ is a union relation} \\ -1 & \text{if edge } i \text{ is a subtraction relation} \\ 0 & \text{otherwise (side-to relation)} \end{cases}$$

Relation no.

	1	2	3	4	5	6	7	8	9	10	11	12	13	
	0	0	0	0	0	0	0	0	1	1	0	0	0	1
	1	1	0	0	0	0	0	0	0	1	0	0	0	2
	0	1	1	0	0	0	0	0	0	0	0	0	0	3
	0	0	1	1	0	0	0	0	0	0	0	0	0	4
	0	0	0	1	1	0	0	0	0	0	0	0	0	5
C = Feature no.	0	0	0	0	0	0	0	0	0	0	1	1	0	6
	0	0	0	0	1	1	0	0	0	0	0	1	1	7
	1	0	0	0	0	0	1	1	1	0	0	0	0	8
	0	0	0	0	0	0	0	0	0	0	0	0	1	9
	0	0	0	0	0	0	0	1	0	0	0	0	0	10
	0	0	0	0	0	1	1	0	0	0	0	0	0	11

The weight vector for the graph in Figure 3.12 is

$$W = [1\ -1\ 0\ 0\ 1\ 0\ 0\ 1\ 0\ 0\ 0\ 0\ 1]$$

Using both C and W, the geometry relations can be completely described.

The geometry relation of features can also be represented by a three-edge colored graph simply by assigning three different colors to the relations of union, subtraction, and side-to. The corresponding matrix can be easily constructed similar to that of explicit function relations.

3.5 REPRESENTATION OF FUNCTION–FEATURE RELATIONS

There are at least two scenarios where the function–feature relations are considered. First, one needs to select suitable features according to the functional requirements. Second, it is required to check whether candidate features are necessary and sufficient to meet the functional requirements of the part. This leads to further checking of feature-related functions. In such a case, the manufacturability of features must be evaluated.

In fact, the feature set and the function set of a part can be partitioned into two distinct groups. Although there exist relationships among the elements in the set of functions and features (see Sections 3.3 and 3.4), such a relationship is not considered here. The feature and function sets can be considered as two all-inclusive but mutually exclusive subsets. This allows one to represent the function–feature relations with a bipartite graph.

A bipartite graph of a function–feature relation for a component design is denoted by $G = (F_i, B_{pd}, E)$, where F_i and B_{pd} is the *feature set of interest* and *design function set of a part*, respectively. The feature set of interest F_i is a collection of features of interest to a designer. The design function set of a part B_{pd} is a collection of intended functions of the part. Each set of edges E indicates the relationship between features and functions.

A distinction should be made between the process-related functions and performance-related functions (as mentioned in Section 3.2.3) in certain applications. This implies incorporating the orientation to the edges of process-related functions in a bipartite graph, that is, directing an edge from a process-related function to the corresponding feature. The relations between performance-related functions and features are represented by undirected edges.

Figure 3.14 shows the bipartite graph of a function–feature relation for the part in Figure 3.8. Only the features of interest are shown in this graph.

The *relation matrix* $R = [r_{ij}]$ with entries $p \times l$ is used to represent relations between p features and l functions, where

$$r_{ij} = \begin{cases} -1 & \text{relation between feature } i \text{ and process-related function } j \\ 1 & \text{relation between feature } i \text{ and performance-related function } j \\ 0 & \text{no relation between feature } i \text{ and function } j \end{cases}$$

It should be emphasized that different features may have relations with the same function and one feature may have multiple functions. A feature that is not of interest, such as the primary feature 5 shown in Figure 3.8, can be deleted.

The representation approach discussed in this chapter forms a foundation for future research topics, such as:

- The inheritance among design objects in a given model and their extensions to new designs

- The integration of process models with product models

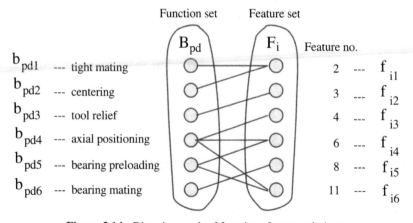

Figure 3.14. Bipartite graph of function–feature relations.

- The extension of feature validation procedures to include the checking of functionality
- The automation of the design process driven by form-feature-based information
- The application of representation libraries of design objects with defined inputs and outputs

3.6 SUMMARY

In this chapter, representation schemes of function relations, feature relations, and function–feature interrelations were discussed. These schemes are important in automated design of parts, especially at the detail design stage. A clear, complete, and simple representation of these relations is a prerequisite for feature selection, manufacturability evaluation, and process planning. To date, limited research has been done in the representation area, especially for function relations and function–feature relations.

The classification of feature-related functions were presented. The functions were classified into performance-related functions, process-related functions, and ergonomics-related functions. A detailed decomposition of feature-related functions was also presented.

It was proposed that the function relations be divided into explicit and implicit relations. Further classification of both explicit and implicit function relations was also provided. Colored graphs were applied to represent explicit function relations and implicit geometry relations. The implicit precision relations facilitate design of tolerances and evaluation of candidate designs. These relations were represented by a weighted graph.

The feature relation problem is important in the selection of features and manufacturability evaluation at the detail design stage. Three relations—union, subtraction, and side-to—were proposed in the chapter. The graphs and matrices were used to represent the feature relations. The precision relations of features were also considered. The representation scheme for feature relations proposed in this chapter is complete and simple.

Besides feature and function relations, interrelations between features and functions exist. It appears that such interrelations are useful in feature selection. A bipartite graph was used to represent the interrelations of functions and features.

The main advantage of the approaches presented in this chapter is that they are appropriate for computer implementation as they are matrix and graph based.

REFERENCES

Bronsvoort, W. F. and F. W. Jansen (1993), Feature modeling and conversion—key concepts to concurrent engineering, *Computers in Industry,* Vol. 21, No. 1, pp. 61–86.

Chen, Y. M., R. A. Miller, and K. R. Vemuri (1991), A framework for feature based part modeling, in *Computers in Engineering,* Vol. 1, ASME, New York, pp. 357–365.

Feng, C. H., C. C. Huang, A. Kusiak, and P. G. Li (1996), Representation of functions in detail design, *Computer-Aided Design,* Vol. 28, No. 12, pp. 961–971.

Henderson, M. R., and L. E. Taylor (1993), A meta-model for mechanical products based upon the mechanical design process, *Research in Engineering Design*, Vol. 5, Nos. 3/4, pp. 140–160.

Hodgson, B. A., and G. Pitts (1991), Designing for CNC manufacture, in J. Corbett, M. Dooner, J. Meleka, and C. Pym (Eds.), *Design for Manufacture,* Addison-Wesley, Wokingham, England, pp. 56–68.

Kannpan, S. M., and R. M. Marshek (1990), An algebraic and predicate logic approach to representation and reasoning in machine design, *Mechanism and Machine Theory,* Vol. 25, No. 3, pp. 335–353.

Kusiak, A., and E. Szczerbicki (1992), A formal approach to specifications in conceptual design, *ASME Journal of Mechanical Design,* Vol. 114, No. 4, pp. 659–666.

Pahl, G., and W. Beitz (1988), *Engineering Design,* Springer-Verlag, New York.

Rosen, D. W. (1993), Feature-based design: Four hypotheses for future CAD systems, *Research in Engineering Design*, Vol. 5, Nos. 3/4, pp. 125–139.

Shah, J. J., and M. T. Rogers (1988), Functional requirements and conceptual design of the feature-based modeling system, *Computer-Aided Engineering Journal,* Vol. 5, No. 1, pp. 9–14.

Welch, R. V., and J. R. Dixon (1992), Representing function, behavior and structure during conceptual design, in D. L. Taylor and L. A. Stauffer (Eds.), *Design Theory and Methodology,* Vol. DE-42, ASME, New York, pp. 11–18.

Xue, D., and Z. Dong (1993), Automated concurrent design based on combined feature, tolerance, production process and cost models, in *Proceedings of Advances in Design Automation Conference*, Vol. 2, ASME, New York, pp. 199–210.

QUESTIONS

3.1. What are some typical activities of the detail design stage of mechanical components?

3.2. What is a requirements space?

3.3. What is meant by mapping functional requirements into the functions of a part?

3.4. What is an example of an explicit relation?

3.5. How are the function relations represented?

3.6. What feature relations do you know?

3.7. How are the function–feature relations represented? Construct an example.

PROBLEMS

3.1. Draw a mechanical part and specify its functions.

3.2. Draw a mechanical part and construct a relation graph.

3.3. Draw a mechanical part, define features and geometry relations, and represent the two as a matrix.

CHAPTER 4

REASON MAINTENANCE IN PRODUCT MODELING

4.1 INTRODUCTION

Three major groups of activities involved in the development of a product computational model include (Pahl and Beitz, 1988):

1. Problem definition, definition of design requirements and constraints, and identification of functional requirements
2. Product modeling that involves model synthesis, rough physical implementation of a product model, and evaluation of the functional behavior of the model
3. Selection of the best product model

The focus in this chapter is on searching for a product model that best satisfies functional requirements. A design solution meets a number of constraints such as cost, weight, and shape, but the most important is that it is capable of performing the function for which the product is aimed. Usually, a number of product models satisfy functional requirements and should be considered as candidate design solutions. The problem is to develop an efficient search procedure to identify all feasible product models, that is, product models that provide accomplishment of the overall design function by properly defining relationships among design subfunctions. No feasible model should be overlooked since we do not know which functionally valid model will lead to the best design solution subject to other design requirements and objectives.

This research was concerned with the development of a product modeling procedure that (a) minimizes search activities in the identification of feasible models, (b) captures model structure and topological connections among components needed for the evaluation of model behavior, and (c) supports modular product design.

This chapter describes an approach to product modeling by using a reason maintenance system to reduce the search activity required for the identification of feasible models. An assumption-based truth maintenance system and multiple worlds are used to discover and store information about feasible designs and to avoid further investigation of infeasible design alternatives.

The product modeling procedure draws from the following approaches: systematic product design (Pahl and Beitz, 1988), a model base concept (Oren et al., 1984), hierarchical, modular discrete-event modeling (Zeigler, 1987), truth maintenance, and qualitative reasoning (de Kleer and Brown, 1984).

Pahl and Beitz (1988) established the theoretical foundation for the process of breaking down the overall design function, derived from design requirements, as the determination of subfunctions that facilitate the search for physical solutions and the combination of these subfunctions into a simple and unambiguous function structure. A functional object, assigned to a functional description, is a symbolic representation described by a set of parameters characterizing input and output quantities in a function, a set of parameters characterizing the input–output transformation, and a set of constraints between parameters from the first two sets. A functional model consists of a number of interconnected functional objects. Hundal (1990) implemented the Pahl and Beitz (1988) systematic design methodology into a general-purpose system for computer-aided conceptual design. The system is aimed at assisting the designer in applying a methodological approach to design, suggesting a set of alternatives at various stages, and evaluating alternative design solutions.

A conceptual design methodology must also describe the process of transforming functional requirements into mechanical components (design building blocks). For that purpose, a scheme for the representation of mechanical components and their interactions must be applied. Such representation schemes may be developed using graph-based languages (Rinderle and Finger, 1990), shape and structure grammars (Longenecker and Fitzhorn, 1990), or functional trees (Pahl and Beitz, 1988).

However, these approaches do not succeed in capturing the hierarchical and modular nature of a conceptual design process. The representation scheme presented in this chapter assumes that a design model consists of individual components. Each component represents a specific elementary design function or functions and is autonomous. Its description contains relationships between input and output variables. The system-theoretic synthesis requires that a designed object has a distinct boundary, within which it is decomposed into a number of hierarchically structured models and components at various levels of abstraction. This approach is directly derived from hierarchical object-oriented model-based concepts presented in Oren et al. (1984) and Zeigler (1987).

A truth maintenance system is applied to provide consistency and efficiency during a product modeling process and for detecting problems due to the erroneous input data or the definition of constraints that cannot be satisfied. A truth maintenance system alone does not provide mechanisms for modeling actions and state changes present in the incremental development of a product model. An approach that follows the implementation of truth-maintained multiple worlds in the IntelliCorp Knowledge

Engineering Environment (KEE) (Filman, 1988) has been applied to provide those capabilities.

Multiple worlds is a computational intelligence technique for simultaneous exploration of alternative assumptions leading to feasible problem solutions. An assumption-based truth maintenance system allows the problem solver to work simultaneously with multiple contexts and find all possible solutions, which makes it suitable for product modeling.

Multiple worlds is particularly well suited for problems with a number of feasible solutions that should be evaluated (a number of product models that should be evaluated) and problems that involve evaluation of solution states during the solution process. Feature-based product design can be modeled using multiple worlds by storing information about various stages of a product model in different worlds. These worlds can then be used for various analyses such as cost, assemblability, manufacturability, and maintainability. The use of truth maintenance multiple worlds in a design-for-producibility system is presented in Ranta et al. (1989).

A qualitative reasoning approach has been increasingly used to model mechanical devices and analyze their functional behavior. Joskowicz (1987) developed an algorithm for qualitative analysis of mechanical devices. The possible relative motions of all pairs of objects that are in contact can be determined given a geometric description of the objects and their initial position. Then, the actual motion of each object is defined based on the pairwise relative motions and an input motion. Bradshaw and Young (1991) combined the knowledge of purpose and the knowledge of structure to predict a device's possible behavior and evaluate its actual behavior. A knowledge-capturing scheme for qualitative modeling of mechanical devices is presented in Pu (1990). The scheme is implemented in a system that allows for the automatic qualitative simulation of the device's functional behavior. An object-oriented system for qualitative reasoning that can be applied for a wide range of engineering tasks such as design, analysis, diagnosis, monitoring, and inspection is presented in Biswas et al. (1989)

4.2 PRODUCT MODELING

A modeling procedure discussed in Vujosevic et al. (1995) for development of a product model that meets requirements defined in the previous section is presented in Figure 4.1. The main activities involved in the product modeling include development of a model base of abstract components that correspond to design functions, synthesis of abstract product models, and analysis of the functional behavior of product concepts developed based on abstract models.

A product model is constructed from a number of abstract components representing product functions generated from the design requirements. Each abstract component becomes a physical component. A model synthesis algorithm generates abstract models by reasoning about the relationships among abstract components. For that purpose, input and output variables are defined for each abstract component.

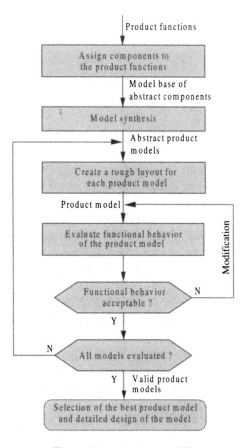

Figure 4.1. Product modeling.

The product modeling approach is illustrated with the design of a holding device. Assume that the following product functions have been defined based on the design requirements and constraints (Kusiak et al., 1991):

F_1: Apply human force.

F_2: Apply human force by hand and amplify the force.

F_3: Apply force rotationally.

F_4: Hold the material.

F_5: Hold the material by a fixed part of the device.

F_6: Hold the material by a movable part of the device.

F_7: Transform radial force.

F_8: Transform force by one screw.

F_9: Change radial force into horizontal force.

F_{10}: Change radial velocity into horizontal velocity.

F_{11}: Fix the unmovable part of the device.

F_{12}: Guide the motion of a movable part,

F_{13}: Fix the guide of a movable part.

The above functions are identified and assigned to the abstract components at an earlier stage of product modeling. The model base of abstract components is shown in Figure 4.2. Abstract components, representing a function or group of functions, are later matched to corresponding mechanical components to generate a physical product layout.

Larger rectangular boxes in Figure 4.2 depict component boundaries. The arrows on the left of each box represent inputs and the arrow on the right represent outputs. Small dark squares mean that an input/output is external, while white squares mean that the input/output is internal. Variables F (force), M (moment), Ω (radial velocity), and V (linear velocity) are input/output variables.

The component C_1 in Figure 4.2 represents functions F_1, F_2, and F_3, that is, apply human force by hand rotationally and amplify the force. The functions F_9 and F_{10} (change radial force and velocity into horizontal force and velocity) are represented by component C_5. Component C_1 accepts input coming from the outside of the system boundary. Components C_6 and C_8 send the output to the outside of the system boundary.

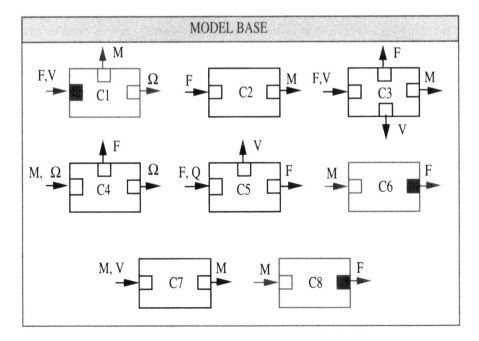

Figure 4.2. Abstract components for modeling a holding device.

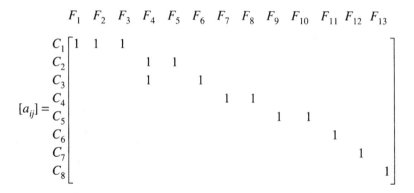

Figure 4.3. Component-function incidence matrix for a holding device.

The component–function incidence matrix $[a_{ij}]$ for the components and functions above is presented in Figure 4.3, where an entry a_{ij} equals 1 if component i represents function j (Kusiak et al., 1991).

Figures 4.2 and 4.3 describe the representation scheme used to convert functional requirements into abstract components that represent mechanical components at this early stage of conceptual design (first activity in Figure 4.1).

Each component can be described by its attributes, dynamic behavior, and functions. The important attributes of a component are input and output variables. Input and output variables are defined based on function(s) represented by the component. For example, in component C_5, F and Ω (force and radial velocity) are the input variables, and F and V (force and linear velocity) are the output variables. The connection of two components C_1 and C_5 from the model base in Figure 4.2 is illustrated in Figure 4.4.

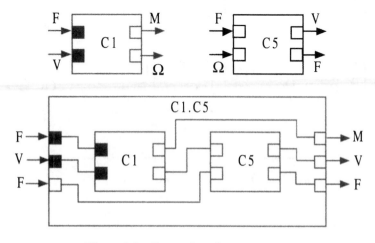

Figure 4.4. Connection of two components.

Model C_1, C_5 has three input variables: F, V (the original inputs of C_1), and F, the unmatched input variable of C_5. The output variables of the model are the original output of component C_5 (F, V) and the unmatched output of component C_1 (M).

A product model must satisfy both structural and functional requirements. The structural requirements are considered in the model synthesis that produces a topology of the interconnected components.

Two major activities involved in the product modeling phase involve synthesis of abstract components that correspond to one or more elementary design functions into a set of product models and analysis of functional behavior of the overall model.

A rough physical layout has to be created for an abstract model in order to be able to evaluate the functional behavior of the model.

When the best overall product model is selected, the detailed design of the product can be performed. A widely used approach for the development of a physical product model is based on a set of interconnected geometric features. Expressing a design in terms of features supports the human way of reasoning about a design. A feature-based design captures the design information necessary for performing a number of activities that could be encountered in the life cycle of a product.

4.3 TRUTH-MAINTAINED MULTIPLE WORLDS

A basic characteristic of truth maintenance systems is the use of dependency-directed backtracking (Stallman and Sussman, 1977). When an inconsistency is detected, the algorithm backtracks to the most recent choice that caused inconsistency, in contrast with chronological backtracking, which backtracks to the most recent choice the search has made.

An assumption-based truth maintenance system (ATMS) avoids dependency-directed backtracking by making a distinction between assumptions and other data (or facts). Assumptions are data that are presumed to be true, unless there is evidence to the contrary. The problem solver to the ATMS submits as a justification the derivation of a datum from other data or assumptions. Each datum is assigned a label indicating a list of environments. Environments represent a set of assumptions under which a datum holds. An environment is inconsistent if it allows deriving a datum representing the contradiction and in that case is called *no good*. The ATMS ensures that no node will be considered to follow from a set of assumptions if a contradiction node also follows from that set of assumptions (de Kleer and Forbus, 1991). When such an environment is discovered, it has to be removed from all the labels. The context of a consistent environment is the set of facts that can be derived from assumptions of that environment.

A dependency network is created during the problem-solving procedure. It consists of nodes assigned to each problem solver's datum and justifications (de Kleer and Forbus, 1991). A node can be a premise (holds universally), a contradiction (never holds), or an assumption (whose belief may be changed). A justification consists of a consequent (node assigned to a problem solver's datum), antecedents (nodes of the data used as antecedents to the inference rule), and an informant aimed to explain inference made. Figure 4.5 illustrates the concepts defined with an example of

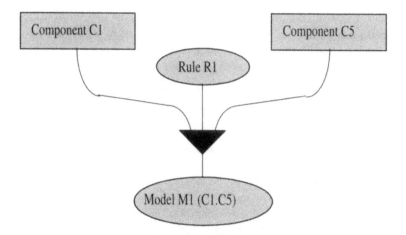

Figure 4.5.   Dependency network.

components C_1 and C_5 from Figure 4.4. It is a simple dependency network consisting of two assumption nodes capturing information about components C_1 and C_5, a justification created by executing production rule R_1, and a nonassumption node storing the problem-solving inference—a complex model M_1 (C_1, C_2).

Rule 1

IF a component has at least one input variable that is the same as an output variable of a component (or an evolving product model)

AND neither the matched input(s) nor output(s) are boundary

AND there are no other rules that prevent coupling of these components

THEN create a model by coupling the two components

AND define its input and output variables based on the input and output variables of its components.

An ATMS alone does not provide mechanisms for modeling the actions and state changes. It can be coupled with the concept of multiple worlds to provide these characteristics (see Filman, 1988).

A consistent world captures a set of related facts that are true in that world and do not violate the constraints established. For example, it can store the information about an evolving product model at a particular point in the model synthesis. Facts can be contradictory across worlds but must be internally consistent within the world. The facts that are true in every world are called background facts.

A world is defined by a set of assumptions. A fact is believed in a world if its set of assumptions is a subset of the world's assumptions.

Each world is associated two ATMS entities (Morris and Nado, 1986): (1) world assumption, representing the action modeled by the world, and (2) world environment that defines the state of the world and consists of world assumptions of a particular world and its ancestors. A world environment is calculated by ATMS using nodes (ATMS entities) assigned to each world.

Typically, a problem-solving procedure involves creation of a tree of worlds, where each world inherits facts true for a parent world.

4.4 MODEL SYNTHESIS

Model synthesis is a constraint satisfaction problem of creating a product model that meets design requirements. A simple enumeration algorithm is able to find all solutions, if any, by exploring the search space and evaluating each feasible design solution. However, an enumeration algorithm can be inefficient, especially when the number of components to be synthesized is large. This is due to the fact that whenever a contradiction is encountered, the algorithm backtracks to the most recent choice of the search procedure instead to the most recent choice that actually contributed to the contradiction.

A model synthesis algorithm should also generate a structural description of the model and the description of topological connections between components needed to evaluate functional behavior of the model.

A problem solver that incorporates a depth-first search and truth-maintained multiple worlds has been developed in order to meet the requirements mentioned above. Multiple worlds are used for storing points of the search space (intermediate and final models).

The model synthesis procedure begins with the creation of the root world (W_1), which is assigned the component with only boundary inputs (C_1). The assumption representing this action is: "component C_1 is added to the world W_1." The datum captured by the world is an object that captures the information about the abstract component C_1.

Each time a world is created, the constraint-definition rules generate and assign to the world a list of components that can be connected to the intermediate model captured in the world by matching output from the intermediate model with inputs of remaining components. Such a list for world W_1 is as follows:

$$W_1: \ C_1 \ \rightarrow \ (C_4, \ C_5, \ C_6, \ C_7, \ C_8)$$

The depth-first search algorithm then selects the first component from the interconnection list and creates a world that captures that particular point of the search space. The world creation rules create the world W_2 to represent the coupling of components C_1 and C_4 into a model M_1: C_1, C_4 shown in Figure 4.6.

The world W_2 captures the following information:

1. A fact inherited from the ancestor world—component C_1.
2. An addition to the world—the assumption "component C_4 added to the world W_2" is stored in the world assumption.

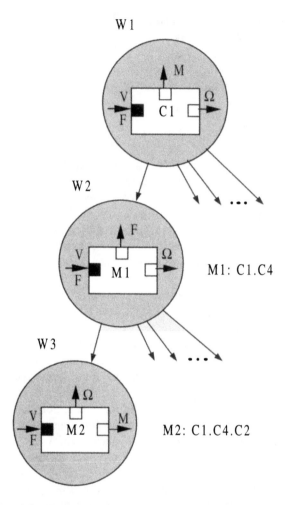

Figure 4.6.   Truth-maintained multiple worlds in the model synthesis.

3. A deduction from the primitive facts of the world—a justification that represents the deduction (model M_2: C_1, C_4) from the primitive facts (components C_1 and C_4) of the world W_2:

$$C1 \; \wedge \; C4 \; \rightarrow \; M2$$

   The world environment is computed by the ATMS and consists of world assumptions added to the world and world assumptions from its ancestors. Each world has a single environment stored in the label of the world node, assigned to each world in order to facilitate the truth maintenance across the worlds. The world node (N_{w2}) of the world W_2 is given a single justification:

$$N_{w1} \quad ^{\wedge} \quad A_{w2} \; \rightarrow \; N_{w2}$$

where N_{w1} is the world node of the world W_1 and A_{w2} is the world assumption of the world W_2.

Again, the interconnection list for the intermediate model captured in the world W_2 is created and the first component from the list selected and captured in the world W_3. The problem-solving procedure continues until all product models have been generated.

The contradiction–detection rules analyze the world and mark it as "no good" if it stores a contradiction. A world is a contradiction if the model captured by the world includes only boundary outputs but does not include all components, at least one internal output that cannot be connected to any of the remaining abstract components from the model base, or all abstract components that have one or more internal outputs.

The ATMS ensures that no world is considered to follow from a set of assumptions if a contradiction world also follows from that set of assumptions. An ATMS algorithm (de Kleer and Forbus, 1991) propagates label changes throughout the three worlds.

The output of model synthesis of the holding device is shown next.

```
FEASIBLE PRODUCT MODELS:

World #953
Number of Components: 8
World Assumption: 'C6'
World Node Environment: (1 7 4 2 3 8 5 6)

World #955
Number of Components: 8
World Assumption: 'C8'
World Node Environment: (1 7 4 2 3 6 5 8)

World #958
Number of Components: 8
World Assumption: 'C6'
World Node Environment: (1 6 4 2 3 8 5 7)

World #959
Number of Components: 8
World Assumption: 'C8'
World Node Environment: (1 6 4 2 3 7 5 8)

World #969
Number of Components: 8
World Assumption: 'C8'
World Node Environment: (1 6 4 2 3 5 7 8)
```

```
World #978
Number of Components: 8
World Assumption: 'C6'
World Node Environment: (1 6 4 2 3 8 7 5)

World #985
Number of Components: 8
World Assumption: 'C6'
World Node Environment: (1 5 4 2 3 8 7 6)

World #990
Number of Components: 8
World Assumption: 'C8'
World Node Environment: (1 5 4 2 3 6 7 8)

World #993
Number of Components: 8
World Assumption: 'C6'
World Node Environment: (1 5 4 2 3 8 6 7)

World #994
Number of Components: 8
World Assumption: 'C8'
World Node Environment: (1 5 4 2 3 7 6 8)

MODEL SYNTHESIS RESULT:

Total number of models generated: 10
Total number of worlds created: 2513
Input data:
  1 → 'force' → Boundary
  2 → 'linear velocity' → Boundary
Output data:
  1 →- 'force' → Boundary
  2 → 'force' → Boundary
```

The worlds displayed are considered consistent since they represent feasible models that capture all eight components from Figure 4.2 and ensure the overall design function. The output shows that the total of 10 feasible models have been identified. The algorithm created 2513 worlds in searching for feasible models. All feasible models have force and linear velocity as input variables and two forces as output variables.

Using the notation from Figure 4.6, the feasible models are represented as follows:

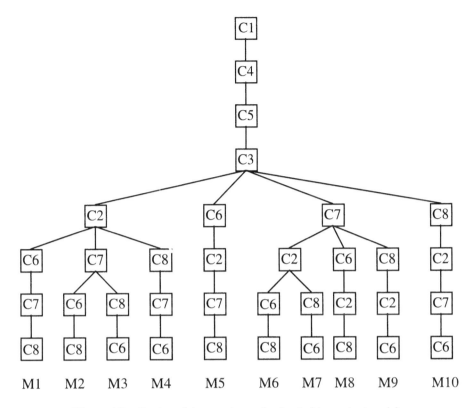

Figure 4.7. Portion of the search tree that has led to product models.

```
M1:   C1.C4.C5.C3.C2.C6.C7.C8
M2:   C1.C4.C5.C3.C2.C7.C6.C8
M3:   C1.C4.C5.C3.C2.C7.C8.C6
M4:   C1.C4.C5.C3.C2.C8.C7.C6
M5:   C1.C4.C5.C3.C6.C2.C7.C8
M6:   C1.C4.C5.C3.C7.C2.C6.C8
M7:   C1.C4.C5.C3.C7.C2.C8.C6
M8:   C1.C4.C5.C3.C7.C6.C2.C8
M9:   C1.C4.C5.C3.C7.C8.C2.C6
M10:  C1.C4.C5.C3.C8.C2.C7.C6
```

The portion of the search tree that has led to the generation of overall product models is shown in Figure 4.7.

The next step in product modeling involves the development of a product layout based on the models generated. This phase can be performed by following the approach given in Pahl and Beitz (1988). Each abstract component is assigned a physical solution that consists of physical effects assigned to elementary design functions represented by that component.

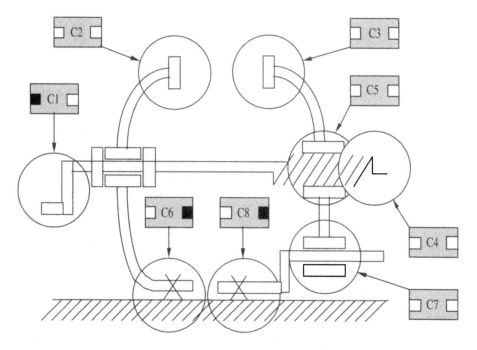

Figure 4.8.   Physical representation of model M_6.

As an example, consider component C_1 in Figure 4.2, which represents elementary functions F_1, F_2, and F_3, described earlier. Physical effects of these functions can be described quantitatively by using the following formulas:

$$M = Fr \qquad \Omega = V/r$$

The same approach can be used to define physical solutions for each abstract component captured in the product models. Various physical solutions can be assigned to a single component; however, they must be compatible with each other.

A physical representation of model M_6 from Figure 4.7 is shown in Figure 4.8.

To be accepted as a valid product concept, the functional behavior of a product layout must be evaluated. An approach for performing this task activity is presented next.

4.5 MODEL ANALYSIS

The framework for model analysis is based on ideas derived from qualitative reasoning. A product model, which consists of a number of interconnected abstract components, maintains the *no-function-in-structure* principle. This principle requires that the behavior of an individual component should in no way refer to the overall behavior of the device (de Kleer and Brown, 1984). This leads to the development of a set of models from a database of primitive components. Each component is assigned information that is valid locally for that component only. The topology of the model supports the bottom-up analysis of functional behavior of the product layout, developed by assigning physical solutions to the components captured by the model.

To be able to simulate the functional behavior of a product model, a domain-dependent hierarchy of processes must be defined. Biswas et al. (1989) defined three types of processes: (1) processes that define the primitive functionality of objects, (2) processes that define the functionality of interconnections between components, and (3) processes that define causal parameter relations that usually correspond to first principles in the domain. The interconnections between components allow for the propagation of model variables, for example, moment, force, and linear and angular velocity.

Based on the three types of processes defined earlier, two types of causalities can be provided (Pu, 1990): (1) transformation of model variables (force, moment, linear and angular velocity) within a component and (2) transmission of model variables along interconnections between components.

A forward-constraint propagation algorithm is developed for the evaluation of the behavior of a product model:

Step 1.   Load into the memory the model selected for simulation and define its input variables.

Step 2.   Select an intermediate model (IM) according to the structural topology of the model stored at the world node of the world representing that model. Initially, select the root component (component C_1 from Figure 4.2), represented by the world W_1, which has only the boundary inputs defined in the previous step.

Step 3.   Calculate the output variables of an IM based on input variables and physical processes (effects) represented by the model. For example, for the model IM_2 (C_1, C_4) in Figure 4.6, the following activities are performed: (a) variable transformations in C_1 ($M = Fr$, $\Omega = V/r$), (b) variable transmission between component C_1 and C_4, (c) variable transformation in C_4 ($V = M/r$), and (d) definition of output variables, which in this case are output variables of C_4.

Step 4.   If all IMs included in the model have been simulated, display the output and stop; otherwise, go to Step 2. The topology of model M_6 from Figure 4.7 used to illustrate the algorithm is shown in Figure 4.9.

The output from the simulation of this model is presented below.

```
SIMULATION OF THE MODEL FROM WORLD #959

Simulation of root component C1
  Component Input:
    force = 100
    linear velocity = 2
  Processes executed:
```

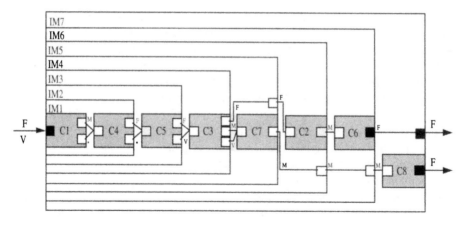

Figure 4.9.   Topology of model M_6.

```
 moment calculated
 radial velocity calculated
Component Output:
 radial velocity = 10.0
 moment = 30.0
Simulation of intermediate model IM1
 Model Input:
  force = 100
  linear velocity = 2
 Processes executed:
  radial velocity propagated
  force calculated
 Model Output:
  force = 75.0
  radial velocity = 10.0
Simulation of intermediate model IM2
 Model Input:
  force = 100
  linear velocity = 2
 Processes executed:
  force propagated
  linear velocity calculated
 Model Output:
  force = 75.0
  linear velocity = 7.0
Simulation of intermediate model IM3
 Model Input:
  force = 100
```

```
  linear velocity = 2
 Processes executed:
  force propagated
  linear velocity propagated
  moment calculated
 Model Output:
  force = 75.0
  linear velocity = 7.0
  moment = 37.5
Simulation of intermediate model IM4
 Model Input:
  force = 100
  linear velocity = 2
 Processes executed:
  force propagated
  moment propagated
 Model Output:
  force = 75.0
  moment = 37.5
Simulation of intermediate model IM5
 Model Input:
  force = 100
  linear velocity = 2
 Processes executed:
  moment propagated
  moment calculated
 Model Output:
  moment = 30.0
  moment = 37.5
Simulation of intermediate model IM6
  Model Input:
  force = 100
  linear velocity = 2
 Processes executed:
  moment propagated
  force calculated
 Model Output:
  force = 100.0
  moment = 37.5
Simulation of intermediate model IM7
 Model Input:
  force = 100
  linear velocity = 2
 Processes executed:
```

```
  force propagated
  force calculated
Model Output:
  force = 93.75
  force = 100.0

SIMULATION OUTPUT:
 Model Input:
  force = 100 (N)
  linear velocity = 2 (m/s)
 Model Output:
  force = 93.75 (N)
  force = 100.0 (N)
```

The simulation output shows that the overall design function is achieved and that the model produces the desired output.

All parameters that cannot be calculated during the simulation are entered interactively by the designer. Those are the geometric characteristics of a product, defined in the layout, needed for calculation of physical effects.

The simulation results should be compared with the desired values of the output variables in order to evaluate the functional behavior of the model. If the desired output has not been generated, the designer may rerun the simulation by changing some of the design parameters. The computer implementation of the approach discussed in this chapter is presented in Vujosevic et al. (1995).

4.6   DISCUSSION

The model synthesis algorithm presented here combines depth-first search and truth-maintained multiple worlds to minimize the number of explored alternatives in the search space and find feasible solutions. The algorithm uses the assumption-based truth maintenance system in a different manner than was used by Ranta et al. (1989). They used multiple worlds to capture particular design decisions, and the result is a directed graph of worlds representing changes of the design model from the initial to the final state of the product design. The number of worlds used is limited. The assumption-based truth maintenance system is used not to reduce the search required to find a feasible solution but rather to detect and maintain dependency relationships between design and manufacturing information. Due to the different application, Ranta et al. (1990) used a different set of rules to detect contradictions and mark worlds as "no good."

An application of a truth maintenance system incorporated in the Proteus AI programming language in the development of Design Advisor to design semicustom very large scale integrated (VLSI) circuits is presented in Steele (1987). The model synthesis algorithm and Design Advisor use a truth maintenance system for its basic intended purpose to determine the current set of beliefs from a current set of justifications. The difference is that Design Advisor uses a justification-based truth maintenance system, which is the simplest form of truth maintenance system upon which all others are based. However, the model synthesis algorithm uses an assumption-based truth maintenance system that allows for making inferences in multiple contexts at once. For more discussion about the differences between these two truth maintenance systems see de Kleer and Forbus (1991).

4.7 SUMMARY

A computer-based approach to mechanical design was presented. The approach assumes that the functional requirements have been previously defined and begins with the definition of abstract components assigned to one or more functional requirements. The designer who defines its input creates an abstract component and output variables and the physical process performed by the component.

The procedure for model synthesis is applicable to mechanical system design and performs the synthesis on a set of defined abstract components. The use of an assumption-based truth maintenance system increases the efficiency of the synthesis algorithm by significantly reducing the number of worlds evaluated in the search process. For the example considered in the chapter, the algorithm has created 2513 worlds to identify 10 feasible solutions, compared to 109,500 worlds that would be needed to explore all possible alternatives using a pure depth-first search algorithm.

The forward-constraint propagation algorithm for model analysis can also be used for a broad range of mechanical systems. However, the calculation of physical processes performed by components may require the input of certain design (geometric) characteristics of the component. The nature of the design process dictates how the system variables are to be modeled (qualitative, quantitative, or binary). For the holding device, quantitative relations among model variables in processes that defined the functionality of each component are the most appropriate. For some other cases, a hierarchy of qualitative processes could be created and assigned to the corresponding components.

REFERENCES

Biswas, G., W. J. Hagins, and K. A. Debelak (1989), Qualitative modeling in engineering applications, in *Proceedings of the 1989 IEEE Conference on Man, Systems, and Cybernetics*, IEEE, New York, pp. 997–1002.

Bradshaw, J. A., and R. M. Young (1991), Evaluating design using knowledge of purpose and knowledge of structure, *IEEE Expert*, April, pp. 33–40.

de Kleer, J., and J. Brown (1984), A qualitative physics based on confluences, *Artificial Intelligence*, Vol. 24, pp. 7–83.

de Kleer, J., and K. D. Forbus (1991), Truth maintenance systems, Tutorial on the Ninth National Conference on Artificial Intelligence, Los Angeles, CA, July, pp. 14–19.

Filman, R. E. (1988), Reasoning with worlds and truth maintenance in a knowledge-based programming environment, *Communications of the ACM*, Vol. 31, No. 4, pp. 382–401.

Hundal, M. S. (1990), A systematic method for developing functional structures, solutions and concept variants, *Mechanisms and Machine Theory*, Vol. 25, No. 3, pp. 243–256.

Joskowicz, L. (1987), Shape and function in mechanical devices, *Proceedings of the Sixth National Conference on Artificial Intelligence*, July 13–17, pp. 611–615.

Kusiak, A., E. Szczerbicki, and R. Vujosevic (1991), Intelligent design synthesis: An object-oriented approach, *International Journal of Production Research*, Vol. 29, No. 7, pp. 1291–1308.

Longenecker, S., and P. A. Fitzhorn, (1990), A shape grammar for modeling solids, in P. A. Fitzhorn (Ed.), *Proceedings of the First International Workshop on Formal Methods in Engineering Design, Manufacturing, and Assembly*, Colorado Springs, CO, pp. 70–68.

Morris, P. H., and R. A. Nado (1986), Representing actions with an assumption-based truth maintenance system, in *Proceedings of the 5th National Conference on Artificial Intelligence*, Philadelphia, pp. 13–17.

Nardi, B. A., and E. A. Paulson (1986), Multiple worlds with truth maintenance in AI applications, in B. du Boulay and D. Hogg (Eds.), *Proceedings of the 7th European Conference on Artificial Intelligence*, North-Holland, Amsterdam, pp. 563–572.

Oren, T. I., B. P. Zeigler, and M. S. Elzas, Eds. (1984), *Simulation and Model-Based Methodologies: An Integrative View*, North-Holland, Amsterdam.

Pahl, G., and W. Beitz (1988), *Engineering Design*, Springer-Verlag, New York.

Pu, P. (1990), Intelligent computer-aided design systems: A synergic approach of artificial intelligence and engineering, *Computing Intelligence*, Vol. 6, pp. 81–90.

Ranta, M., M. Inui, and F. Kimura (1989), A process planning systems for producibility feedback to designers, in *Proceedings of the Third International IFIP Conference on Computer Applications in Production and Engineering*, Tokyo, Japan, Elsevier Science, New York, pp. 373–382.

Rinderle, J. R., and S. A. Finger (1990), Transformational approach to mechanical synthesis, in *Proceedings of the NSF Design and Manufacturing Systems Conference*, Society of Manufacturing Engineers, Dearborn, MI, pp. 67–75.

Stallman, R. and G. J. Sussman (1977), Forward reasoning and dependency-directed backtracking in a system for computer-aided circuit analysis, *Artificial Intelligence*, Vol. 9, pp. 135–196.

Steele, R. L. (1987), An expert system application in semicustom VLSI design, in *Proceedings of the 24th ACM/IEEE Design Automation Conference*, pp. 679–686.

Vujosevic, R., A. Kusiak, and E. Szczerbicki (1995), Reason maintenance in product modeling, *ASME Transactions: Journal of Engineering for Industry*, Vol. 117, No. 2, pp. 223–231.

Zeigler, B. P. (1987), Hierarchical, modular discrete-event modeling in an object-oriented environment, *Simulation*, Vol. 49, No. 5, pp. 219–230.

QUESTIONS

4.1. What are multiple words?

4.2. What is a component?

4.3. What is a product model?

4.4. What is a model base in product modeling?

4.5. What is model synthesis?

PROBLEMS

4.1. Draw models of four components, define inputs and outputs for each of them, and synthesize them into a product model.

4.2. Sketch a product, define its components, define inputs and outputs for each of them, and synthesize them into a product model. Compare the product model with your initial sketch.

CHAPTER 5

PROCESS PLANNING

5.1 INTRODUCTION

Process planning for a mechanical part determines the processing route, operations, machines, fixtures, and tools required to manufacture the part at the lowest cost and best quality (Kusiak, 1991). Human experts often perform this complex task. Numerous efforts have been undertaken to automate process planning. Many of these efforts are surveyed in Alting and Zhang (1989). Automating process planning offers various benefits, for example, reduced lead time and increased consistency of process plans.

There are two basic approaches to automated process planning:

- Variant approach
- Generative approach

In the variant approach, each part is classified based on its attributes and coded using a classification and coding system. The code and the process plan of each part are stored in a database. When a process plan is required for a new part, the part is coded and the most similar process plan is retrieved from the database. The retrieved process plan is modified when necessary. The main advantage of the variant approach is its simplicity. The effectiveness of the variant approach depends on the quality of the process plans stored in the database, user experience, as well as the quality of the classification and coding system.

The generative process planning system builds a process plan using the information about parts, machines, tools, and so on. The existing generative process planning systems generate process plans for parts with rather simple geometry. The advantage of using a generative approach versus a variant approach is that the former produces process plans for parts that are not necessarily of similar shape or belong to the same part family.

Using a computational intelligence (CI) framework, Tsang and Lagoude (1986) formulated the process planning problem as a sequence of actions (operations) and resources (machines, tools, etc.) that enable the goal state (producing of a finished part) to be reached given the initial state (raw material or stock). On the basis of the above formulation, a number of intelligent process planning systems have been developed.

A process planner often extracts manufacturing details in the form of individual elements of the part, known as manufacturing features. To automate the task of part interpretation, feature recognition and extraction become issues.

The degree of automation of manufacturing hardware and software appears to be positively correlated. It is not possible to justify an automated manufacturing system without efficient control software. The concept of concurrent engineering places a great emphasis on process planning. An automated process planning system bridges design with manufacturing. The design of products and components in a concurrent engineering environment emphasizes the availability of the data related to manufacturability, cost, and so on at design stages. In order to provide the data required by a designer in real time, at least some phases of process planning have to be automated. The time to generate a process plan at different levels of abstractions becomes a crucial issue in modern manufacturing.

Numerous models and algorithms discussed throughout this chapter apply to processes other than machining, for example, mechanical and electrical assembly.

5.2 PHASES OF PROCESS PLANNING

Due to the diversity of process planning tasks, it is difficult to apply a uniform approach for its automation. In this chapter, process planning for parts to be machined is considered. The overall process planning task can be partitioned into a number of phases. First of all, it is imperative that the part geometry and topology are completely understood before a process plan is generated. Following this phase, appropriate manufacturing processes are selected. Next, the machines, cutting tools, and fixtures have to be selected. Depending on the value of manufacturing parameters (e.g., the depth of cut, feed rate), the material volume to be removed from the stock is decomposed into smaller manufacturing features. Next precedence constraints are generated that determine the order of manufacturing. Then the manufacturing features are sequenced to generate a final process plan.

Table 5.1 lists eight phases of process planning for parts to be machined with a recommended solution approach for each phase. Each phase in Table 5.1 has been assigned a dominant solution approach (Kusiak, 1990b). For example, all functions in phase 1 are basically performed by a knowledge-based subsystem, while in phase 4 optimization algorithms are complemented with a knowledge-based system. In the next sections of this chapter the eight phases are discussed in detail.

TABLE 5.1. Phases of Process Planning

Phase Number	Phase Name	Solution Approach[a]
1	Interpretation of part design data	KB
2	Selection of manufacturing processes	KB
3	Selection of machines, tools, and fixtures	KB
4	Process optimization	OPT/KB
5	Decomposition of the material volume to be removed	KB
6	Selection of manufacturing features	OPT/KB
7	Generation of precedence constraints	KB
8	Sequencing manufacturing features	OPT/KB

5.3   INTERPRETATION OF PART DESIGN DATA

Interpretation of the part design data depends on the part representation. There are different ways of representing a part for the purpose of process planning. A description of features and their representation and recognition is presented next.

5.3.1   Feature-Based Part Modeling

Traditional CAD (computer-aided design) systems have been successfully used as a tool to increase productivity of the design process. A significant research effort has been directed toward the development of an interface between CAD and CAM (computer-aided manufacturing) systems. A way of thinking about the relationship between the design and manufacturing activities has evolved in different directions. Developmental efforts in the CAD area aim at the design of an intelligent system that would provide a designer with the necessary manufacturing information at the early stages of the design process. A product should be designed to simplify process planning, scheduling, quality control, assembly, and so on (Kusiak, 1992).

Since traditional CAD systems do not meet the requirements of a modern design environment, some alternative approaches have been considered. A widely used approach for representing parts is based on a set of interconnected geometric features. There are many definitions of features, and one suggested by Dixon et al. (1990) is as follows: "A feature is any geometric form or entity uniquely defined by its boundaries, or any uniquely defined geometric attribute of a part, that is meaningful to any life-cycle activity."

Expressing a part design in terms of features supports the human way of reasoning about the design. A feature-based design captures the design information necessary for performing numerous activities that could be encountered in a product life cycle. Although the number of these activities is large, this chapter is concerned with process planning. Even with a limited knowledge about manufacturing, one can conclude that design features are not directly usable in process planning. One needs to find a way, by reasoning about design features, to identify manufacturing features that would be appropriate for process planning.

The term "form feature" as used in this chapter was introduced by Pratt and Wilson (1985). Over the years numerous attempts have been made to clarify the feature concept, establish relationships between different types of features, define a general architecture for a feature-based design system, and develop feature-based application for the life-cycle activities.

The features (material volumes) used in this chapter are defined next (Kusiak, 1992); for an in-depth discussion of these features, see Chapter 3):

1. *Form feature*: a geometric feature of interest to the designer, such as a slot, hole, and so on (see Table 5.2).
2. *Basic form feature*: the basic geometric feature of a part, usually of the same type as the raw material.
3. *Subfeature*: a form feature defined for a basic feature, such as a slot or a hole
4. *Machining feature*: a volume of material to be removed to produce a form feature.
5. *Basic machining feature*: a volume of material to be removed from the raw material to obtain a basic form feature.

TABLE 5.2. Sample Form Features and Corresponding Machining Features

Geometry	Form Feature	Machining Feature
	Plane	Block
	Slot	Block
	Hole	Cylinder

6. *Elementary machining feature*: a volume of material obtained by decomposing the basic machining feature, or a volume of material that corresponds to a particular subfeature that is not decomposable.

7. *Machining feature*: one or more elementary machining features that are removable in a single tool pass.

A relationship between form features and machining features needs to be established. Table 5.2 illustrates such relationship with the example of three machining features. Note that a machining feature is a manufacturing feature confined to a machining domain. The methods that have been used to represent and recognize features can be categorized as follows (Joshi and Chang, 1990):

- Syntactic pattern recognition
- State transition diagrams
- Decomposition approach
- Knowledge-based approach
- Constructive solid approach
- Graph-based approach

Each of these methods is briefly described next.

5.3.2 Syntactic Pattern Recognition

In this recognition method, a set of semantic primitives represents a picture (Liu and Srinivasan, 1984). Here, a pattern is defined with a grammar and a parser is used to apply this grammar to the picture. If the syntax of the picture matches the grammatical specifications, then that picture can be classified as a particular pattern.

An extended context-free grammar

$$G = (V_n, V_t, P, S)$$

and an input string

$$w = a_1, a_2, \ldots, a_n$$

form the input to the schema, where

$V_n =$ set of nonterminal (i.e., defined in the input string) symbols
$V_t =$ set of terminal (i.e., forming the solution) symbols
$P =$ finite set of production rules denoted by $a \Rightarrow b$, where a and b are strings over $V_n \cup V_t$ and with string a involving at least one symbol of V_n
$S =$ a subset of V_n, the starting symbol of a sentence

For extensive discussion of context-free grammars see Hopcroft and Ullman (1969). The syntactic pattern recognition appears to fit the most two-dimensional patterns.

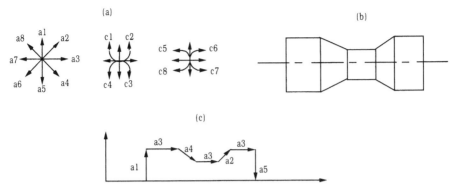

Figure 5.1. Syntactic pattern recognition: (*a*) basic pattern primitives; (*b*) part to be represented; (*c*) part representation.

Figure 5.1 shows a set of basic primitives and a part representation using these primitives. Liu and Srinivasan (1984) used syntactic pattern recognition to select machine tools to manufacture a part. Kyprianou (1980) applied a set of structural primitives such as a convex edge, a concave edge, and a smooth edge against line primitives for feature recognition.

5.3.3 State Transition Diagrams

A state transition diagram is similar to the context-free grammar. In this approach, part geometry is described with the sweep operators and/or the union of swept volumes. Machining features are identified using a state transition. For an input string, the classification of a feature is performed with the state transition diagram.

5.3.4 Decomposition Approach

In this approach, the geometry of a part is decomposed into several volumes. These volumes are generally either design or manufacturing features. A recognition step is necessary after the volume decomposition has taken place. Two different decomposition methods are used: the cut-and-collect method and the use of macros for machining features. The latter method directly uses the results of feature decomposition, and hence further description of the use of macros becomes redundant. The cut-and-collect method determines the volume of material to be removed by identifying the part and its solid model. This volume is further broken down into machining features. That is, if P is the final part and S is the initial stock, then the material V to be removed by machining is obtained from a simple equation, $V = S - P$. The volume V is further cut into smaller pieces. Another strategy in the cut-and-collect approach is to divide this volume into cells. The dimensions of each cell are related to the cutter diameter and the depth of cut.

5.3.5 Knowledge-Based Approach

In the knowledge-based approach, features are recognized using the coded human knowledge. It is believed that a human identifies a feature with a set of predefined notions. For example, a blind hole is recognized with the following production rule (Henderson and Chang, 1988):

Rule R$_1$

　　IF　a hole entrance face exists

AND　the face adjacent to the entrance is cylindrical

AND　the cylindrical face is concave

AND　the next adjacent face is a plane adjacent only to the cylinder

THEN　the feature is a simple blind hole

5.3.6 Constructive Solid Geometry Approach

This approach is also referred to as the set-theoretic approach. Despite the fact that boundary representation (B-rep) of solids is favored, constructive solid geometry (CSG) has been used for extracting feature information. Woo (1977) used a set of algebraic expressions for the primitives in creating the part to extract volumes for NC (numerically controlled) machining. In this approach, the description of an algebraic volume has a nested structure, with each set of parentheses corresponding to a construct at an intermediate stage of design. The constructs are then evaluated one at a time. Woodwark (1988) presented two ways of using the CSG approach for feature recognition: first, restricting the scope of the set-theoretic model by imposing restrictions on the primitives and, second, exploring the possibility of using the information from the way in which the model was created to trigger templates to match them with the model. The basic idea of the CSG approach is illustrated in Figure 5.2, where the volume of material V_f corresponding to the final part is obtained from the volume of the stock V_s.

5.3.7 Graph-Based Approach

In the graph-based approach, a boundary representation of the part is converted to graph $G = (N, A, T)$, where N is the set of nodes, A is the set of arcs, and T is the set of attributes assigned to the arcs. Joshi (1987) proposed an attribute adjacency graph to represent features. Here, each face of the part is represented as a node and each edge or face adjacency is represented as an arc. The attributes take values 0 and 1 if the two adjacent faces are concave and convex, respectively. The features are recognized by analysis of the attribute adjacency graph. Figure 5.3 illustrates the attribute adjacency graphs for two features, a step and a pocket. It is assumed that the features are made of depressions or cavities. A production rule is formulated for each feature type based on the properties of the corresponding attribute adjacency graph. An example rule for the recognition of a pocket is shown next:

Rule R$_2$

　　IF　the graph is cyclic

$V_f = V_s - V_1 - V_2$

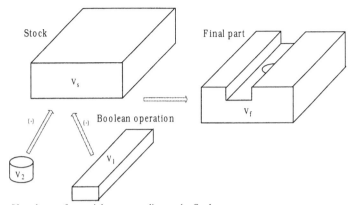

V_f volume of material corresponding to the final part
V_s volume of material corresponding to the stock
V_1, V_2 design features defined in the CAD system

Figure 5.2. Constructive solid geometry.

AND   it has exactly one node with the number of incident 0 arcs equal to the total number of nodes -1

AND   all other nodes are of degree 3

THEN the corresponding feature is a POCKET

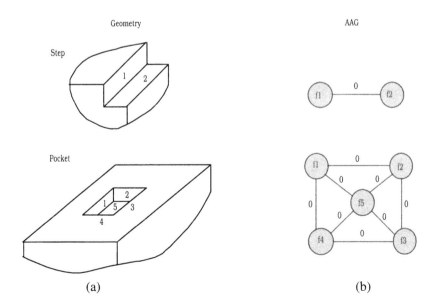

Figure 5.3. (*a*) Two features and (*b*) corresponding attribute adjacency graph.

Sakurai and Gossard (1988) used feature graphs by letting the user interactively define the features from a three-dimensional (3D) boundary representation (B-rep) model. To define a feature, the information relevant to part geometry and topology is extracted. Recognition is established by matching the edges, faces, and geometric information from the B-rep to determine the subset that corresponds to the features.

De Floriani (1987) converted the B-rep to a face-oriented relational model, called the generalized edge face graph. The features to be recognized are classified into depressions and protrusions, holes, and handles. A compound feature is expressed as a combination of simple depressions, protrusions, and hole features.

To fully represent a part, other features including dimensions and tolerances are used. Dimensions and tolerances are typically available in a CAD file. To illustrate manufacturing features jointly with dimensions and tolerances, consider the pocket illustrated in Figure 5.4. The manufacturing features, dimensions, and tolerances of the pocket in Figure 5.4 can be represented by the following frame (Mantyla et al., 1987):

```
Frame 1
 (Pocket-1
 (Type Rectangular_pocket)
 (Length 2.5 Tol .01)
 (Width 1.4 Tol .01)
 (Depth 0.75 Tol .01)
 (Corner_radius 0.25)
 (Side_wall_thickness .250)
 (Bottom_wall_thickness .250)
 (Side_surface_finish 125)
 (Bottom_surface_finish 125)
 (Axial_clearance .75)
```

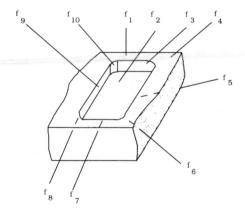

Figure 5.4. Pocket.

```
(Radial_clearance .25)
(Connecting_face f₁)
(Bottom_face f₂)
(Side_faces f₃ f₅ f₇ f₉)
(Fillet_faces f₄ f₆ f₈ f₁₀))
```

The frame above provides information on topology, dimensions, and tolerances of the pocket. If required, additional information may be incorporated in the frame.

5.4 SELECTION OF PROCESSES

Consider the part P in Figure 5.5 obtained by the removal of volume V of material from the stock. The material volume V can be segmented into machining features. A machining feature (machinable volume) is a volume of material that is removed in one tool path. The process of selection of machines, tools, and fixtures is based on the part design features. It is typically performed in two stages. In the first stage one selects a process (e.g., drilling, milling) and in the second stage specific machines, cutting tools, and fixtures are selected. In this section, the process selection procedure is described. Section 5.5 describes the second stage of machine tool and fixture selection.

A manufacturing process is selected for the removal of machining features. It becomes imperative to select processes that can machine the parts according to their design specifications. The rules may include the information about shape and size limitations for the process selected, tolerances that can be obtained, and surface finish capability; for example:

Rule R_3
  IF feature is a POCKET
 AND tolerance = +0.010 in.
 AND surface finish ≤ 94
 THEN machining_process is END_MILLING
 AND machining_direction is Z_Axis

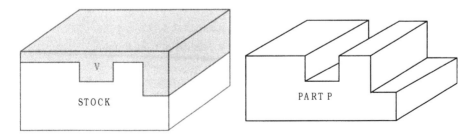

Figure 5.5. Stock and part P.

Similarly, production rules for selection of preparatory processes are formulated; for example:

Rule R₄
    IF machining_process is REAMING
  THEN preparatory_process is DRILLING
    OR preparatory_process is BORING
  AND tolerance = +0.001 in.
  AND surface finish < 63

Alternatively, frames can be used to represent the relationship between processes and features as shown next:

```
Frame 2
 ((BROACHING
  (AKO (Value MACHINING_PROCESS))
  (FRAME_LABEL (Value BROACHING))
  (FEATURES_PRODUCED (Value THROUGH_SLOT,
    Value KEY_WAY))
  (BATCH_QTY (Value 150_PER_HOUR))
  (PREPARATORY_PROCESS (Value DRILLING))
  (TOLERANCE (Value + 0.002 in))
  (FINISH (Value < 94))))
```

5.5 SELECTION OF MACHINES, TOOLS, AND FIXTURES

In this section, it is assumed that selection of manufacturing equipment (i.e., machines, tools, and fixtures) is performed based on the equipment existing in a manufacturing system, rather than procuring the equipment, which is a task considered during design of the manufacturing system. The latter topic is discussed in the next chapter. Since the selection of machines, tools, and fixtures is qualitative, a knowledge-based approach is likely to be used. The production rule and two frames presented next illustrate the nature of knowledge at this process planning phase:

Rule R₅
    IF drilling process p_3 is recommended for feature f_5
  AND drilling machine M_1 that can perform process p_3 is available
  THEN use machine M_1 to derive the feature f_5

As an alternative to production rules, frames can be used; for example:

```
Frame 3
(DRILLING
(AKO (Value DRILLING))
```

```
(FRAME_LABEL (Value DRILLING))
(TOOL_SELECTED (Value DRILL_BIT))
(SMALLER_SIZE (Value D1 diameter))
(LARGER_SIZE (Value D2 diameter))
(MACHINE SELECTED (Value MACHINE_M₁)))
Frame 4
(MACHINE_M₁
(AKO (Value DRILLING_MACHINE))
(FRAME_LABEL (Value DRILLING_MACHINE))
(MAX_AXIAL_LOAD (Value X))
(MAX_POWER_ATTAINABLE (Value Y)))
```

5.6 PROCESS OPTIMIZATION

A basic issue of concern in machining optimization is tool life. It is defined as the cut time of a new tool before a certain flank wear is reached (Chang et al., 1991). Taylor's tool life equation is given as

$$t = \frac{\lambda C}{V^{\alpha_T} f^{\beta_T} a_p^{\gamma_T}} \tag{5.1}$$

where

$$
\begin{aligned}
t &= \text{tool life} \\
\lambda, C &= \text{constants for a specific tool/workpiece combination} \\
\alpha_T, \beta_T, \gamma_T &= \text{exponents for a specific tool/workpiece combination} \\
V &= \text{cutting speed} \\
f &= \text{feed rate} \\
a_p &= \text{depth of cut}
\end{aligned}
$$

Numerical examples with Eq. 5.1 may be found in Ludema et al. (1987). Machining optimization models are classified as single-pass and multipass models. In a single-pass model, one assumes that only one pass is needed to produce the required geometry. In this case the depth of cut is fixed. In a multipass model, this assumption is relaxed and the depth of cut becomes a decision variable.

5.6.1 Single-Pass Model

Before presenting the single-pass model, the following notation is introduced:

$$
\begin{aligned}
t_{pr} &= \text{total (machining, handling, and tool change) time for a prismatic part} \\
t_m &= \text{machining time} \\
t_h &= \text{material handling time} \\
t_t &= \text{tool changing time} \\
t &= \text{tool life}
\end{aligned}
$$

C_{pr} = production cost per part
C_b = setup cost for a batch of parts
C_m = total machine and operator rate
C_r = tool cost
N_b = batch size

Based on the above notation, the objective of the single-pass model for machining a prismatic part is formulated as follows (Chang et al., 1991):

$$\min t_{pr} = t_m + t_h + \frac{t_m}{t} \tag{5.2}$$

or in terms of cost

$$\min C_{pr} = \frac{C_b}{N_b} + C_m \left[t_m + t_h + \frac{t_m}{t} \left(t_t + \frac{C_r}{C_m} \right) \right]$$

The typical constraints included in the model are:

Spindle-speed constraints:

$$n_{w,min} < n_w < n_{w,max} \qquad \text{(for the part)} \tag{5.3}$$

$$n_{t,min} < n_t < n_{t,max} \qquad \text{(for the tool)} \tag{5.4}$$

Feed constraint:

$$f_{min} < f < f_{max} \tag{5.5}$$

Cutting force constraint:

$$F_c < F_{c,max} \tag{5.6}$$

Power constraint:

$$P_m < P_{max} \tag{5.7}$$

Surface finish constraint:

$$R_a < R_{a,max} \tag{5.8}$$

The model (5.2)–(5.8) does not reflect the total complexity of the process planning problem, for example the relationship between the feed rate and surface finish is not considered. Each of the variables n_w, n_t, F_c, P_m, and R_a is a function of some of the variables included in the objective function as well as some other constants. The formulas for n_w, n_t, F_c, P_m, and R_a are provided in the literature (e.g., see Chang et al.,

1991). A formulation of the single-pass model for the turning process is presented in Philipson and Ravindran (1979), and multipass turning is presented in Mesquita et al. (1995) and Tan and Creese (1995). The same authors also discussed linear, geometric, and goal programming approaches for solving this model.

5.6.2 Multipass Model

In addition to the previously defined symbols, the following notation is used:

n = number of machining passes
a_p^i = depth of cut in machining pass i
t_m^i = time required for machining pass i
C_{pr}^i = production cost per part for machining pass i

The multipass model can be formulated as follows:

$$\min t_{pr} = t_h + \sum_{i=1}^{n} \left(t_m^i + \frac{t_m^i}{t} t_t \right) \tag{5.9}$$

or in terms of cost per part:

$$\min C_{pr} = \frac{C_b}{N_b} + C_m t_h + \sum_{i=1}^{n} C_{pr}^i$$

where

$$C_{pr}^i = C_m \left[t_m^i + \frac{t_m^i}{t} \left(t_t + \frac{C_T}{C_m} \right) \right] \tag{5.10}$$

and the subscript i represents the ith pass. The model includes constraints (5.3)–(5.8) and

$$a_{p,min} < a_p^i < a_{p,max} \tag{5.11}$$

$$a_p = \sum_{i=1}^{n} a_p^i \tag{5.12}$$

Optimization of the machining process is a complex task involving various mathematical programming models, formulas, and algorithms as well as empirical parameters and procedures. A knowledge-based system integrated with mathematical models, formulas, and algorithms seems the most suitable alternative for optimizing the machining process. A hybrid solution approach for such a problem is presented in Kusiak (1987).

A rule for selecting a model for machining optimization is given next:

Rule R_6

 IF the depth of cut of feature f_2 is ≤ 2 mm

AND surface finish class is sf_3

AND tolerance class is t_4

THEN use the single-pass model for machining optimization

The two models for machining optimization are not likely to be used in a factory environment. Tables and simple formulas are often used. The formulas for calculation of machining parameters (e.g., feed rate, depth of cut) and process planning parameters (e.g., machining time) can also be selected by production rules; for example:

Rule R_7

 IF the process is turning

THEN calculate machining time from the formula $t_m = l_w/fn_w$

where

 $l_w =$ length of surface to be turned

 $f =$ feed rate

 $n_w =$ rotational frequency of the part (rpm)

5.7 DECOMPOSITION OF MATERIAL VOLUME TO BE REMOVED

The volume of material to be removed (volume V in Figure 5.5) can be decomposed into volumes $v_1, \ldots, v_9$ shown in Figure 5.6. The latter can be accomplished by drawing planes adjacent to all surfaces of the part. Solving the multipass machining model discussed in the previous section results in one or more tool passes. This might also be viewed as generating additional planes for the volume to be removed. The planes decompose the previously generated volumes. The machining optimization phase produces a plane p_1 and other planes resulting in the decomposition of volumes $v_1, \ldots, v_9$ into elementary machining features (elementary volumes) $e_1, \ldots, e_{18}$ in Figure 5.7.

 The plane p_1 and the two vertical planes decomposed the volumes $v_3, \ldots, v_7$ into elementary machining features $e_1, \ldots, e_{10}$. The remaining horizontal planes

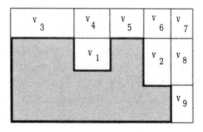

Figure 5.6. Decomposition of volume V to be removed into volumes $v_1, \ldots, v_9$.

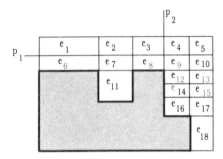

Figure 5.7. Segmentation of volumes $v_1, \ldots, v_9$ in Figure 5.6 into elementary machining features $e_1, \ldots, e_{18}$.

resulted in the elementary machining features $e_{12}, \ldots, e_{17}$. Volume v_3 in Figure 5.6 was renamed feature e_{18} in Figure 5.7. Each elementary machining feature $e_i, i = 1, \ldots, 18$, in Figure 5.7 is not greater than the corresponding volume $v_k, k = 1, \ldots, 9$, in Figure 5.6. The elementary machining features can be combined into machining features. A machining feature V_j is a set of elementary machining features that can be removed in one tool pass [e.g., see columns $V_j, j = 1, \ldots, 11$, in matrix (5.22)]. The composition of each machining feature V_j depends mostly on the type of machine and tool (standard versus special) suggested for removal of this feature. For example, the set $\{e_1, e_9, e_{15}\} = V_{12}$ of elementary features is not a machining feature as no tool and technology exist that would remove V_{12} in one tool pass. Therefore, V_{12} is not included as a column in matrix (5.22).

5.8 SELECTION OF MANUFACTURING FEATURES

Before the model for optimal selection of machining features is presented, the following notation needs to be defined:

I = set of all elementary machining features of the part

J = set of all machining features V_j

$$a_{ij} = \begin{cases} 1 & \text{if elementary machining feature } e_i \text{ corresponds to machining feature } V_j \\ 0 & \text{otherwise} \end{cases}$$

T = set of available tools for machining the part

F = set of available fixtures

N_t = upper limit on the number of tools to be used for machining the part

J_t = subset of machining features for which tool $t \in T$ applies, $\Sigma_{t \in T}(\cup J_t = J)$

J_f = subset of machining features for which fixture $f \in F$ applies $(\cup_{f \in F} J_f = J)$

c_j = removal cost of machining feature $V_j, j \in J$

p_t = utilization cost of tool $t, t \in T$

k_f = utilization cost of fixture $f, f \in F$

N_f = upper limit on a number of fixtures to be used

$$x_j = \begin{cases} 1 & \text{if machining feature } V_j \text{ is selected, } j \in J \\ 0 & \text{otherwise} \end{cases}$$

$$y_t = \begin{cases} 1 & \text{if tool } t \text{ is selected, } t \in T \\ 0 & \text{otherwise} \end{cases}$$

$$z_f = \begin{cases} 1 & \text{if fixture } f \text{ is selected, } f \in F \\ 0 & \text{otherwise} \end{cases}$$

The objective function of the model presented next minimizes the total cost of removal of machining features and tool and fixture utilization costs:

$$Z = \min \sum_{j \in J} c_j x_j + \sum_{t \in T} p_t y_t + \sum_{f \in F} k_f z_f \tag{5.13}$$

such that

$$\sum_{j \in J} a_{ij} x_j \geq 1 \qquad \text{for all } i \in I \tag{5.14}$$

$$\sum_{t \in T} y_t \leq N_t \tag{5.15}$$

$$\sum_{j \in J_t} x_j \leq |J_t| y_t \qquad \text{for all } t \in T \tag{5.16}$$

$$\sum_{f \in F} z_f \leq N_f \tag{5.17}$$

$$\sum_{j \in J_f} x_j \leq |J_f| z_f \qquad \text{for all } f \in F \tag{5.18}$$

$$x_j = 0, 1 \qquad \text{for all } j \in J \tag{5.19}$$

$$y_t = 0, 1 \qquad \text{for all } t \in T \tag{5.20}$$

$$z_f = 0, 1 \qquad \text{for all } f \in F \tag{5.21}$$

Note that $|\cdot|$ in constraints (5.16) and (5.18) denotes the cardinality of a set.

Constraint (5.14) ensures that each elementary machining feature is included in at least one machining feature. Inequality (5.15) imposes an upper limit on a number of tools to be used for machining the part. Constraint (5.16) ensures that a machining feature can be removed only if a tool required has been used. The number of fixtures to be used is limited to N_f by constraint (5.17). Constraint (5.18) ensures that a machining feature can be removed only if the fixture required has been used. The integrality of variables x_j, y_t, and z_f is imposed by constraints (5.19), (5.20), and (5.21), respectively.

It should be stressed that not all constraints (5.15)–(5.21) need to be used in a particular industrial application. For example, constraint (5.15) may be imposed only when the tool magazine capacity is critical.

Example 5.1. For the part in Figure 5.7 the following incidence matrix has been constructed:

$$T = [t_1 \quad t_2 \quad t_3 \quad t_2 \quad t_2 \quad t_3 \quad t_1 \quad t_1 \quad t_2 \quad t_4 \quad t_2]$$

$$F = [f_1 \quad f_1 \quad f_2 \quad f_2 \quad f_2 \quad f_1 \quad f_1 \quad f_1 \quad f_3 \quad f_2 \quad f_1]$$

	V_1	V_2	V_3	V_4	V_5	V_6	V_7	V_8	V_9	V_{10}	V_{11}
e_1	1		1								
e_2	1			1							
e_3	1				1						
e_4	1				1		1				
e_5	1									1	
e_6		1	1								
e_7		1		1							
e_8		1			1						
e_9		1			1		1				
e_{10}		1				1				1	
e_{11}				1	1						
e_{12}									1		
e_{13}						1			1		1
e_{14}									1		
e_{15}						1			1		1
e_{16}									1		
e_{17}						1			1		1
e_{18}						1	1				1

(5.22)

$$C = [5 \quad 5 \quad 2 \quad 3 \quad 1 \quad 4 \quad 5 \quad 1 \quad 5 \quad 6 \quad 6]$$

Each machining feature $V_j, j = 1, \ldots, 11$, in matrix (5.22) has been defined as a combination of elementary machining features that can be removed in one path of a standard cutting tool. Of course, other machining features are possible; however, their number has been restricted to 11 to make the model (5.13)–(5.21) easy to present and solve.

For each machining feature V_j in matrix (5.22) the following data have been specified:

(a)   Vector $C = [c_j]$ of machining costs:

$$C = [5, 5, 2, 3, 1, 4, 5, 1, 5, 6, 6]$$

(b)   Vector T of tools:

$$T = [t_1, t_1, t_3, t_2, t_2, t_3, t_1, t_1, t_2, t_4, t_2]$$

(c)   Vector F of fixtures:

$$F = [f_1, f_1, f_2, f_2, f_2, f_1, f_1, f_1, f_3, f_2, f_1]$$

(d)   The maximum number of tools to be used: $N_t = 3$.
(e)   The maximum number of fixtures to be used: $N_f = 2$.

The vectors of tool and fixture utilization costs $[p_t] = [k_f] = 1$. Solving the model (see the Appendix) for the data provided above produces the following solution:

1. $x_1 = x_2 = x_5 = x_8 = x_{10} = 1$, with the following corresponding machining features:

$$V_1 = \{e_1, e_2, e_3, e_4, e_5\} \qquad V_2 = \{e_6, e_7, e_8, e_9, e_{10}\}$$

$$V_5 = \{e_{11}\} \qquad V_8 = \{e_{18}\} \qquad V_{10} = \{e_{12}, e_{13}, e_{14}, e_{15}, e_{16}, e_{17}\}$$

Note that the five machining features $V_j, j = 1, 2, 5, 8, 10$, include all the elementary machining features $e_1, \ldots, e_{13}$ as shown in matrix (5.22).
2. $y_1 = y_2 = y_4 = 1$, with the corresponding tools $t_1, t_2,$ and t_4 selected.
3. $z_1 = z_2 = 1$, with the corresponding fixtures f_1 and f_2 selected.
4. The value of the objective function is $Z = 23$.

The solution generated includes the minimum number of tools and fixtures and has the lowest cutting cost. Experience with the model (5.13)–(5.21) for a set of five industrial parts indicates that the cutting cost can be reduced as much as 8% and the number of fixtures and tools can also be reduced.

5.9   GENERATION OF PRECEDENCE CONSTRAINTS

Solving the model (5.13)–(5.21) results in a set of machining features. It is obvious that some precedence constraints exist among these features. The constraints can be generated using production rules; for example:

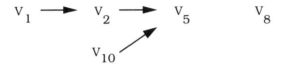

Figure 5.8. Precedence graph for the part in Figure 5.5.

Rule R_8

  IF   machining feature V_i can be accessed only after machining feature V_k is removed

THEN   generate a precedence constraint $V_k \rightarrow V_i$

Rule R_8 is illustrated based on the solution of Example 5.1. Looking at Figure 5.7, it is clear that for the features $\{e_1, e_2, e_3, e_4, e_5\} = V_1$ and $\{e_6, e_7, e_8, e_9, e_{10}\} = V_2$ the following precedence constraint can be generated:

$$V_1 \rightarrow V_2$$

Rule R_9

  IF   plane $p \in V_k$ is a tolerance base for machining feature V_i

THEN   precedence $V_k \rightarrow V_i$ is generated

Assuming that plane p_2 in Figure 5.7 is a tolerance base for the slot feature $e_{11} \in V_5$, the precedence constraint presented below is generated:

$$V_{10} \rightarrow V_5$$

Recall that $V_{10} = \{e_{12}, e_{13}, e_{14}, e_{15}, e_{16}, e_{17}\}$.

5.10   SEQUENCING MANUFACTURING FEATURES

Given the graph of precedence constraints among manufacturing features, an optimal sequence of manufacturing features can be generated by Algorithm 5.1, known as the topological ordering algorithm (Horowitz and Sahni, 1983).

Algorithm 5.1

  Step 1. For $i = 1, \ldots, n$ output the manufacturing features (vertices).

  Step 2. If every manufacturing feature has a predecessor, then the precedence graph has a cycle and is infeasible. Stop.

Step 3. Select a manufacturing feature V_i with no predecessors.

Step 4. Output V_i.

Step 5. Delete the manufacturing feature V_i and all edges leading out of V_i from the precedence graph.

Step 6. End.

The topological ordering algorithm does not give any consideration to the sequence of manufacturing resources used by the features. Process plans generated by Algorithm 5.1 may require an excessive number of toll, fixture, or machine changeovers. The minimization of the number changeovers can be easily incorporated in Algorithm 5.1 by replacing Step 3 with Step 3′.

Step 3′. Select a machining feature V_i that has no predecessors and requires manufacturing resources most similar to the resources required by V_{i-1}.

The topological ordering algorithm is illustrated in Example 5.2.

Example 5.2. Order topologically the directed precedence graph with the five machining features shown in Figure 5.8. In fact, the digraph in Figure 5.8 has been obtained using the machining features generated in Example 5.1.

One can easily check that there are no cycles in the digraph of Figure 5.8. Step 2 of the topological sorting algorithm is then not relevant. In Step 3, any of the three vertices V_1, V_{10}, or V_8 can be selected. Beginning with vertex V_3 the topological sorting algorithm produces the following sequence of machining features: $\{V_1, V_2, V_{10}, V_5, V_8\}$. Assigning each machining feature, tools, fixtures, and machines required results in the following process plan: $\{(V_1, t_1, f_1, \text{MC-5}), (V_2, t_1, f_1, \text{MC-5}), (V_{10}, t_4, f_2, \text{MC-5}), (V_5, t_2, f_2, \text{MC-5}), (V_8, t_1, f_1, \text{MC-5})\}$.

This process plan is presented in an expanded form in Table 5.3 and specifies machining features (V_j), tools (t_t), fixtures (f_f), and the machining center (MC-5). The original topological sorting algorithm (Algorithm 5.1) has generated it. This process plan requires two fixture changes and three tool changes, as the changeover aspect of fixtures and tools was not considered. Sorting the graph in Figure 5.8 with the modified topological sorting algorithm (Step 3 replaced with Step 3′) results in a different sequence of machining features, in fact with a smaller number of changeovers. For the data in Example 5.2, the following process plan is generated with the modified topological sorting algorithm: $\{(V_8, t_1, f_1, \text{MC-5}), (V_1, t_1, f_1, \text{MC-5}), (V_2, t_1, f_1, \text{MC-5}), (V_{10}, t_4, f_2, \text{MC-5}), (V_5, t_4, f_2, \text{MC-5})\}$. This process plan obeys all precedence constraints in Figure 5.8 and requires only one tool change and one fixture change.

Depending upon the objective and constraints of the feature selection model (5.13)–(5.21), the topological sorting algorithm can be modified to fit that problem. In particular, the total sequence-dependent fixture setup cost can be minimized. In this case the fixture setup cost would not have to be included in the model (5.13)–(5.21).

TABLE 5.3. Process Plan for Part *P* in Figure 5.5

Operation Number	Machine Number	Tool Number	Fixture Number	Feature Removed	Graphical Representation
1	MC-5	t_1	f_1	V_1	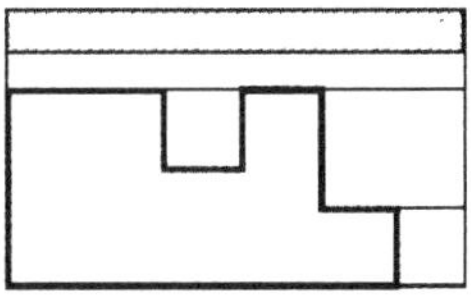
2	MC-5	t_1	f_1	V_2	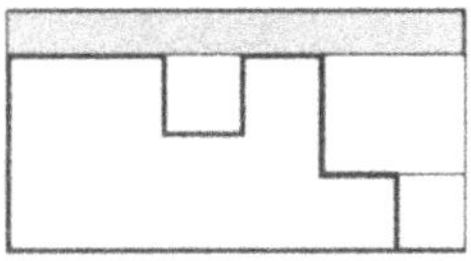
3	MC-5	t_4	f_2	V_{10}	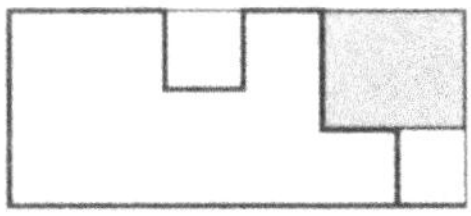
4	MC-5	t_2	f_2	V_5	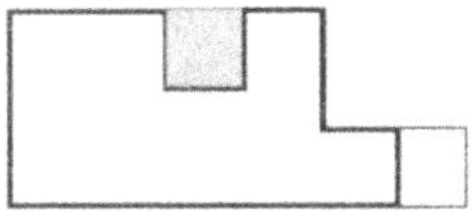
5	MC-5	t_1	f_1	V_8	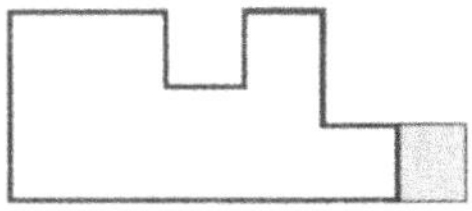

5.11 OBJECT-ORIENTED SYSTEM FOR PROCESS PLANNING

In this section, implementation of phases 5–8 from Table 5.1 is presented. The system was developed in Smalltalk-80 on a Macintosh computer (Vujosevic and Kusiak, 1992). The Smalltalk-80 class hierarchy implemented in this system is shown next:

Object
    Model
        PartModel
    Controller
        MouseMenuController

        FBPPController
  View
    Canvas
      FBPPView
  Feature
    Block
    Cylinder
    Face
      PlanarFace
    Hole
    LBracket
    Slot
      OpenSlot
      BlindSlot
    Surface

The feature-based information about a part is contained in the class **PartModel**, which is a subclass of the class **Model**. The class **FBPPController** controls the user–model interaction. The class **FBPPView** displays a two-dimensional view of the part being modeled as well as a set of elementary manufacturing features created during the part modeling phase. The class **Feature** is a superclass for a number of classes representing specific feature types. It contains data and methods common to all types of features. Classes representing feature types contain data and methods specific for a particular feature type.

The user interface includes a set of buttons for performing feature-based part modeling and selection of machining features, a window for display of a feature-based part model and a set of elementary volumes generated, and a window for textual explanation of the activities performed and display of results. A part model is stored in the model base.

5.11.1 Part Modeling and Generation of Elementary Manufacturing Features

In this section, the procedure for selection of machining features for the part shown in Figure 5.9 is presented. Designing a part begins with the definition of the basic part characteristics, for example, name, type, and material. Then, the part dimensions and part position with respect to the predefined coordinate system are defined. This is followed by the definition of the basic form features of the part. The reasoning process about machining features begins at this point. The basic machining feature (BMF), that is, the overall volume of material to be removed, is generated. This process is performed by production rules generating an object representing the BMF. Having defined the part in Figure 5.9, a block object representing raw material with dimensions (200, 200, 200) and origin (0, 0, 0), and another block object representing

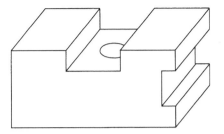

Figure 5.9. Mechanical part.

the basic form feature with dimensions (180, 180, 200) with the origin $(-10, -10, 0)$, the following production rule is executed:

Rule R_{10}
> IF   raw material length is greater than the length of the basic form feature length
> AND   raw material height is greater than the height of the basic form feature height
> AND   the width of the raw material and basic form feature are the same
> AND   the x coordinate of the basic feature is: (raw material length – basic form feature length)
> AND   the y coordinate of the basic feature is: (raw material height – basic form feature height)
> THEN   the basic manufacturing feature is an L-bracket

An object representing the L-bracket feature is created by instantiating the corresponding **L-Bracket** class. Parameters of the L-bracket BMF are determined based on the dimensions of the raw material and basic part feature.

If the BMF has a complex geometric form, it is automatically decomposed into a number of elementary machining features. The generated L-bracket type of BMF (see Figure 5.10a) is decomposed into the three elementary machining features shown in Figure 5.10b. The elementary machining features generated are displayed on the screen.

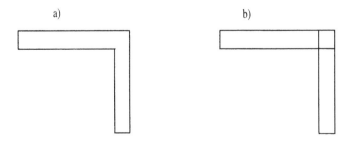

Figure 5.10. L-bracket: (a) basic manufacturing feature; (b) three elementary manufacturing features.

In general, a BMF includes one or more elementary manufacturing feature stored in the instance variable *elementaryManufacturingFeature* of the class **PartModel**.

Next, the set of subfeatures is defined and the corresponding elementary manufacturing features are generated. For example, for a slot the corresponding manufacturing feature (an instance of the class **Block**) is created and placed in *elementaryManufacturingFeature*. For each newly created manufacturing feature, Algorithm 5.2 (the extended edges algorithm) for the decomposition of manufacturing features is applied (for details see Vujosevic and Kusiak, 1992).

Algorithm 5.2

Step 1. From the list of existing manufacturing features, determine a feature that is adjacent to the newly created manufacturing feature. If no such feature exists, stop.

Step 2. Decompose the manufacturing feature selected and/or the new manufacturing feature using geometric reasoning about the relationship between them. For example, for two adjacent block features, 11 different cases may occur (for the two-dimensional case), as shown in Table 5.4.

Step 3. Remove the decomposed manufacturing feature(s) from the list, and place the elementary manufacturing features obtained by decomposition in the same list.

Step 4. Go to Step 1.

TABLE 5.4. Examples of Relationships between Two Block Elementary Features

Spatial Relationship	Relationship Type	Decomposition Outcome
A / B	A and B do not overlap	A / B
A / B	A partially overlaps B with no edges aligned	A_1 A_2 / B_1 B_2
A / B	A partially overlaps B with left edges aligned	A / B_1 B_2
A / B	A completely overlaps B with left edges aligned	A_1 A_2 / B
A / B	A partially overlaps B with no edges aligned	A / B_1 B_3 B

Algorithm 5.2 defines a slot subfeature with dimensions (100, 20, 200) and the origin in (0, 70, 0) for the part in Figure 5.9.

An instance of the class **Slot** is created and placed in the instance variable *subfeature* of the class **PartModel**. The corresponding manufacturing feature that represents a material volume to be removed in order to produce the slot is also generated. After completion of this step, four material volumes (one manufacturing feature and three elementary manufacturing features obtained from decomposition of the basic manufacturing feature) of the block type are stored in the instance variable *elementaryManufacturingFeature* (see Figure 5.11a). The **PartModel** method is then executed and Algorithm 5.2 produces the result shown in Figure 5.11b. The system does not allow for the decomposition of features with different types, for example, block- and cylinder-type features.

Algorithm 5.2 is executed for each newly created machining feature. The adjacency relationships between elementary machining features are generated. To accomplish this, a representation scheme for different types of elementary machining features must be established, for example, the representation scheme for the faces of a block is shown in Figure 5.12.

5.11.2 Grouping Elementary Manufacturing Features

Elementary machining features that can be removed in a single tool path are grouped to form a machining feature. For example, all block-type elementary machining features are grouped separately from the cylinder-type ones. This simplifies reasoning about the formation of machining features. The algorithm for grouping elementary manufacturing features of the same type along a coordinate axis is presented next.

Algorithm 5.3

Step 1. Form list L_1 of elementary manufacturing features included in a manufacturing feature and list L_2 of manufacturing features to be generated.

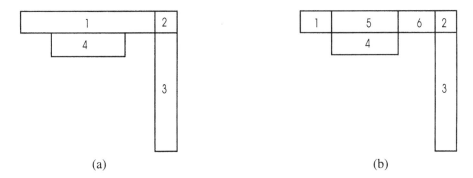

(a)   (b)

Figure 5.11. Decomposition of machining features: (*a*) geometry before decomposition; (*b*) geometry after decomposition.

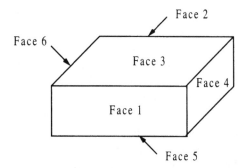

Figure 5.12. Representation of faces of the elementary machining feature "block."

Step 2.   Select an elementary manufacturing feature e from list L_1 of the existing elementary manufacturing features.

Step 3.   If V is not adjacent to any other elementary manufacturing feature, add e to the list L_1, delete from the list of existing elementary manufacturing features, and go to Step 2; otherwise, go to Step 4.

Step 4.   Add the elementary manufacturing features that are adjacent to e in a particular direction (x, y, or z) and are not included in the lists L_1 and L_2, to the list L_1, and delete them from the list of existing elementary manufacturing features.

Step 5.   Add the lists L_1 and L_2 and repeat Step 2 until no adjacent elementary manufacturing features in a particular direction exist.

The implementation of Step 4 is different for each group of elementary machining features. Algorithm 5.3 is further applied for each of the three directions in order to obtain all feasible groups of elementary machining features. Only feasible machining features are generated, and this process is controlled by production rules. Algorithm 5.3 produces a set of machining features stored in the instance variable *machiningFeature* of the class **PartModel**.

Grouping elementary machining features into machining features is performed after the modeling process has been completed. For the part in Figure 5.9, the

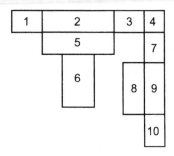

Figure 5.13. Elementary machining features of the part in Figure 5.9.

feature-based modeling process produced a set of elementary machining features, shown in Figure 5.13. The features include one cylinder-type (volume 6) and nine other block elementary machining features.

5.11.3   Selection of Manufacturing Features

The set-covering model for selection of machining features is considered.

Define

I = set of elementary machining features e_i

J = set of machining features V_j

$$a_{ij} = \begin{cases} 1 & \text{if elementary machining feature } e_i \text{ corresponds to machining feature } V_j \\ 0 & \text{otherwise} \end{cases}$$

c_j = removal cost of machining feature V_j

$$x_j = \begin{cases} 1 & \text{if machining feature} V_j \text{ is selected, } j \in J \\ 0 & \text{otherwise} \end{cases}$$

Then

$$\min \sum_{j \in J} c_j x_j \tag{5.23}$$

subject to

$$\sum_{j \in J} a_{ij} x_j \geq 1 \qquad \text{for all } i \in I \tag{5.24}$$

$$x_j = 0, 1 \qquad \text{for all } j \in J \tag{5.25}$$

An efficient heuristic algorithm for solving the model (5.23)–(5.25) was developed by Chvatal (1979).

To present Chvatal's algorithm, denote

$|V_j|$ = cardinality of vector V_j (number of nonzero elements in vector V_j)

Algorithm 5.4 (Chvatal, 1979)

Step 0.   Set the solution set $J^* = \varnothing$.

Step 1.   If $V_j = \varnothing$ for all j, stop; J^* is a solution. Otherwise, find a subscript k maximizing the ratio $|V_j|/c_j$ and proceed to Step 2.

Step 2.   Add k to J^*, replace each V_j with $V_j - V_k$, and go to Step 1.

This algorithm is illustrated in Example 5.3.

Example 5.3. Consider the part in Figure 5.14 with four elementary features removed. Given the incidence matrix $[a_{ij}]$ (5.26) and vector $[c_j]$ of machining costs, solve the set-covering model (5.23)–(5.25):

$$[a_{ij}] = \begin{array}{c} \\ e_1 \\ e_2 \\ e_3 \\ e_4 \end{array} \begin{bmatrix} V_1 & V_2 & V_3 & V_4 & V_5 & V_6 & V_7 \\ 1 & & & & 1 & & 1 \\ & 1 & & & 1 & 1 & 1 \\ & & 1 & & 1 & & \\ & & & 1 & & 1 & \end{bmatrix} \qquad (5.26)$$

$$[c_j]\; [0.4 \quad 0.4 \quad 0.4 \quad 0.3 \quad 0.8 \; 0.7 \quad 0.6]$$

In iteration 1 and Step 1 of Algorithm 5.4 the value

$$\max |V_j|/c_j = \max\ \{1/0.4,\ 1/0.4,\ 1/0.4,\ 1/0.3,\ 3/0.8,\ 2/0.7,\ 2/0.6\} = 3/0.8 = 3.75$$

with the corresponding $k = 5$ is computed.

In Step 2, the set $J^* = \{5\}$ is updated and each V_j is replaced with $V_j - V_5$, which corresponds to the removal of rows e_1, e_2, and e_3 from matrix (5.26).

In iteration 2 and Step 1 of Algorithm 5.4, the value $\max|V_j|/c_j = \max\ \{1/0.3,\ 1/0.7\}$ $= 1/0.3 = 3.33$ with the corresponding $k = 4$ is computed. The final solution is $J^* = \{4, 5\}$, that is, machining features $V_4 = \{e_4\}$ and $V_5 = \{e_1, e_2, e_3\}$ have been selected.

This algorithm is implemented as an instance method of the class **PartModel** and is applied to the machining features stored in the instance variable *machiningFeature* of the same class.

For the set of elementary machining features in Figure 5.13, the following incidence matrix is formed:

$$\begin{array}{c} \\ e_1 \\ e_2 \\ e_3 \\ e_4 \\ e_5 \\ e_6 \\ e_7 \\ e_8 \\ e_9 \\ e_{10} \end{array} \begin{bmatrix} V_1 & V_2 & V_3 & V_4 & V_5 & V_6 & V_7 & V_8 \\ 1 & & & & & & & \\ 1 & 1 & & & 1 & & & \\ & 1 & & & & & & \\ & 1 & & & 1 & & & \\ 1 & & & 1 & & & & \\ & & 1 & & & & & \\ & & & & 1 & & 1 & \\ & & & & & 1 & & \\ & & & & & 1 & 1 & \\ & & & & & 1 & & 1 \end{bmatrix}$$

$$[c_j]\ [\ 14 \quad 4 \quad 1 \quad 6 \quad 6 \quad 12 \quad 3 \quad 2]$$

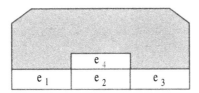

Figure 5.14. Part with elementary features e_1, e_2, e_3, and e_4 removed.

The final solution generated by Algorithm 5.4 is the set of machining features $\{V_2, V_3, V_4, V_7, V_8\}$.

5.11.4 Generation of Precedence Constraints and Sequencing Machining Features

Depending on the nature of precedence among machining features, various objectives and constraints are considered. The constraint concerned with geometric relationships among machining features is one of the most important ones.

Examples of two production rules for generating precedence constraints are given next:

Rule R_{11}
  IF the selected machining feature has no adjacent machining features
  OR the selected machining feature is external
THEN precedence constraints do not exist.

For example, machining feature 2 in Figure 5.15 is external and there is no machining feature that must be machined prior to removal of feature 2.

Rule R_{12}
  IF the machining feature selected is internal
 AND the volume selected has one or more adjacent features
THEN the adjacent feature(s) must be machined first.

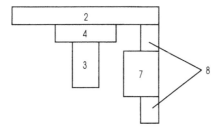

Figure 5.15. Machining features selected for part in Figure 5.9.

To machine feature 4 in Figure 5.15, feature 2 must be machined first, resulting in the following precedence constraint:

$$V_2 \rightarrow V_4$$

Another constraint may imply that a selected surface be a tolerance base for a feature. The corresponding constraint is generated by executing the previously stated Rule R_9:

IF plane p of machining feature V_k is a tolerance base for feature V_i
THEN the precedence $V_k \rightarrow V_i$ is established

One may also consider the sequence-dependent fixture setup cost. An approach addressing this problem is presented in Hayes and Wright (1989).

Given the precedence constraints among machining features, an optimal sequence of machining features can be generated by Algorithm 5.1, the topological ordering algorithm (Horowitz and Sahni, 1983). The topological ordering algorithm modified to handle the objects representing machining features is presented next.

Algorithm 5.5

Step 1. From a set of machining features stored in the instance variable *machiningFeature*, select one that has no predecessors. If every machining feature has a predecessor, then there is no feasible sequence of machining features. Stop.

Step 2. Store the feature selected in the array *sequenceOfmachiningFeatures*; delete this feature from *machiningFeature*; delete the feature number from the predecessor array of each volume for which the feature selected is a predecessor.

Step 3. Go to Step 1 until *machiningFeature* is empty.

For the part in Figure 5.9, Algorithm 5.5 produces the sequence of machining features $\{V_2, V_3, V_4, V_7, V_8\}$.

5.12 PROCESS PLANNING SHELL

In this section, a process planning shell that can be used for various manufacturing processes is presented, in particular, the machining process. Most existing automated processes planning systems are implemented around the knowledge in a specific domain; that is, the system is restricted to a limited domain of parts and processes. The conventional expert system shells appear to be too restrictive for effective building of knowledge-based process planning systems. These two major difficulties have led to the development of the process planning shell presented in this section. The shell allows the user to define the elements involved in process planning, for example, product features, processes, machine tools, and the knowledge required.

5.12.1 Process Planning Domain

Process planning knowledge can be classified as the knowledge about part, manufacturing facility, and operations planning. Product knowledge is concerned with part characteristics such as geometry, material, tolerances, surface finish, and production volume. Manufacturing facility knowledge is concerned with manufacturing processes, machine tools, materials, standards, and so on. Finally, operations planning knowledge is concerned with the order in which the manufacturing processes are used to manufacture the part. Figure 5.16 presents an entity–relationship model of process planning. The nodes in the model are entities (objects) involved in process planning, and the arcs are relations linking the entities based on the underlying semantics. For example, the entity "process" is related to the entity "feature" based on the relation "generates," that is, *a process may generate a feature*.

The model in Figure 5.16 is at a high level of abstraction. The details of relations between processes, features, machines, and parts as well as the type of relationship (one-to-many, many-to-one, and many-to-many) are shown in Figure 5.17.

Each entity in the model is a class representative. Each class includes a hierarchy of members (objects). Each object contains structural, declarative, procedural, and control knowledge about an entity. As an example, consider the exploded view of entity "feature" that has been instantiated with the object "Cyl-block" (see Figure 5.18).

5.12.2 Part Description Language

Due to the diversity of parts and manufacturing operations, it is not likely to define a universal set of features able to satisfy the needs of all types of parts. For this reason,

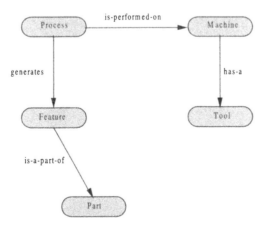

Figure 5.16. Process planning entities.

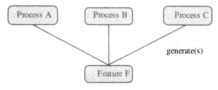

A "feature" may be generated by
many processes

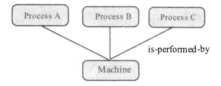

A "process" is performed by a machine
A "machine" may perform many "processes"

A "process" can be performed
by numerous machines

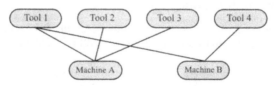

A "tool" is used by a "machine"
A "tool" may be used by several "machines"
A "machine" uses one or more "tool"

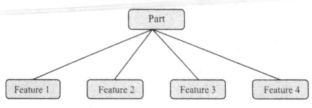

A "part" consists of one or more "features"

Figure 5.17. Relationship between manufacturing entities.

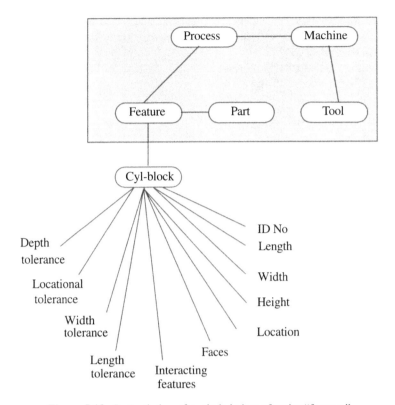

Figure 5.18. Instantiation of exploded view of entity "feature."

it is important for a feature representation structure to be both flexible and generic so that it can be tailored to a specific application.

In this section, primitives for defining a part with its manufacturing features are defined. The part is defined using a set of commands that constitute a feature-based language, named the part description language (PDL), which facilitates automatic transformation of CAD design data to a desired form. The PDL uses a set of primitives (commands) to describe different aspects of a part, for example:

assembly: Identifies a product consisting of several subassemblies.

part: Identifies a part.

material: Identifies the material used for a part.

remove: "REMOVE f, t" removes feature f of type t.

finish: Identifies the surface finish for the specified surface.

reference: Identifies the reference surface.

volume: Identifies a feature of the part to be removed.

intersect: Identifies intersections of two parts.

To illustrate the PDL, consider a part and the stock used to manufacture it (see Figure 5.19*a*). The material volume to be removed to generate the part in Figure 5.19*a* is shown in Figure 5.19*b*.

The required set of PDL primitives to define the part in Figure 5.19 is presented next:

PART	ABC-X45
MATERIAL	Aluminum x23
REMOVE	H1, cyl-hole
REMOVE	H2, cyl-hole
REMOVE	H3, cyl-hole
REMOVE	H4, cyl-hole

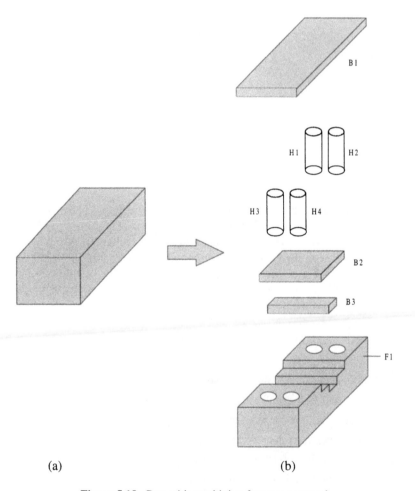

(a)   (b)

Figure 5.19. Part with machining features removed.

REMOVE	B1, rect-block
REMOVE	B2, through-slot
REMOVE	B3, through-slot
REFERENCE	F1
FINISH	F1, 34

5.12.3 System Architecture

The process planning shell is implemented with four major modules: user interface, object definition, product description interpreter, and planning (Figure 5.21). The user interface module allows for the interaction between a user and the system. The planning module generates process plans using the information from the product/manufacturing database. The input to this module is the part data, knowledge of processes, machines, tools, and features. The output of the planning module is a process plan.

The shell operates in two different modes: the generation mode and the application mode. The user selects an option from the *function* menu. In the generation mode, the system generates new objects (e.g., processes, machines, features) or modifies or deletes an existing object. In this mode, the system begins by asking questions regarding the object under consideration and finally generates (or modifies) an object. The module responsible for acting in the generation mode is the object definition module (Figure 5.20).

An example of the object definition session creating a new feature is illustrated in Figure 5.21.

The user defines a product using objects created in the generation mode. The objects are arranged in libraries within the database (object base). However, if an object needed is not available in the object base, the user can switch to the generation mode to create an object and then use the newly created object in the application mode. The selection of objects from the libraries is supported by a set of menus. If the object selected needs to be instantiated, the user should provide the system with the values of uninstantiated attributes within the object. The part description module (Figure 5.20) handles the application mode.

5.12.4 Knowledge Organization

In the process planning shell, the static knowledge (e.g., processes, machines, tools, materials) and the dynamic knowledge (e.g., planning procedures, constraints) are combined as structured objects. A structured object is similar to a frame. Figure 5.22 illustrates an object of the entity "process." Each slot is a pointer to another object, a control knowledge, or a procedure.

Objects may have vertical and horizontal connections. The vertical connections constitute the taxonomy of objects related to each other based on the generalization–specialization relationship. Such taxonomy facilitates the task of managing the objects within the system and assists the user in organizing his or her knowledge at different levels of abstraction. Figure 5.23 illustrates the object

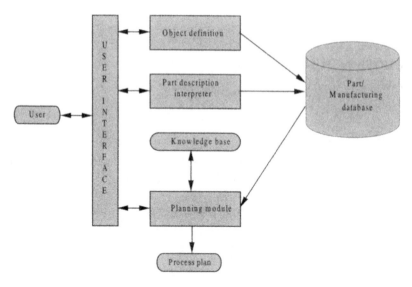

Figure 5.20. System architecture.

```
What is the name of the feature?
>>> rect_block
Is the feature "rect_block" a subset of another feature? [y/n]
>>> n
Does feature "rect_block" include other features? [y/n]
>>> y
Specify the name of subfeatures for "rect_block"
>>> rect_face
>>>
What are the attributes of feature "rect_block"?
>>> id_no
>>> length
>>> width
>>> height
>>> critical faces
>>> reference face
>>>
What process(es) generates the feature "rect_block"?
>>> milling
>>> broaching
>>> laser cutting
>>>
Do you want to introduce any control rules for this feature? [y/n]
>>> n
```

Figure 5.21. Session of creating a new feature.

Name: ....

SuperFrame:
....
Specialization:
....
Preparatory processes:
....
Feature:
....
Machine: ....
Surface finish: ....
Tolerances: ....
Material:
....
Constraints:
....

Figure 5.22. Structure of the object "process."

"process" taxonomy. It should be noted that the system allows the user to define the content of each object and to create a taxonomy of objects. This is done through the user interface module and object definition module. In particular, the pointer slots allow constructing a complex network representing the domain knowledge.

5.12.5 Reasoning Mechanism

Reasoning takes place by the activation of structural objects. The activation of objects is controlled by local and global control mechanisms. The local control knowledge is referred to the control rules inside an object, whereas the global control knowledge is in the form of metarules controlling the activation of objects or making conclusions

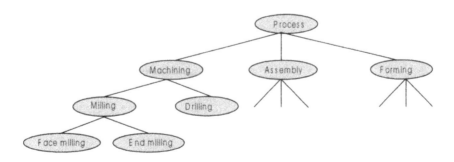

Figure 5.23. Taxonomy of objects.

based on a global condition. The final outcome of the inference process is a process plan.

5.13 SUMMARY

In this chapter, knowledge-based and optimization perspectives to process planning were discussed. The process planning task was decomposed into eight phases, each with a general solution approach proposed. The knowledge-based approach solves the qualitative problems while the optimization approach tackles problems of a quantitative nature. Integrating the two approaches appears to be a viable alternative for developing intelligent process planning systems. Future research and development efforts should focus on closing the gap between design and manufacturing, which implies understanding the underlying science.

An object-oriented approach for selection of machining features was presented. The goal of the methodology presented in the chapter was to increase the degree of concurrency between the part modeling and machining feature selection activities.

The process planning knowledge is heuristic in nature, product and manufacturing facility dependent, and prone to dynamic changes. Most existing processes planning systems are implemented around a fixed set of knowledge in a specific domain of products, parts, and processes. The diversity of products and manufacturing operations demands the implementation of domain-specific process planning systems.

The conventional expert system shells are too restrictive to be effectively used for building knowledge-based process planning systems. The shell presented provides a generic environment allowing the user to obtain a working process planning system in a domain of interest.

A generic model of process planning was suggested. This model was used to implement various modules of the system. The issue of major concern is the language for product description, the formalism for representing the domain objects, the inference strategy for generating process plans, and the user interfaces. The process planning shell uses the part description language (PDL) to define parts.

APPENDIX: MODEL LISTING FOR EXAMPLE 5.1

```
MIN
5X1+5X2+2X3+3X4+1X5+4X6+5X7+1X8+5X9+6X10+6X11
+Y1+Y2+Y3+Y4
+Z1+Z2+Z3
SUBJECT TO:
X1+X3>=1
X1+X4>=1
X1+X6>=1
X1+X6+X9>=1
X1+X11>=1
```

```
X2+X3>=1
X2+X4>=1
X2+X6>=1
X2+X6+X9>=1
X2+X7+X11>=1
X4+X5>=1
X10>=1
X7+X10+X11>=1
X10>=1
X7+X8+X11>=1
Y1+Y2+Y3+Y4<=3
X1+X2+X7+X8<=Y1
X4+X5+X9+X11<=4Y2
X3+X6<=2Y3
X10<=Y4
Z1+Z2+Z3<=2
X1+X2+X6+X7+X8+X11<=6Z1
X3+X4+X5+X10<=4Z2
X9<=Z3
END
INTEGER 18
```

REFERENCES

Alting, L., H. Zhang (1989), Computer aided process planning: The state-of-the-art survey, *International Journal of Production Research*, Vol. 27, No. 4, pp. 553–585.

Chang, T. C., R. A. Wysk, and H. P. Wang (1991), *Computer Integrated Manufacturing*, Prentice-Hall, Englewood Cliffs, NJ.

Chvatal, V. (1979), A greedy heuristic for the set-covering problem, *Mathematics of Operations Research*, Vol. 4, No. 3, pp. 233–235.

de Floriani, L. (1987), Graph based approach to object feature recognition, *Proceedings of the Third ACM Symposium on Computational Geometry*, Waterloo, Ontario, Canada, pp. 131–137.

Dixon, J. R., E. C. Libardi, and E. H. Nielsen (1990), Unresolved research issues in development of design-with-features systems, in M. J. Wozny, J. U. Turner, and K. Preiss (Eds.), *Geometric Modeling for Product Engineering*, Elsevier, New York, pp. 183–197.

Hayes, C. and P. K. Wright (1989), Setup planning in machining: An expert system approach, paper presented at the NSF Conference on Advances in Manufacturing System Integration and Process, SME, Dearborn, MI, pp. 441–443.

Henderson, M. R., and G. J. Chang (1988), FRAPP: Automated feature recognition and process planning from solid model data, in *Proceedings of the ASME International Computers in Engineering Conference*, Vol. 1, San Diego, CA, pp. 529–536.

Hopcroft, J. E., and J. D. Ullman (1969), *Formal Languages and Their Relationship to Automata*, Addison-Wesley, Reading, MA.

Horowitz, E., and S. Sahni (1983), *Fundamentals of Data Structure*, Computer Science Press, Rockville, MD, pp. 312–313.

Joshi, S. (1987), CAD interface for automated process planning, Ph.D. Thesis, School of Industrial Engineering, Purdue University, West Lafayette, IN.

Joshi, S., and T. C. Chang (1990), Feature extraction and feature based design approaches in the development of design interface for process planing, *Journal of Intelligent Manufacturing*, Vol. 1, No. 1, pp. 1–15.

Kusiak, A. (1987), Artificial intelligence and operations research in flexible manufacturing systems, *Information Processing and Operational Research (INFOR)*, Vol. 25, No. 1, pp. 2–12.

Kusiak, A. (1990a), Selection of machinable volumes, *IIE Transactions*, Vol. 22, No. 2, pp. 151–160.

Kusiak, A. (1990b), *Intelligent Manufacturing Systems*, Prentice-Hall, Englewood Cliffs, NJ.

Kusiak, A. (1991), Process planning: A knowledge-based and optimization perspective, *IEEE Transactions on Robotics and Automation*, Vol. 7, No. 3, pp. 257–266.

Kusiak, A., Ed. (1992), *Intelligent Design and Manufacturing*, John Wiley, New York.

Kyprianou, L. K. (1980), Shape classification in computer aided design, Ph.D. Thesis, University of Cambridge, United Kingdom.

Liu, C. R., and R. Srinivasan (1984), Generative process planning using syntactic pattern recognition, *Computers in Mechanical Engineering*, Vol. 2, No. 5, pp. 63–66.

Ludema, K., R. M. Caddell, and A. G. Atkins (1987), *Manufacturing Engineering: Economics and Processes*, Prentice Hall, Englewood Cliffs, NJ.

Mantyla, M., J. Opas, and J. Puhakka (1987), A prototype system for generative process planning of prismatic parts, in A. Kusiak (Ed.), *Modern Production Systems,* Elsevier, New York, pp. 599–611.

Mesquita, R., E. Krasteva, and S. Doytchinov (1995), Computer-aided selection of optimum machining parameters in multipass turning, *International Journal of Advanced Manufacturing Technology*, Vol. 10, No. 1, pp. 19–26.

Philipson, R. H., and A. Ravindran (1979), Application of mathematical programming to metal cutting, *Mathematical Programming*, Vol. 11, pp. 116–134.

Pratt, M. J., and P. H. Wilson (1985), Requirements for support of form features in a solid modeling system, Report R-85-ASPP-01, Consortium for Advanced Manufacturing International CAM-I, Arlington, TX.

Sakurai, H., and D. C. Gossard (1988), Shape feature recognition from 3-D solid models, in *Proceedings of the ASME International Computers in Engineering Conference*, Vol. 1, San Diego, CA, pp. 515–519.

Tan, F. P., and R. C. Creese (1995), A generalized multi-pass machining model for machining parameters selection in turning, *International Journal of Production Research*, Vol. 33, No. 5, pp. 1467–1487.

Tsang, J. P., and Y. Lagoude (1986), Process plan representation and manipulation in generic expert planning systems, paper presented at the 1986 IEEE International Conference on Robotics and Automation, April 7–10, San Francisco, CA.

Vujosevic, R., and A. Kusiak (1992), Selection of machinable volumes: An object-oriented approach, *Expert Systems with Applications*, Vol. 4, No. 3, pp. 273–283.

Woo, T. C. (1977), Computer aided recognition of volumetric designs, in D. McPherson (Ed.), *Advances in Computer-Aided Manufacturing*, North-Holland, Amsterdam, pp. 121–135.

Woodwark, J.R. (1988), Some speculations on feature recognition, *Computer Aided Design*, Vol. 20, pp. 189–196.

QUESTIONS

5.1. What is a process plan in metal cutting?

5.2. What is an assembly plan?

5.3. What is the relationship between a design feature and a machining feature? Draw an example of the two features.

5.4. What methods of representing features do you know?

5.5. Do the representation methods of design features differ from the methods used to represent manufacturing features?

5.6. What are the basic phases involved in the process planning of machined parts?

5.7. Why do we limit the number of elementary machining features in the decomposition of a mechanical part?

5.8. What are the objective function and constraints of the problem solved by the Chvatal heuristic?

5.9. What is a process planning shell?

5.10. How do you envision using the part description language in process planning for machining and assembly?

PROBLEMS

5.1. For the part (shown in two dimensions) in Figure 5.24 with 18 elementary features and two tolerance bases b_1 and b_2:

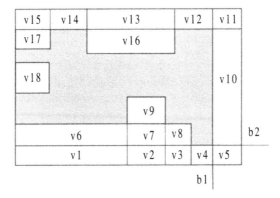

Figure 5.24. Prismatic part with 18 elementary machining features.

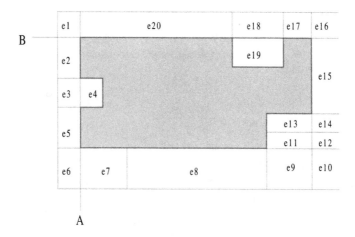

Figure 5.25. Prismatic part with 21 elementary features.

(a) List five machining features (machinable volumes).

(b) Show two precedence constraints among some of the machining features in (a).

5.2. For the part in Figure 5.25:

(a) Set up an elementary machining feature–machining feature (machinable volume) incidence matrix.

(b) Assume tools and fixtures and assign them to the machining features.

(c) Assume the maximum number of tools and fixtures to be selected.

(d) Assume the cost (in cents) of removing each machining feature to be proportional to the number of elementary features included in the machining feature.

(e) Formulate a model for the selection of machining features that minimizes the total cost of removing all machining features, the total number of tools, and the total number of fixtures.

(f) Solve the model with an integer programming software (e.g., LINDO).

(g) Discuss the results.

5.3. The surfaces A and B of the part in Figure 5.25 serve as tolerance bases. Consider the surfaces in the development of a precedence graph. Sequence the machining features selected with the topological ordering algorithm to minimize:

(a) The total number setups (fixture changes)

(b) The total number of tool changes

Assuming that the fixture setup cost is $3 and the tool setup cost is $1, compare the total cost (machining and setup) of solutions (a) and (b).

5.4. Consider the part in Figure 5.26.

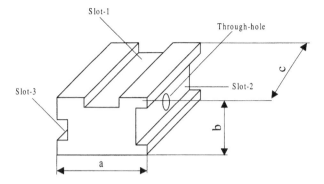

Figure 5.26. Part with three slots and a through-hole.

(a) The dimensions of the stock are $a + 4$, $b + 2$, c (all dimensions are given in millimeters). The depth of each slot is 2 mm. The required surface quality can be accomplished with milling.

(b) The following tools have been selected: slot 1, t_1; slot 2, t_2; slot 3, t_3; through hole, t_4; and tool t_5 to remove all other volumes of material.

(c) Decompose the volume of the material to be removed into elementary machining features. Draw the decomposition in two dimensions.

(d) Write down an elementary feature–machining feature incidence matrix.

(e) Formulate a model for selection of machining features that minimizes the total number of machining features.

(f) Solve the model formulated in (e) with Chvatal's heuristic.

(g) Generate precedence constraints for the machining features generated in (f).

(h) Generate a feasible process plan.

(i) Formulate a model for selection of machining features that minimizes the total number of machining features and the number of tools.

(j) Find the lowest possible number of tools necessary to machine the part in Figure 5.26 by solving the model in (i) with an integer programming software (e.g., LINDO). Is this solution different from the solution obtained in (f)? Why?

5.5. Given the precedence graph among machining features in Figure 5.27, determine a process plan that minimizes the number of tool changes. What is the minimum number of tool changes?

5.6. The precedence graph in Figure 5.28 represents 11 assembly features. Features F_1, F_2, F_7 are assembled with tool T_1; features F_4, F_9, and F_{10} with tool T_2; and the remaining features are assembled with tool T_3. Generate an assembly plan that minimizes the total number of tool changes.

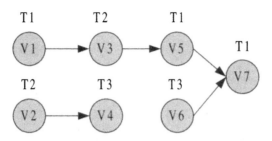

Figure 5.27. Precedence graph for a part.

5.7. Consider the rotational part in Figure 5.29. The diameter of the stock is $D = 65$ mm and the length of the stock is $L = 220$ mm. The dimensions of the part are: $l_1 = 200$ mm, $l_2 = 70$ mm, $l_3 = 50$ mm, $d = 55$ mm, $d_1 = 28$ mm, and $d_2 = 45$ in.

- Use three different cutting tools: T_1 to remove the material corresponding to the length l_1 and T_2 for removing the material corresponding to the length l_2, and tool T_3 to remove the features corresponding to the remaining surfaces.

- Assume that removal of the material over the length l_1 is done in two cuts (rough and fine).

- Assume that the cost of each feature removal is equal to its length (measured along the part horizontal axis).

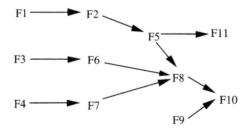

Figure 5.28. Precedence graph.

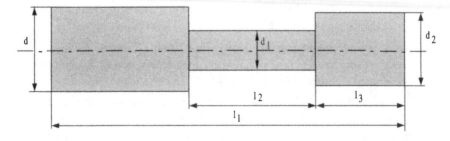

Figure 5.29. Rotational part.

(a) Decompose the volume of material to be removed into elementary manufacturing features.

(b) Set up an elementary machining feature–machining feature incidence matrix.

(c) Assign the cost of removing each machining feature and assign tools to the features.

(d) Ignore the tooling and determine the lowest cost manufacturing features with the Chvatal heuristic algorithm.

(e) For the set of features determined in (d), find a process plan with the topological ordering algorithm.

CHAPTER 6

SETUP REDUCTION

6.1 INTRODUCTION

Modern manufacturing aims at reducing the product time to market by improving design and manufacturing activities. Another effective approach for reducing the time to market is through shortening of setup time. Activities preceding actual manufacturing of an item might be considered as setup activities. The following benefits of setup time reduction have been indicated in the literature (Black, 1991; Gallego and Moon, 1992; Hall, 1983; Jordan and Frazier, 1993; Shingo, 1985; Steudel and Desruelle, 1991; Trevino et al., 1993):

- Reduced work-in-process inventory
- Reduced floor space occupancy
- Reduced material handling
- Reduced lead time
- Improved efficiency and flexibility of manufacturing systems (manufacturing can be adjusted to the actual demand)
- Reduced waste (e.g., of product, equipment, time)
- Reduced manufacturing cost
- Increased worker productivity (reduction and elimination of unproductive tasks)
- Increased sense of equipment ownership (operators are responsible not only for production but also for setups)
- Progressive implementation of new attitudes and culture (workers get involved in improvement of work conditions)

The reduction of setup time can be achieved in three basic ways:

- Designs of products
- Designs and optimization of processes and operations (Dobson, 1992; Flynn, 1987; Shingo, 1985; Steudel and Desruelle, 1992; Trevino, et al. 1993)
- Designs of manufacturing systems (Burgess et al., 1993; Dahel and, Smith, 1993; Jordan and Frazier, 1993).

The literature has dealt with the setup time reduction approach. As mentioned above, setup time can be reduced from three perspectives. Hitomi and Ham (1977), Foo and Wager (1983), Flynn (1987), and Dobson (1992) have discussed the reduction of setup time based on the sequence dependence of setups. Chand (1989), Chand and Sethi (1990), and Karwan et al. (1988) have considered setup time reduction in the context of worker learning. There are two major reasons for reducing the internal setup time defined in Section 6.2. First, the internal setup time lowers the rate of equipment utilization, which may create bottlenecks in a manufacturing system. Second, a longer internal setup time implies a higher internal setup cost. According to economic lot size theory, a higher internal setup cost requires larger lot sizes, which works against the market requirements imposing small lot sizes. Therefore, the reduction of internal setup time is critical in a market-driven manufacturing environment. The importance of the reduction of internal setup time has been addressed in the literature. For example, Gallego and Moon (1992), Inman et al. (1991), and Trevino et al. (1993) have discussed the impact of externalizing setup activities on scheduling of economic lots. Methods for reduction of the internal setup time by improving processes are provided in Shingo (1985). The reduction of setup time by sequencing setup activities and making a trade-off between time and cost of each activity has not been discussed in the literature.

There are two types of project scheduling approaches: the critical path method (CPM) and the project evaluation and review technique (PERT). A brief comparison of CPM and PERT is provided in Hillier and Lieberman (1990). Since the models presented in this chapter are deterministic (based on CPM), only the CPM literature is reviewed next. Two basic concepts were introduced to the classical CPM approach in Kelley and Walker (1959) and Kelly (1961). The first one is the concept of precedence that represents the ordering among activities. The other one is the concept of time–cost trade-off between the duration of an activity and its cost. Such a trade-off assumes that the duration of an activity can be shortened from its "normal" duration at a certain cost. Reducing the activity time by using more resources, such as operators, machines, and overtime, is called crashing. The time of an activity that has crashed is called crash time as opposed to the normal time. Accordingly, the cost of achieving this crash time from its normal time is called a crash cost. The literature on resource-constrained scheduling can be classified based on two criteria: (1) the objective function and (2) the resource category. Basically, there are two types of objective functions (see Talbot, 1982): (1) the cost-based objectives for maximizing the net present value of the project (the classical critical path scheduling model presented in Section 6.3 is of this type) and (2) the time-based objectives for minimizing project duration (the models presented in this chapter are of this type). Resources are classified into three categories: (1) renewable, (2) nonrenewable, and (3) double constrained. A brief review of the literature on project scheduling with respect to these three categories of resources is provided in Ulusoy and Özdamar (1994). The literature on project scheduling is quite extensive, for example, Elmaghraby (1977), Weglarz (1979), Christofides et al. (1987), Patterson et al. (1990), Elmaghraby and Kamburowski (1992), Leachman and Kim (1993), and Premachandra (1993). Its review is provided in Davis and Patterson (1975) and Russel (1986). However, scheduling setup activities has not been discussed in the literature.

This chapter focuses on improving the setup process by optimizing the sequence of setup activities. Sequencing setup activities can be modeled as a project scheduling problem with an objective of minimizing the duration of the setup process. The difference between the standard project scheduling and the scheduling of setup activities is discussed. Computational results are provided. Factorial analysis is used to verify the models for scheduling setup activities and to investigate the effects of critical factors on setup time reduction.

6.2   CHARACTERISTICS OF SETUP ACTIVITIES

Before providing models and methods for minimization of the setup time, characteristics of setup activities are identified. The following definitions are provided (Feng et al., 1997):

Definition 1.   Setup activities required for manufacturing a new lot of items or a new item in the same lot can be classified as (a) interlot and (b) intralot setup activities. The interlot setup activity is performed in order to manufacture a new lot of items (e.g., acquiring materials, equipment, tools, and fixtures), and the intralot setup activity is performed in order to manufacture a new item in the same lot (e.g., changing of tools on a machine tool, reorientation of the item on a machine tool, transportation of an item from one machine tool to another).

Definition 2.   Setup activities required to manufacture a new lot of items can be also classified as (a) internal and (b) external setup activities. The internal setup activity is performed when the manufacturing facility (e.g., machine tools, robots, CMMs) is not in operation (see class I in Figure 6.1; e.g., installation and adjustment of tools and fixtures on a machine tool) and the external setup activities are the remaining setup activities (see classes E and F in Figure 6.1).

Definition 3.   External setup activities can be classified into classes E and F. A class E external setup activity (e.g., generation of production reports, training of operators)

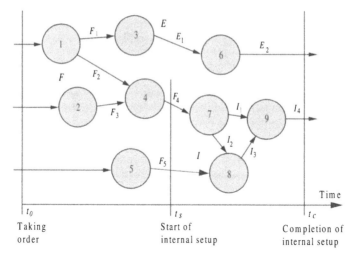

E : External setup activities that do not affect internal setup activities
F : External setup activities that affect internal setup activities
I : Internal setup activities

Figure 6.1. Classification of interlot setup activities.

is the one that its completion does not affect the start of the internal setup activity, and a class F external setup activity (see Figure 6.1; e.g., process planning, production planning, NC programming, item scheduling, preparing tools and fixtures) impacts the start of the internal setup activity.

Definition 4. The external setup activity that is completed last (e.g., E_2 in Figure 6.1) is called a final external setup activity.

Based on the above definitions, the setup activities can be characterized as follows:

1. To schedule setup activities, one should partition them into three classes: E, F, and I (see Figure 6.1). Resource (e.g., machine tool, operator) and cost constraints differ for each of the three classes.

2. The internal setup activities can be performed on one machine tool, but the independent setup activities can be conducted simultaneously, which is analogous to scheduling jobs on multiple machines. The start and completion times of an internal setup activity are usually constrained. An external setup activity can be normally performed by multiple resources. The start and completion times of an external setup activity of classes E and F are constrained.

3. Although the focus of this chapter is to minimize the internal setup time, the models apply to minimization of the duration of the entire setup process as well.

6.3   SCHEDULING MODEL

6.3.1   Example of Scheduling Setup Activities

Two notations are used for activity networks (Elmaghraby, 1977): (1) the activity-on-arc network (AOA) and (2) the activity-on-node (AON) network. In this chapter, the AOA convention is adopted. In an AOA network, a directed arc denotes an activity and a node denotes an event. The directed arc also represents the duration of the activity and a precedence relation between activities. The first (last) node represents the beginning (completion) of the entire project. Activities represented by arcs leaving a node cannot be started until the event represented by that node has occurred. An event occurs only when all activities entering it have been completed. A dashed line represents a dummy activity, which imposes precedence constraints with zero time duration. In this chapter, it is assumed that no cycles exist in the AOA network of setup activities. The networks of setup activities presented in Figures 6.1 and 6.3 follow the AOA convention.

An example part to be machined on a CNC machine is shown in Figure 6.2. The network of setup activities for manufacturing this part is shown in Figure 6.3.

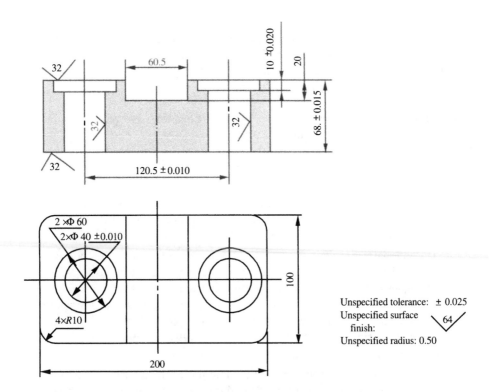

Figure 6.2. Example part to be machined on a CNC machine.

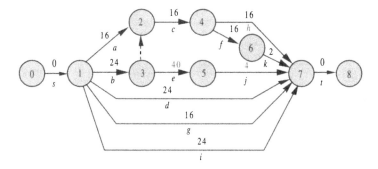

Figure 6.3. Setup activity network for machining the part in Figure 6.2.

For the network of interlot setup activities in Figure 6.3, the data and the normal schedule of activities are presented in Table 6.1 and Figure 6.4, respectively. From Table 6.1, one can see that the latest beginning time of the internal setup is 52 (activity h or g), and the earliest completion time is 68 (activity h, j, k or g). Thus, the internal setup time $t_{is} = t_c - t_s = 68 - 52 = 16$ (time units), where t_c is the completion time and t_s the start time. During the internal setup time of 16 time units, the machine is not available for production.

TABLE 6.1. Data for Setup Activities

Activity Label	Description	Duration	Predecessors	EB	EF	LB	LF	Slack
a	Production planning	16	None	0	16	8	24	8
b	Process planning	24	None	0	24	0	24	0
c	Scheduling	16	a, b	24	40	34	40	10
d	Material acquisition	24	None	0	24	44	68	44
e	Tool/fixture delivery	40	b	24	64	24	64	0
f	NC programming	16	c	40	56	50	68	10
g	CMM calibration	16	None	0	36	52	68	52
h	Robot calibration	16	c	40	80	52	68	12
i	Production report generation	24	None	0	24	44	68	44
j	Loading and adjusting tools and the fixture	4	e	64	68	64	68	0
k	Loading NC program	2	f	56	58	66	68	10

EB = earliest begin time
EF = earliest finish time
LB = latest begin time
LF = latest finish time

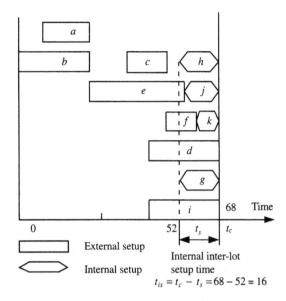

Figure 6.4. Gantt chart of interlot setup activities.

Suppose that the internal setup time is to be reduced based on the due date of the part. Let the available time for internal setup of the machine be $T_{is} = 4$ time units rather than 16 time units. The machine is not available until $T_s = 40$, and the production must start for the new lot of items no later than $T_c = 44$. Assume the design of the item cannot be modified. No change of the layout of the manufacturing system is permitted. The only way to reduce the internal setup time is by improving the setup process.

The principles for reducing the internal setup time by improving processes and operations are provided in Shingo (1985). Without going into details of setup reduction, Table 6.2 provides the crash cost per time unit and the allowable crash time

TABLE 6.2. Crash Time and Crash Cost

Activity	Crash Cost/Time Unit C_i	Minimum Crash Time R_i
a	20	8
b	20	16
c	20	8
d	10	8
e	15	12
f	15	8
g	15	4
h	15	4
i	15	8
j	20	2
k	15	1

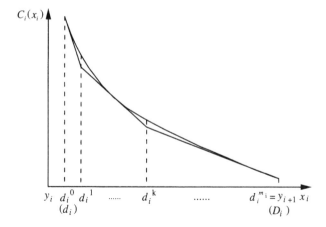

Figure 6.5. Illustration of the activity duration time and crash cost.

for each activity. Although the crash cost with respect to the reduced time in Table 6.2 is linear for simplicity, models 6.1 and 6.2 apply to nonlinear cases. For illustration of the nonlinearity of the crash cost, see Figure 6.5. Table 6.3 shows the classification of setup activities based on Definitions 3 and 4.

6.3.2   Project Scheduling Model

In this chapter, it is assumed that only one machine is to be set up. For a part to be manufactured in a manufacturing system with multiple machines, one can identify a bottleneck machine and then apply models 6.1 and 6.2. The single machine scheduling assumption is being frequently used in the literature based on the above fact.

Assume a common start activity s and a common terminal activity t in the network of setup activities; otherwise create an artificial one (see Figure 6.3). Let y_s be the start time of the start activity and y_t the start time of terminal activity. The additional variables and parameters used in the model are defined next:

TABLE 6.3.   Classification of Setup Activities in Figure 6.3

Internal Setup Activities	
Class I	g, h, j, k
External Setup Activities	
Class E	i
Class F	a, b, c, d, e, f

$x_i =$ the crash time of activity i, $x_i = D_i - \sum_{k=1}^{m_i} t_i^k$

$y_i =$ (unknown) start time of activity i, $i = 1, 2, \ldots, n$

$t_i^k =$ reduced time in the range of d_i^{k-1} and d_i^k, $k = 1, 2, \ldots, m_i$

$D_i =$ normal duration of activity i (see Figure 6.5)

$d_i =$ minimum allowable crash time for activity i, $d_i \le x_i \le D_i$

$$d_i^k = \begin{cases} d_i & \text{if } k = 0 \\ D_i & \text{if } k = m_i \end{cases}$$

$C_i(x_i) =$ crash cost for activity i

$C_i^k =$ crash cost per unit time for activity i in the range of d_j^{k-1} and d_i^k, $k = 1, 2, \ldots, m_i$

A standard project scheduling model for minimizing the duration of the entire setup process regardless of the class of the activities with the crashing budget constraint is presented next.

Model 6.1

$$\min Z = y_t - y_s \tag{6.1}$$

such that

$$\sum_{i=1}^{n} C_i(x_i) \le B \tag{6.2}$$

$$y_i = y_{i-1} - x_{i-1} \ge 0 \qquad i = 2, 3, \ldots, n-1 \tag{6.3}$$

$$d_i \le x_i \le D_i \qquad i = 1, 2, \ldots, n \tag{6.4}$$

$$x_i, y_i \ge 0 \qquad i = 1, 2, \ldots, n \tag{6.5}$$

The objective of the model (6.1)–(6.5) minimizes the duration of the entire setup process. Constraint (6.2) ensures the allowable crash budget, the left-hand side of which might be nonlinear. Inequality (6.3) expresses precedence relationships among activities. Constraint (6.4) imposes limit on the crash time for each activity. The nonnegativity of x_i and y_i is ensured by constraint (6.5).

The crashing cost function in Model 6.1 is nonlinear, continuous, and convex. The convexity of the crashing cost function reflects the fact in project crashing that the shorter the crash time, the more the resources. In practice, a piecewise linear approximation to the continuous nonlinear crashing cost function is often sufficient (see Figure 6.5). For a linear crash cost function, constraint (6.2) is replaced with

$$\sum_{i=1}^{n} C_i(D_i - x_i) \leq B \tag{6.6}$$

and Model 6.1 can be applied. Rewriting the nonlinear continuous cost function in a piecewise linear form, the following linear programming formulation is obtained for minimizing the duration of the entire setup process.

Model 6.1 (a)

$$\min Z = y_t - y_s \tag{6.7}$$

such that

$$y_i - y_{i-1} - x_{i-1} \geq 0 \; i = 2, 3, \ldots, n-1 \tag{6.8}$$

$$\sum_{i}^{n} \sum_{k=1}^{m_i} C_i^k t_i^k \leq B \tag{6.9}$$

$$x_i = D_i - \sum_{k=1}^{m_i} t_i^k \qquad i = 1, 2, \ldots, n \tag{6.10}$$

$$0 \leq t_i^k \leq d_i^k - d_i^{k-1} \qquad \begin{aligned} k &= 1, 2, \ldots, m_i, \\ i &= 1, 2, \ldots, n \end{aligned} \tag{6.11}$$

$$d_i \leq x_i \leq D_i \qquad i = 1, 2, \ldots, n \tag{6.12}$$

$$x_i, y_i \geq 0 \qquad i = 1, 2, \ldots, n \tag{6.13}$$

For a linear crash cost function, constraints (6.9)–(6.11) are replaced with (6.6). The computational results obtained from the model (6.7)–(6.13) solved with LINDO (Schrage, 1984) for three different values of crash budget B [see constraint (6.2) or (6.9)] are presented in Table 6.4. One can see that the completion time T_c affects the project duration Z and the start time t_s of the internal setup activities.

TABLE 6.4. Computational Results from Model 6.1

Constraint B	t_s	t_c	t_{is}	Z
100	40.0	61.3	21.3	61.3
300	32.5	50.5	18.0	50.5
500	28.3	44.3	16.0	44.3
600	25.4	41.4	16.0	41.4
700	22.9	38.9	16.0	38.9
900	18.4	34.4	16.0	34.4

6.4 MINIMIZING INTERNAL SETUP TIME

Model 6.1 can be extended to solve the internal setup time minimization (ISTM) problem due to the following similarities between standard project scheduling and setup activity scheduling:

- Deals with the time (duration) of activities
- Makes a trade-off between cost and time
- Considers precedence constraints

However, Model 6.1 cannot be directly applied for minimizing the internal setup time due to the following differences between standard project scheduling and setup activity scheduling:

- The objective of Model 6.1 is to minimize the duration of the entire setup process, while the ISTM problem minimizes the internal setup time.
- Model 6.1 is concerned with the completion time of the entire setup process. The ISTM problem restricts the earliest start time and the latest completion time of the internal setup activities and the external setup activities. The two constraints are imposed by a manufacturing facility.

In modeling the setup activity scheduling problem, the above differences must be considered.

6.4.1 Setup Scheduling Model

In a market-driven manufacturing environment, the reduction of internal setup time is often more important than that of the entire setup process. Before presenting the scheduling model for minimization of the internal setup time, the following definitions and notation are introduced:

Figure 6.6 illustrates the calculation of the objective function, that is, the internal setup time $t_{is} = t_c - t_s$. Activities I_1, I_2, and I_3 are called the *initial activities* (set I_s) since they begin the internal setup (see Figure 6.6). Similarly, activities I_8 and I_9 are called

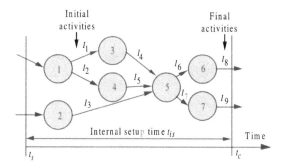

Figure 6.6. Calculation of the internal setup time.

the *final activities* (set I_c) since they complete the internal setup (see also Figure 6.6). The completion time (t_c) of the last final setup activity terminates the setup process. The start time (t_s) of the first initial setup activity begins the internal setup process.

x_i = crashed time of setup activity i of one of the following classes: external setup activity of class E for $i \in E$, external setup activity of class F for $i \in F$, internal setup activity of class I for $i \in I$

y_i = start time of activity i of one of the following classes: external setup activity of class E for $i \in E$, external setup activity of class F for $i \in F$, internal setup activity of class I for $i \in I$

z = completion time of the latest completed setup activity (activities)

y_{lc}^q = completion time of the final external setup activity of class E, $i \in E_c$, which is completed last, $y_{lc}^E = \max\{y_{n_E} + x_{n_E} : n_E \in E_c\}$

y^r = start time of the initial internal setup activity which starts first

y_{lc}^p = completion time of the internal setup activity which is completed last, $y_{lc}^I = \max\{y_{n_I} + x_{n_I} : n_I \in I_c\}$

I_S = set of initial internal setup activities (see I_1, I_2, I_3 in Figure 6.6)

I_c = set of final internal setup activities (see I_8, I_9 in Figure 6.6)

E_c = set of final external setup activities of class E (see E_2 in Figure 6.1)

B = total allowed crash budget

T_s = earliest allowable start time for internal setup activities

T_c = latest completion time of entire setup project

The basic model for minimizing the internal setup time is presented next.

Model 6.2

$$\min Z = \max\{y_{lc}^E, y_{lc}^I\} - y^r \qquad (6.14)$$

such that

$$\sum_{i=1}^{n} C_i(x_i) \le B \tag{6.15}$$

$$y_i - y_{i-1} - x_{i-1} \ge 0 \qquad i = 2, 3, \ldots, n-1 \tag{6.16}$$

$$d_i \le x_i \le D_i \qquad i = 1, 2, \ldots, n \tag{6.17}$$

$$y^r \ge T_s \qquad \text{for all } r \in I_S \tag{6.18}$$

$$y_{lc}^l \le T_c \qquad \text{for all } n_l \in I_c \tag{6.19}$$

$$y_{lc}^E \le T_c \qquad \text{for all } n_E \in E_c \tag{6.20}$$

$$x_i, y_i \ge 0 \qquad i = 1, 2, \ldots, n \tag{6.21}$$

The objective of (6.14)–(6.21) minimizes the duration of internal setup. Constraint (6.15) imposes a limit on the total crash budget of the setup process. Constraint (6.16) ensures the precedence relations between setup activities. Constraint (6.17) limits the allowable crash time for each activity. Constraint (6.18) implies the earliest start time of an internal setup activity is equal to or greater than T_s when the previous lot of products has finished and the manufacturing facility has stopped for setup. Constraints (6.19)–(6.20) limit the latest completion time of all setup activities of classes I and E, respectively. Constraint (6.21) ensures nonnegativity of all decision and intermediate variables.

The duration of the longest path among all possible paths in Figure 6.6 should be used in the objective function. In Figure 6.6, the internal setup time T_{is} relates to the duration of the following 6 paths of activities:

(P$_1$) 1–3–5–6
(P$_2$) 1–3–5–7
(P$_3$) 1–4–5–6
(P$_4$) 1–4–5–7
(P$_5$) 2–5–6
(P$_6$) 2–5–7

For example, consider the duration of the first path P$_1$:

$$T_{P_1} = y_8^I - y_1^I + x_8$$

Using the notation for the first and last column activities, the following equation for calculating the duration of any path of internal setup activities is obtained:

$$T_{P_k} = y_{n_I} - y^r + x_{n_I} \qquad \text{for all } n_I \in I_c \qquad (6.22)$$

Although there are six internal setup paths in Figure 6.6, the number of paths used to calculate the duration of the longest internal setup paths depends only on the number of the initial and final setup activities, that is, there are $3 \times 2 = 6$ equations in Figure 6.6 from Eq. (6.22). Let z denote the maximum duration of all the possible paths of internal setup activities,

$$z = \max\{T_{P_k} : \text{for all } P_k\}$$

where T_{P_k} is the duration of path P_k ($k = 1, 2, \ldots, 6$ in this case). Therefore, objective function (6.14) becomes

$$\min Z = \max\{T_{P_k} : \text{for all } P_k\} \qquad \text{or} \qquad \min Z = z \qquad (6.23)$$

The computational complexity of finding all possible paths in a digraph with V vertices is exponential (Horowitz and Sahni, 1984). However, in our case, the value V is small due to the following two reasons: (1) the number of internal setup activities is limited (nine in Figure 6.6) and (2) the number of initial and final internal activities (five out of nine in Figure 6.6) is smaller than that of internal setup activities. The remaining question is how to directly incorporate the definition of z in Model 6.2. The definition of z implies that

$$z \geq T_{P_k} \qquad \text{for } k \in I$$

$$(y_{n_I} + x_{n_I} - y^r) - z \leq 0 \qquad \text{for all } n_I \in I_S \qquad (6.24)$$

If $y_{lc}^E > y_{lc}^I$, the following constraint should be considered:

$$(y_{n_E} + x_{n_E} - y^r) - z \leq 0 \qquad \text{for all } n_E \in E_c \qquad (6.25)$$

Since z is nonnegative, a nonnegativity constraint is required in Model 6.2 as well, that is

$$z > 0 \qquad (6.26)$$

Substituting expressions (6.23)–(6.26) into Model 6.2, the scheduling problem of minimizing the internal setup time is expressed as follows:

Model 6.2(a)

$$\min Z = z \tag{6.27}$$

such that

$$y_i - y_{i-1} - x_{i-1} \geq 0 \qquad i = 2, 3, \ldots, n-1 \tag{6.28}$$

$$\sum_{i}^{n} \sum_{k=1}^{m_i} C_i^k t_i^k \leq B \tag{6.29}$$

$$x_i = D_i - \sum_{k=1}^{m_i} t_i^k \qquad i = 1, 2, \ldots, n \tag{6.30}$$

$$0 \leq t_i^k \leq d_i^k - d_i^{k-1} \qquad \begin{aligned} k &= 1, 2, \ldots, m_i \\ i &= 1, 2, \ldots, n \end{aligned} \tag{6.31}$$

$$d_i \leq x_i \leq D_i \qquad i = 1, 2, \ldots, n \tag{6.32}$$

$$(y_{n_E} + x_{n_E} - y^r) - z \leq 0 \qquad \text{for all } n_E \in E_c \tag{6.33}$$

$$(y_{n_I} + x_{n_I} - y^r) - z \leq 0 \qquad \text{for all } n_I \in I_c \tag{6.34}$$

$$y^r \geq T_s \qquad \text{for all } r \in E_c \tag{6.35}$$

$$y_{n_E} + x_{n_E} \leq T_c \qquad \text{for all } n_E \in E_c \tag{6.36}$$

$$y_{n_I} + x_{n_I} \leq T_c \qquad \text{for all } n_I \in I_c \tag{6.37}$$

$$x_i, y_i, z \geq 0 \qquad i = 1, 2, \ldots, n \tag{6.38}$$

For a linear crash cost function, constraints (6.29)–(6.31) are replaced with (6.6).

6.4.2 Model Extensions

Model 6.2(a) can be extended to include the following cases:

1. The earliest start time of the internal setup process is fixed.
2. The earliest start time of the internal setup process is free.
3. The objective is to minimize the duration of the entire setup process rather than the duration of the internal setup process.

Scheduling Model with a Fixed Start Time of the Internal Setup Process. It can be shown next that for scheduling with a fixed start time of the internal setup (i.e., $t_s = T_s$), a special case of Model 6.2(a) is applicable. In fact, calculation of the duration of internal setup paths is simplified when the start time is fixed. Looking at Figure 6.6, the longest internal setup duration is obtained by subtracting T_s from the completion time of activities 8 and 9, respectively, that is,

$$T_{P_k} = (y_{n_I} + x_{n_I} - T_s) \quad \text{for all } n_I \in I_c \tag{6.39}$$

The number of internal setup paths is equal to the number of final internal activities (two in Figure 6.6), no longer multiplication of the number of initial and final internal activities (see Figure 6.6). Then, constraint (6.37) becomes (6.40),

$$(y_{n_I} + x_{n_I} - T_s) - z \leq 0 \quad \text{for all } n_I \in I_c \tag{6.40}$$

And constraint (6.17) or (6.35) becomes (6.41),

$$y^r = T_s \tag{6.41}$$

If $y_{lc}^E > y_{lc}^I$, then constraint (6.6) is rewritten as (6.42),

$$(y_{n_E} + x_{n_E} - T_s) - z \leq 0 \quad \text{for all } n_E \in E_c \tag{6.42}$$

Therefore, the scheduling model for a fixed start time of the internal setup is obtained.

Model 6.2(b)

$$\min Z = z$$

such that constraints (6.28)–(6.34), (6.40)–(6.42), and (6.38) hold.

Comparing Model 6.2(b) with Model 6.2(a), the later might generate a start time $t_s > T_s$, meaning a slack. This slack time can be used for preventive maintenance of the manufacturing facility so as to reduce the number of machine breakdowns, which is an important factor.

Scheduling Model with a Free Start Time of the Internal Setup Process. It is clear that minimizing the duration of the internal setup process with a free start time of the internal setup process is a special case of Model 6.2(a) when constraint (6.35) in Model 6.2(a) is removed.

Scheduling a Model for Minimizing the Overall Setup Duration. It can be shown that minimizing the overall duration of the internal and external setup process can be modeled as a special case of Model 6.2(a). Then, the objective becomes

$$\min Z = \max\{y_{nI} + x_{nI} : n_I \in I_c, y_{nE} + x_{nE}: n_E \in E_c\}$$

which is equivalent to

$$\min z \tag{6.43}$$

with the following constraints:

$$y_{nI} + x_{nI} - z \le 0 \quad \text{for all } n_I \in I_c \tag{6.44}$$

$$y_{nE} + x_{nE} - z \le 0 \quad \text{for all } n_E \in E_c \tag{6.45}$$

Therefore, the following model is obtained for minimizing the entire duration of the internal and external setup process:

Model 6.2(c)

$$\min Z = z$$

such that constraints (6.28)–(6.32), (6.44)–(6.45), and (6.38) hold.

Comparing Model 6.2(c) with Model 6.1(a), one observes that they are similar. The difference is that the latter requires a common terminal activity as well as a common start activity, while the former does not. Thus, Model 6.1 [including Model 6.1(a)] can be considered as a special case of Model 6.2(c), and this in turn is a special case of Model 6.2(a).

6.5 COMPUTATIONAL EXPERIENCE

6.5.1 Numerical Example

The data presented in Figure 6.3 and Tables 6.1 and 6.2 are used to investigate the effects of three critical factors on the internal setup time and the completion time of the entire setup process obtained from Model 6.2(a). The three critical factors are: (a) the total crash budget (B), (b) the start time of the internal setup (T_s), and (c) the allowed completion time of the internal (entire) setup process (T_c). Three levels are selected for each of the three factors. Therefore, a 3^3 factorial design is required

(Montgomery, 1991), which implies 27 experiments with no replicates (see Table 6.5). In the design of experiments, the three levels of crash cost are 500, 600, and 700; the three levels of start time of the internal setup are 20, 25, and 30; and the three levels of allowed completion time of the entire setup process are 45, 55, and 65 (see Table 6.5). Using the data provided in Figure 6.3 and Tables 6.1 and 6.2, the computational results obtained from Model 6.2(a) are presented in Table 6.5.

Based on the experimental results in Table 6.5, plots of internal setup time with respect to three critical factors are presented in Figures 6.7, 6.8, and 6.9. In Figure 6.7, $t_{is}(M1)$ and $t_{is}(M2)$ is the internal setup duration from Model 6.1 and Model 6.2(a), $t_c(M1)$ the completion time from Model 6.1, and $t_{c,arg}(M2)$ and $t_{c,min}(M2)$ is the average and minimum completion time from Model 6.2(a). The analysis of variance of the experiments with the response t_c is presented in Table 6.6, with the response t_{is} is presented in Table 6.7 (where DF means degrees of freedom, SS sum of squares, and MS mean square).

Based on the results of Tables 6.6 and 6.7, the following observations are made in light of the range of the three parameters used in the experiments:

- For the same value of total crash budget B and start time T_s of the internal setup process, a later allowed completion time T_c of the entire setup process leads to a shorter internal setup time and an earlier completion time of the entire setup process, both at a 99% confidence interval. This is due to the fact that a later completion time of the entire setup process allows us to use less budget on the external setup activities of class E and more budget on the internal setup, thus resulting in a shorter internal setup time.

- For the same value of start time and the allowed completion time of the setup process, a higher crash budget leads to a shorter internal setup time and an earlier completion time of the entire setup process, both at a 97.5% confidence interval.

TABLE 6.5. Computational Results of 3^3 Experimental Design for Model 6.2(a)

		$T_c = 45$			$T_c = 55$			$T_c = 65$		
B	T_s	t_s	t_c	t_{is}	t_s	t_c	t_{is}	t_s	t_c	t_{is}
500	20	30.0	45.0	15.0	48.7	55.0	6.3	55.7	59.7	4.0
	25	30.0	45.0	15.0	48.7	55.0	6.3	55.7	59.7	4.0
	30	30.0	45.0	15.0	48.7	55.0	6.3	55.7	59.7	4.0
600	20	33.7	45.0	11.3	50.0	54.0	4.0	50.0	54.0	4.0
	25	33.7	45.0	11.3	49.5	53.5	4.0	49.5	53.5	4.0
	30	33.7	45.0	11.3	49.5	53.5	4.0	49.5	53.5	4.0
700	20	37.0	45.0	8.0	45.0	49.0	4.0	55.7	49.7	4.0
	25	37.0	45.0	8.0	45.0	49.0	4.0	45.0	49.0	4.0
	30	37.0	45.0	8.0	45.0	49.0	4.0	45.0	49.0	4.0

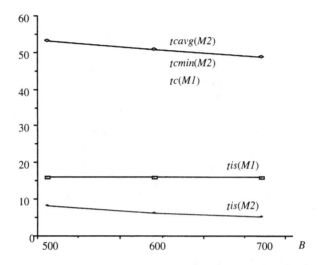

Figure 6.7. Plot of t_c and t_{is} vs. the allowed total crash budget (B).

This is obvious since a higher crashing budget reduces more effectively the activity duration.

- For the same value of total crash budget B of the entire setup process and allowed completion time of the entire setup process, the value of the start time of the internal setup process does not have a significant impact on the internal setup time and the completion time of the entire setup process within the range

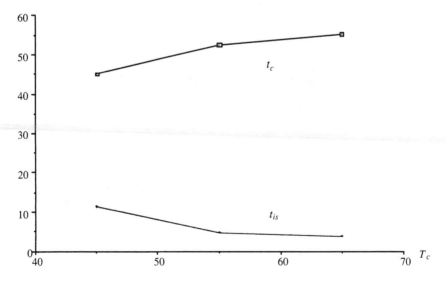

Figure 6.8. Plot of t_{is} vs. the allowed completion time of the entire setup process (T_c).

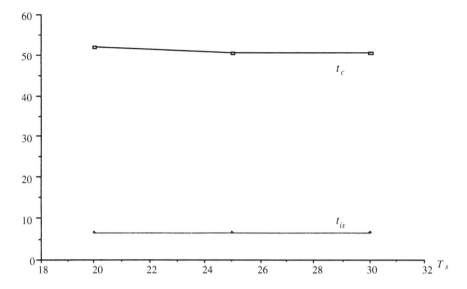

Figure 6.9. Plot of t_{is} vs. the allowed start time of the internal setup (T_s).

of the values of T_s used in the experiment. This implies that the constraint on T_s in Model 6.2(a) is redundant in a certain range of change.

- The preceding observations imply that to minimize the internal setup time, the best combination of the three factors is to set the crash cost at the highest level ($B = 600$) and the allowed completion time of the internal setup time to the highest level ($T_{is} = 40$) with a free T_s in the range of 20–30.
- The effect of the allowed completion time of the internal setup (T_{is}) is significant at a 99% confidence interval (the most important factor). The allowed crash budget (B) is significant at a 97.5% confidence interval (the second important factor), and the least important factor is the start time of internal setup activities (T_s).

TABLE 6.6. Analysis of Variance (ANOVA) with Response t_c

Source	DF	SS	MS	F
T_s	2	7.61	3.81	0.46
T_c	2	478.44	239.22	28.92[a]
B	2	86.24	43.13	5.21[b]
Error	20	165.42	82.71	
Total	26			

[a] Significant at 1%.
[b] Significant at 2.5%.

TABLE 6.7. Analysis of Variance (ANOVA) with Response t_{is}

Source	DF	SS	MS	F
T_s	2	0.00	0.00	0.00
T_c	2	248.65	124.32	26.70[a]
B	2	43.25	21.62	4.64[b]
Error	20	93.13	4.66	
Total	26			

[a] Significant at 1%.

6.5.2 Comparative Analysis of Models

The values of t_{is} and t_c calculated from Models 6.1 and 6.2(a) are presented in Figure 6.7. As mentioned previously, Model 6.1 minimizes the overall project duration, while Model 6.2 minimizes the internal setup time, both with the constrained total crash budget. In the "as is" schedule of setup activities with a zero crash budget (Figure 6.4) the internal setup time was 16 time units. One can see from Table 6.4 that the internal setup time generated from Model 6.1 remains 16 time units despite an increased level of setup budget from 500 to 900. However, the internal setup time generated from Model 6.2(a) (see Table 6.5 and Figure 6.7) is reduced from the 16.0 of Model 6.1 to 15.0, 11.3, and 8.0 time units, respectively, when the same crash budgets 500, 600, and 700 are used. A comparison of Models 6.1 and 6.2(a) on the internal setup time t_{is} and entire setup duration t_c is provided in Table 6.8. The percentage values of the reduction of internal setup and increase of the entire setup for Models 6.1 and 6.2(a) are provided. It is shown from Table 6.8 that Model 6.2(a) always results in a greater reduction of the internal setup than an increase in the entire setup, as compared with Model 6.1, due to different objectives of the models.

TABLE 6.8. Impacts of Models 6.1 and 6.2(a) on Internal Setup Time and Entire Setup Duration

B	t_{is}			t_c		
	M1[a]	M2a[a]	Reduction	M1[a]	M2a[a]	Increase
500	16.0	15.0	6.25	44.3	45.0	1.56
600	16.0	11.3	29.38	41.4	45.0	8.00
700	16.0	8.0	50.00	38.9	45.0	13.56

[a]M1 = Model 6.1, M2a = Model 6.2(a)

6.5 SUMMARY

In this chapter, characteristics of the setup activities were discussed. The internal setup time reduction problem was formulated as a minimax linear programming model. This might be the first attempt to identify these characteristics and apply a project scheduling approach to minimize the setup time, the internal setup time in particular, based on these characteristics. Various cases of setup time reduction were investigated, and the scheduling models were extended to incorporate these cases.

A 3^3 factorial design was used to provide further insights for setup time reduction. The factorial analysis approach showed that a higher crash cost, later start time of internal setup activities, and longer allowed completion time of the internal (entire) setup process led to a shorter internal setup time. Ranking these three factors so as to affect the internal setup time in the example problem implies that the allowed duration of the internal setup is the most important factor, the crash budget is of second importance, and the start time of the internal setup process of least importance.

The models presented in this chapter are of practical significance. Providing the data in Tables 6.1 and 6.2 and a limit on the crash budget B, the earliest start time of the internal setup process T_s, and the maximum allowed duration of the internal setup process T_{is} or the latest completion time of the entire setup process T_c, Model 6.2(a) generates an optimal sequence of internal and external setup activities and makes a trade-off between the time and crash cost of the internal and external setup activities. These models can be applied for one machine or multiple machines. In the case of multiple machines, the values of T_s, T_{is}, (T_c) should be used for the bottleneck machine.

REFERENCES

Black, J. (1991), *The Design of the Factory with a Future*, McGraw-Hill, New York.

Burgess, A., I. Morgan, and T. Vollmann (1993), Cellular manufacturing: Its impacts on the total factory, *International Journal of Production Research*, Vol. 31, No. 8, pp. 2059–2077.

Chand, S. (1989), Lot sizes and setup frequency with learning in setups and process quality, *European Journal of Operational Research*, Vol. 42, No. 2, pp. 190–202.

Chand, S., and S. Sethi (1990), A dynamic lot sizing model with learning in setups, *Operations Research*, Vol. 8, No. 4, pp. 644–655.

Christofides, N., R. Alvarez, and J. Tamarit (1987), Project scheduling with resource constraints: Branch and bound approach, *European Journal of Operational Research*, Vol. 29, No. 3, pp. 262–273.

Dahel, N., and S. Smith (1993), Designing flexibility into cellular manufacturing systems, *International Journal of Production Research*, Vol. 31, No. 4, pp. 933–945.

Davis, E., and H. Patterson (1975), A comparison of heuristic and optimum solutions in resource-constrained project scheduling, *Management Science*, Vol. 21, No. 7, pp. 944–955.

Dobson, G. (1992), The cyclic lot scheduling problem with sequence-dependent setups, *Operations Research*, Vol. 40, No. 4, pp. 736–749.

Elmaghraby, S. (1977), *Activity Networks: Project Planning and Control by Network Models*, John Wiley, New York.

Elmaghraby, S. and J. Kamburowski (1992), The analysis of activity networks under generalized precedence relations (GPRs), *Management Science*, Vol. 38, No. 8, pp. 1245–1263.

Feng, C-X., A. Kusiak, and C. C. Huang (1997), Scheduling models for setup reduction, *ASME Transactions: Journal of Manufacturing Science and Engineering* Vol. 119, No. 4A, pp. 571–579.

Flynn, B. (1987), Repetitive lots: The use of a sequence-dependent set-up time scheduling procedure in group technology and traditional shops, *Journal of Operations Management*, Vol. 7, No. 1, pp. 203–216.

Foo, F., and J. Wager (1983), Set-up times in cyclic and acyclic group technology scheduling systems, *International Journal of Production Research*, Vol. 21, No. 1, pp. 63–73.

Gallego, G., and I. Moon (1992), The effect of externalizing setups in the economic lot scheduling problem, *Operations Research*, Vol. 40, No. 3, pp. 614–619.

Hall, R. (1983), *Zero Inventories*, Dow Jones-Irwin, Homewood, IL.

Hillier, F., and G. Lieberman (1990), *Introduction to Operations Research*, 5th ed., McGraw-Hill, New York.

Hitomi, K., and I. Ham (1977), Group scheduling techniques for multiproduct, multistage manufacturing systems, *Transactions of the ASME: Journal of Engineering for Industry*, Vol. 99B, No. 3, pp. 759–765.

Horowitz, E., and S. Sahni (1984), *Fundamentals of Data Structure with Pascal*, Computer Science Press, Rockville, MD.

Inman, R., P. Jones, and G. Gallego (1991), Economic lot scheduling of fully loaded processes with external setups, *Naval Research Logistics*, Vol. 38, No. 5, pp. 699–713.

Jordan, P., and G. Frazier (1993), Is the potential of cellular manufacturing being achieved? *Production and Inventory Management Journal*, Vol. 34, No. 1, pp. 70–72.

Karwan, K., J. Mazzola, and R. Morey (1988), Production lot sizing under setup and worker learning, *Naval Research Logistics*, Vol. 35, No. 2, pp. 159–175.

Kelley, J., Jr. (1961), Critical path planning and scheduling: Mathematical basis, *Operations Research*, Vol. 9, No. 3, pp. 296–320.

Kelly, J., Jr., and J. Walker (1959), Critical path planning and scheduling, *Proceedings of Eastern Joint Computer Conference*, pp. 160–173.

Leachman, R., and S. Kim (1993), A revised critical path method for networks including both overlap relationships and variable-duration activities, *European Journal of Operational Research*, Vol. 64, No. 2, pp. 229–248.

Montgomery, D. (1991), *Design and Analysis of Experiments*, 3rd ed., John Wiley, New York.

Patterson, J., F. Talbot, R. Slowinski, and J. Weglarz (1990), Computational experience with a backtracking algorithm for solving a general class of precedence and resource-constrained scheduling problems, *European Journal of Operational Research*, Vol. 49, No. 1, pp. 68–79.

Premachandra, I. (1993), A goal-programming model for activity crashing in project networks, *International Journal of Operations and Production Management*, Vol. 13, No. 6, pp. 79–85.

Russel, R. (1986), A comparison of heuristics for scheduling projects with cash flows and resource restrictions, *Management Science*, Vol. 32, No. 10, pp. 1291–1300.

Schrage, L. (1984), *Linear, Integer and Quadratic Programming with LINDO*, Scientific Press, Palo Alto, CA.

Shingo, S. (1985), *A Revolution in Manufacturing: The SMED System*, Productivity Press, Stamford, CT.

Steudel, H., and P. Desruelle (1991), *Manufacturing in the Nineties: How to Become a Mean, Lean, World-Class Competitor*, Van Nostrand Reinhold, New York.

Talbot, F. (1982), Resource-constrained project scheduling with time resource tradeoffs: The nonpreemptive case, *Management Science*, Vol. 28, No. 10, pp. 1197–1210.

Trevino, J., B. Hurley, and W. Friedrich (1993), A mathematical model for the economic justification of setup time reduction, *International Journal of Production Research*, Vol. 31, No. 1, pp. 191–202.

Ulusoy, G., and L. Özdamar (1994), A constraint-based perspective in resource constrained project scheduling, *International Journal of Production Research*, Vol. 32, No. 3, pp. 693–705.

Weglarz, J. (1979), Project scheduling with discrete and continuous resources, *IEEE Transactions on Systems, Man, and Cybernetics*, Vol. 9, No. 10, pp. 644–650.

QUESTIONS

6.1. What is a setup?

6.2. What are the benefits from setup reduction?

6.3. What is CPM?

6.4. What is PERT?

6.5. What is an activity on an arc network?

6.6. What is a crash cost and how can it be calculated?

PROBLEMS

6.1. Think of a mechanical part, machine a tool to process this part, define setup activities, and construct a network of activities.

6.2. Define an application scenario for Model 6.1, formulate the model, solve it, and discuss the results.

6.3. Define an application scenario for Model 6.2, formulate the model, solve it, and discuss the results.

CHAPTER 7

PRODUCTION PLANNING AND SCHEDULING

7.1 PRODUCTION PLANNING

7.1.1 Manufacturing Resource Planning

Based on the information regarding the type and quantity of products and parts to be manufactured, the material required to produce the products and parts is ordered with an appropriate lead time to assure the timely availability for manufacturing. This constitutes a basic function of the material requirements planning (MRP) system. The MRP appears to be one of the most widely used computerized systems for planning the acquisition of material and components. It uses a bill of material (BOM) as a key representation of material requirements. The bill of material is represented as a tree with the root representing a product to be produced and the children representing lower level subassemblies or parts. The MRP system uses the fact that the BOM structure allows us to generate requirements for parts and subassemblies based on the demand for the products.

The MRP system essentially involves a two-step procedure (Kusiak et al. 1993):

1. Computing the net requirement for parts:

    (a)   If (gross requirement – inventory level – scheduled receipt) > 0,

    (b)   then the net requirement = (gross requirement – inventory level – scheduled receipt).

    (c)   Otherwise the net requirement = 0.

2. Place a planned order at an appropriate period by backward scheduling from the required date by the lead time to fulfill the parts order.

This procedure begins at the input from the master production schedule and works down the BOM structure until all parts are planned. Figure 7.1 illustrates a BOM structure for product C that includes two subassemblies S1 and S2 with the subassembly S1 including two units of part P1 and one unit of part P2. The MRP system converts the master production schedule for products into a schedule for assemblies and parts, that is, it generates the information indicating what material is needed, when, and how much. The structure of a typical MRP environment is illustrated in Figure 7.2.

The basic MRP record structure is presented in Figure 7.3:

- The top row in Figure 7.3 indicates periods that can vary in length from a day to a week or longer. The number of periods in the record is called a planning horizon. The planning horizon indicates the number of future periods for which plans are made.

- The second row, "gross requirements," is the anticipated future demand or usage for an item. The gross requirement is time phased, rather than aggregated or averaged, that is, 10 items in period 2, 40 in period 4, rather than as the total requirement of 60 over the five periods. A gross requirement in a particular period signifies that a demand is anticipated that will be unsatisfied unless the items are available.

- The "scheduled receipts" row shows the quantities that have already been ordered and when we expect them to be completed. Scheduled receipts represent a source of the item to meet the gross requirements.

- The next row is the number of items "on hand." The timing convention in this row is the end of the period; that is, the row is the projected balance after the replenishment orders have been received and the gross requirement has been satisfied. There is an extra period shown at the beginning of the MRP record to represent the available balance at the present time.

- Whenever the on-hand available balance would imply a quantity insufficient to satisfy the gross requirement (a negative quantity), additional orders must be planned for. This is done by creating a planned order release in time to keep the on-hand available balance from becoming negative. For example, the on-hand balance at the end of period 4 in Figure 7.3 is 4 units. This is insufficient to meet

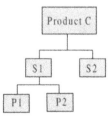

Figure 7.1. The BOM for product C.

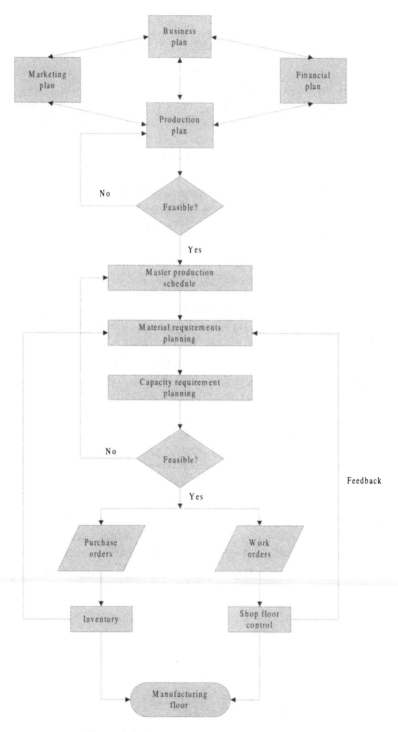

Figure 7.2. Structure of an MRP environment.

Period		1	2	3	4	5
Gross requirements			10		40	10
Scheduled receipts		50				
On hand	4	54	44	44	4	-6
Planned order releases					50	

Lead time = 1 period
Lot size = 50
Safety stock = 4

Figure 7.3. Basic MRP record.

the gross requirement of 10 units in period 5. An order for at least 6 units must be planned for period 4 to avoid a shortage in period 5. Since the lead time is 1 week, the MRP system creates a planned order at the beginning of week 4 providing a lead-time offset of 1 week. As the lot size is 50 units, the quantity of the order to be placed is 50 units, thus resulting in a positive balance of the items on hand of 44 rather than negative 6.

- The safety stock is 4 items.

The basic MRP record provides the information for each part in the system. To manage the flow of parts needed, these single part records must be linked together and the end-item requirements have to be translated into the component requirements, taking existing inventories and scheduled receipts into account. The components must be planned to be available at the time the end-item order is released for production, at which time the components are consumed. This process is called a parts explosion.

The example of a requirements explosion for the subassembly S1 and part P2 from Figure 7.1 is illustrated in Figure 7.4. Again, the order release of 50 items for subassembly S1 replaces the negative balance of the items on hand to 44, and for part P2 the negative balance of 42 changes to positive 58. The linkage of inventory records guides the explosion of requirements from the MPS down the various component levels. Gross requirements for the end items are processed against inventory to determine the net requirements, which are covered by planned orders. The quantity and timing of planned order releases determine, in turn, the quantity and timing of component gross requirements. This process is repetitively performed for items on successively lower levels until a purchased item is reached, at which point the explosion process terminates.

The term material requirements planning (MRP) was established in 1970s to describe the business layer of manufacturing software. Material requirements planning does not consider plant and supplier capacity while generating material requirements and work and purchase orders (Durmusoglu et al., 1996). In the 1980s the traditional MRP system evolved into the manufacturing resource planning (MRP

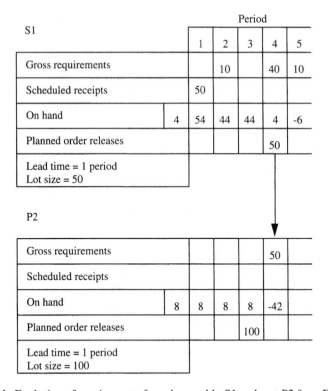

Figure 7.4. Explosion of requirements for subassembly S1 and part P2 from Figure 7.1.

II) system. In addition to the MRP functions, MRP II incorporates additional functions. Some of the most important extensions relate to the incorporation of manufacturing resource capacity. The decade of the 1990s has brought additional extensions to the manufacturing management software that was named Enterprise Resource Planning (ERP). There is widespread agreement that SAP software company has set new standards in the ERP market with its R/3 client/server software (Chase et al., 1998). Leading high-technology companies and small- and medium-size firms have adopted the SAP software and other competing ERP-type solutions developed by Baan, PeopleSoft, and Oracle as well as decision support tools (e.g., Business Objects). One should be aware of potential limitations imposed by single-vendor solutions versus solutions based on broadly accepted standards, e.g., ISO standards. Companies willing to form *virtual enterprises* should follow ERP software solutions based on widely accepted standards [e.g., emerging Internet standards (e.g., XML) and open system architecture standards]. A single-vendor ERP software is more suitable for forming *extended enterprises* where the partners could adopt the requirements of a single software solution.

Unforeseen events occur in production; for example, vendors may not deliver items on time and scrap may become excessive. Consequently, the required and planned order releases change substantially from one MRP run to the next. Whenever the on-hand status and gross requirement changes are reported to the system, the current MRP schedule becomes obsolete. Changes in the planned order releases produce chain reactions downward through the product structures and hence throughout the material requirements plan. Incorporating such changes for upper level inventory items may cause other lower level items to have insufficient or excessive immediate coverage or to require an immediate order release. The degree to which upper level changes are reflected in the lower level exception conditions is called "MRP nervousness." It is commonly defined as significant changes in MRP plans, which are produced even though there are only minor changes in the higher level products or MPS (Vollmann et al., 1988).

In a "nervous" environment, the continuous rescheduling of orders could undermine user's confidence in the system, depriving it of the support needed for successful operation. For that reason, one could implement a "dampening mechanism" to reduce the frequency of replanning and then achieving a stable schedule. However, there is an additional cost associated with the dampening mechanism when it is introduced. Therefore, it is important to have an understanding of the trade-off between achieving a stable MPS at the minimum cost while being responsive to the changes in requirements.

7.1.1.1 *Processing Frequency.*

As the new information becomes available, the MRP records must be updated so plans can be adjusted to reflect these changes. The issue here is how frequently the records should be processed and whether all records should be processed at the same time.

One of the options is to process all records in one computer run, which is called regeneration run. An alternative is called "net change" processing, in which only those records that are affected by the new information are reconstructed. The appropriate frequency for processing the MRP time-phased records depends on the company, its products, and its operations.

The motivation for less frequent processing of new information is the computational cost, which can be especially high with regeneration. The problem with less frequent processing is that the component status expressed in the MRP record becomes increasingly inaccurate. More frequent processing of records results in fewer unpleasant surprises.

The net change approach can reduce computer time enough to make more frequent processing possible. On the other hand, daily processing of part of the MRP records could be computationally more expensive than weekly regeneration.

7.1.1.2 *MRP Nervousness.*

A dampening mechanism can be viewed as a filter that screens out insignificant exception messages. It has been suggested that some "rules" be used to limit the number of exception messages (Steele, 1975). In an MRP system these rules could include constraints on certain variables, usually the quantity and time of planned order releases, to reduce the number of unnecessary exception messages. The goal is to report to the production planner only items requiring immediate response.

The rules that include the constraints on the quantity may act as a "trigger" that determines whether an exception message should be reported if the schedule order receipts or on-hand inventory has changed. A constraint of another type is a time fence. Time fences can be selected so that only certain types of changes occurring in a particular time interval are reported. The rationale is that only "emergency" types of exceptions should be reported.

Several alternative ways of dealing with MRP nervousness have been suggested in the literature. One frequently discussed approach is to use a safety stock of an upper level item to act as a buffer against differences in the actual and forecast requirements. This policy is expected to reduce the order instability in the MPS and thus to minimize the nervousness transmitted to lower levels of the product structure. Yano and Carlson (1987) indicated that in some cases it might be more economical to reschedule infrequently and use safety stock as protection against demand variations.

The approach of freezing the MPS, especially over the cumulative lead time of a product, is frequently used in practice and was described in Berry et al. (1979). Kropp et al. (1983) modified several standard lot-sizing procedures used in a rolling horizon environment to incorporate the cost of changing the MPS. Chung and Krajewski (1986) indicated that under certain cost and demand conditions the master production schedule need not to be updated in every period.

Recent research has focused on design of methods for freezing the MPS as a way of controlling MPS stability under rolling planning conditions (Sridharan and Berry, 1990). They presented a framework for analyzing design parameters of alternative methods for freezing the MPS and compared their performance when the design parameters of these methods were varied.

In addition, the effectiveness of five different strategies for reducing MRP nervousness has been investigated by Blackburn et al. (1986). Their experimental results emphasized the need for considering the trade-off between the costs and benefits of achieving schedule stability. Interaction effects on MRP nervousness of different strategies have also been reported (Minifie and Davis, 1990).

A knowledge-based approach can be applied to deal with the MRP nervousness problem. The user provides the system with input parameters such as the process type, product type, demand type, and performance required. The system recommends a "suitable" mechanism for reducing MRP nervousness.

Production rules illustrating the knowledge-based system are presented next (Kusiak and Yang, 1995).

Rule 1

IF the process type is multistage

AND the product structure is multiproduct

AND the demand is known

AND the measure of performance is cost

THEN the dampening mechanism suggested is the change_cost_procedure

This rule originated from Blackburn et al. (1986).

Rule 2

IF the dampening mechanism suggested is freeze_MPS

    AND the process type is all

    AND the product structure is one

    AND the demand is known

    AND the measure of performance is all

THEN the design parameter of this dampening mechanism suggested is planning_horizon_length

This rule can be traced back to Sridharan and Berry (1990), who indicated that among the MPS freezing parameters, the length of the planning horizon or the type of information considered in freezing the MPS has the smallest impact on the cost and MPS instability measures.

Rule 3

IF the dampening mechanism suggested is freeze_MPS

    AND the design parameter is to use freeze_interval

    AND the process type is all

    AND the product structure is one

    AND the demand is known

    AND the measure of performance is schedule stability

THEN the value of the design parameter is to be adjusted by 50%

This rule is based on Sridharan et al. (1988). They argued that in order to obtain meaningful benefits with respect to schedule stability, the proportion of the planning horizon that is frozen should exceed 50 percent. Furthermore, it had been observed that the order-based freezing method results in more stable schedules than the period-based method.

7.1.2 Optimized Production Technology System

The fundamental principle of the Optimized Production Technology (OPT) system, another production planning tool, is that only the bottleneck operations (resources) are considered critical to scheduling (Vollman et al., 1988). The production output is limited by the bottleneck operations and improving the utilization of the bottleneck facilities can only increase throughput.

The OPT system combines the data in the BOM file with those in the routing file. As a result, a network or extended tree diagram is constructed. The operational data attached to each part in the product structure are then combined with the master production schedule to form what is called a "product network."

Each operation is defined in terms of the resources used and the setup and processing times. The OPT files include the data on the capacity, maximum inventory, minimum batch quantities, order quantities, due dates, alternative routings, and resource constraints. Product network and resource descriptions are then fed into a set of routines called BUILDNET and SERVE that identify the bottleneck resources. The BUILDNET routine combines the product network and resource information to form an engineering network. The SERVE routine determines the backward schedule from the order due dates assuming infinite capacity for the resources.

Average expected loads on machine centers are determined using a rough-cut capacity planning routine. These average loads are arranged in descending order, and the most heavily loaded workcenters are studied. Then, a routine called SPLIT is used to split the OPT product network into two portions. The lower section is called the "SERVE Network," which includes all operations preceding the bottleneck resources. The upper portion is the "OPT Network," which incorporates all of the bottleneck resources and all succeeding operations.

One of the advantages of this split is that one can see where attention should be focused. Bottleneck capacity is used more extensively by finite loading of this small subset of workcenters. When this finite loading through bottleneck resources is completed, the result is a doable master production schedule. In short, OPT conceivably can take any master production schedule as input and determine the extent to which it is doable.

For nonbottleneck resources OPT schedules are based on MRP logic. In such cases OPT reduces batch sizes to the point where some nonbottleneck resources become bottlenecks. This results in less work in process inventory and reduced lead time. This is achieved by overlapping schedules using unequal batch sizes for transferring and processing.

To maximize output from bottleneck operations, larger lot sizes are run. As a result, the percentage of nonproductive time devoted to setups in these workcenters is reduced. On the other hand, smaller batches are made at nonbottleneck workcenters. Calculation of the batch sizes is a part of the OPT procedure.

A transfer batch refers to the lot size that moves from operation to operation. A process batch is the total lot size released to the shop. The OPT distinguishes between a transfer batch and a process batch. Any differences are held in work-in-process inventories. Also, any operation cannot proceed until a transfer batch is built behind it.

The OPT system provides buffers for schedules for critical operations by using both safety stocks and safety lead time. When a sequence of jobs is scheduled on the same machine, safety timing is introduced between subsequent batches. In this way, a cushion is provided against variations adversely affecting the flow of jobs through the same operation.

7.1.3 Just-in-Time System

An MRP system involves some guesswork. Customer demands need to be predicted in order to prepare the schedule and also it is required to guess the amount of time required for production to make the needed parts. Even though this system allows corrections to be made frequently, bad predictions result in excess inventories of some parts (Belhe and Kusiak, 1997).

Manufacturing companies face the difficulty of reducing the production cost and improving the product quality. Three kinds of costs—materials, labor, and overhead—are associated with manufacturing a product. It is most important to use the correct manufacturing strategy to reduce these costs. Just-in-time is a management approach that is focused on integrating and streamlining the manufacturing system into the simplest possible process. It strives to minimize the elements in manufacturing system that restrain productivity of the system.

The beginning of just-in-time can be traced back to the Toyota system. This system was first implemented at Toyota, Japan, which became successful in reducing inventory levels and improving quality. In this system material movement between workcenters follows three main rules:

- Material is moved in a continuous flow rather than in a batch mode.
- Material is moved in the smallest possible quantities.
- Material is moved only when it is required by the next stage.

The Toyota system evolved into the just-in-time system, designed to improve the efficiency of manufacturing organizations with minimum resources. It also improves quality, reduces inventory levels, and provides maximum motivation to solve problems as soon as they occur.

Just-in-time is defined as a production system designed to eliminate waste in the manufacturing environment. Here, waste is described as anything that is not necessary for the manufacturing of the product or is in excess.

Most companies use master schedules and MRP to decide their production schedules and the movement of material in the factory. This system is referred to as a "push system." Another system, called a "pull system," uses bottom-up demand and is driven by the consumption rates of parts in the production process. The goal of the pull system is to pull material required with minimum advance notice of production requirements from the customer. One of the most fundamental changes that just-in-time introduces in a manufacturing organization is the institution of a pull system instead of a push system. The pull system is favored because it eliminates unnecessary elements from the production system. These elements are primarily the cost of material labor diverted into inventory.

Consider a company is building 80 units a day of power supply modules. Assume that a work order for a set of 400 transformers is released to the manufacturing floor on Monday to meet the weekly demand. According to the production schedule, we only need a set of 80 transformers on Monday, but we have released 400, producing an excess of 320 transformers that day, an excess of 240 transformers on Tuesday, and so on. In a just-in-time system, these excess sets of transformers would be considered as waste, because they are not needed to produce the daily quota. In such a system, we would only release 80 transformers each day. For high-volume applications, the frequency of release could be increased. If the production rate increases 10-fold, then we would be better off by releasing 400 transformers every four hours or 100 transformers every hour.

There are some major misconceptions about the just-in-time manufacturing system (Lubben, 1988). A major misconception is that just-in-time is an inventory control system. Although inventory reduction is a key goal of a just-in-time system, it is much broader and affects the operation of many departments in the company. It is important not to reduce buffer inventories until the quality of parts reaches an acceptable level. A company should start solving quality problems with suppliers and the production process long before it starts reducing inventory.

The second misconception is that just-in-time is a method used by the materials function to push inventory back into the supplier's shop, thereby forcing the supplier to carry the customer's inventory. Where the material is stored is irrelevant. The materials and resources required making the part have been committed. The resources that could have been used to produce a needed product have been diverted into nonproductive inventory.

Manufacturing organizations must have a clear procedure to issue materials to the production floor. There are three critical aspects involved in this procedure. First, the procedure must ensure that the materials issued are sufficient for the production build schedule. Second, the procedure must allow the company to track the materials moving through the production process. Third, the procedure must allow analysis of the physical movement of the materials in the factory so as to be able to increase productivity and reduce overhead.

7.1.3.1 *Kanban System Concept.*

The word *kanban* means "visual record" and refers to a manufacturing control system originally developed in Japan. Toyota used the kanban system for many years as a means to communicate material needs between two workcenters. The kanban is a mechanism by which a workstation signals the need for more parts from the preceding station. For example, consider the manufacturing process shown in Figure 7.5. There are three workcenters A, B, and C cascaded together. The requirement for a certain number of finished goods is translated as the requirement of the corresponding number of units from workcenter B. This requirement is transferred backward up to the material stock location. Then only the required number of raw material units is released to workcenter A, which finally is transformed into the exact requirement of the number of finished product units.

In the Toyota kanban system every part number has its own special container design to hold a precise quantity of that part. Usually this quantity is small. There are

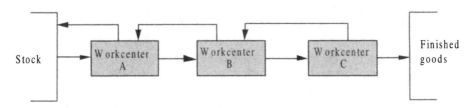

Figure 7.5. Communication needs between workcenters.

two cards, or "kanban," for each container. Each kanban indicates the part number, container capacity, and certain other information. One kanban, called a production kanban, serves the workcenter producing the part, and the other kanban, known as the withdrawal kanban, serves the workcenter using it. Each container cycles from the producing workcenter to the using workcenter and back.

The withdrawal kanban travels between workcenters and is used to authorize the movement of parts from one workcenter to another. A withdrawal kanban must accompany the flow of material from one process to another. Once a withdrawal kanban fetches a part, it will stay with the material all the time. Then, after the subsequent process consumes the last part of the lot, the kanban will travel again to the preceding process to fetch a new set of parts.

The production kanban's job is to release an order to the preceding process to build more parts. The production kanban goes into a queue with other production kanbans at the workcenter. After the workcenter builds the new parts, the production kanban travels back to the wait area until a new withdrawal kanban starts the cycle over again.

As described by the Japan Management Association (1986), the major functions of the kanban system are:

1. Engage in standard operations at any time
2. Give directions based on the actual conditions existing in the workplace
3. Prevent addition of any unnecessary work for those engaged in start-up operations

7.1.3.2 *Kanban Rules.* A kanban is a tool created to manage the workplace effectively. However, this tool should be used appropriately. The rules for operating kanbans are as follows:

1. *Do not send defective products to the subsequent process.* If a defective product is discovered, the highest priority must be given to the measures to prevent its recurrence. This is to make sure that such defects will not be produced again. If the defective products get mixed up with good products, they must be exchanged promptly.

2. *The subsequent process withdraws only what is needed.* The subsequent process must withdraw materials from the preceding process at the time needed and in the quantity needed. A number of concrete steps are required to ensure that these withdrawals are not arbitrary, such as:

    (a) There is no withdrawal without a kanban.

    (b) Items withdrawn cannot exceed the number of kanban submitted.

    (c) A kanban must always accompany each item.

3. *Produce only the exact quantity withdrawn by the subsequent process.* This rule is a logical extension of the second rule. With this rule the process is able to restrict its inventory to the minimum. Not producing more than the number of kanbans and producing in the sequence in which the kanbans are received make

this rule operational. By observing the second and third rules, the production process can function in unison.

4. *Equalize production.* In order to produce the exact quantity of product, adequate facilities and personnel are required. As a result, the processes that cannot deal with the requirements resort to producing material ahead of time, which is a violation of the third rule. This rule insists on load smoothing.

5. *Kanban as a means of fine tuning.* In using a kanban system, it is important to abide by the principle of load smoothing in production. Sudden changes in production demands cannot be handled by the kanban system. A kanban can only respond to the need for fine tuning.

6. *Stabilize and rationalize the process.* The defective parts should not be sent to the subsequent processes, and defective work is the result of not having sufficient standardization and stabilization. Unless the process is rationalized and stabilized, adequate supply and quality cannot be maintained. This in turn supports the load-smoothing system of production.

Details of kanban systems are discussed in Chapter 8.

7.2 CAPACITY BALANCING

The output from an MRP system serves as a basis for detailed balancing of the manufacturing capacity. The manufacturing capacity can be considered at a cell or machine level. The goal of capacity balancing at the machine level is to assign manufacturing operations (which also implies the parts) to machines so that the capacity of machines and other relevant constraints are met. The three models presented in this section illustrate capacity balancing at the machine level. Capacity balancing at the cell level can be accomplished with similar models by using aggregate data.

Capacity balancing is accomplished at the operation level. An operation is a set of tasks (e.g., removal of machining features) performed on a part on one machine or station. The capacity-balancing models discussed in this chapter apply to different manufacturing (e.g., machining and assembly) and service (e.g., assignment of tasks to persons) processes. Terms such as machine, part, and operations can be generalized to other applications.

Model 7.1. Model 7.1 is developed under the assumption that splitting of the batches of parts (or operations) does not take place. The objective function (7.1) minimizes the total processing cost of all batches of parts (operations):

$$\min \sum_{i \in I} \sum_{j \in J} C_{ij} x_{ij} \tag{7.1}$$

$$\sum_{j\in J} x_{ij} = 1 \qquad i \in I \tag{7.2}$$

$$\sum_{i\in I} T_{ij}x_{ij} \le b_j \qquad j \in J \tag{7.3}$$

$$x_{ij} = 0, 1 \qquad i \in I, j \in J \tag{7.4}$$

where

I = set of batches of parts (operations) to be processed

J = set of machines

T_{ij} = time of processing batch i of parts (operations) on machine j

C_{ij} = cost of processing batch i on machine j

b_j = processing time available on machine j (capacity of machine j)

$x_{ij} = \begin{cases} 1 & \text{if batch } i \text{ is processed on machine } j, j \in J \\ 0 & \text{otherwise} \end{cases}$

Constraint (7.2) ensures that each batch is processed on exactly one machine. The capacity limitation of each machine j is expressed by constraint (7.3). The integrality requirement is expressed by constraint (7.4). The model (7.1)–(7.4) is illustrated in Example 7.1.

Example 7.1. Five batches of parts are to be machined in a system with three machines. Given the matrix of batch processing costs $[C_{ij}]$, the matrix of processing of each batch on the three machines $[T_{ij}]$ and the machine capacity vector $[b_j]$ assign optimally the batches to the three machines:

$$[C_{ij}] = \begin{matrix} 1 \\ 2 \\ 3 \\ 4 \\ 5 \end{matrix} \begin{bmatrix} 4 & 7 & 7 \\ 1.5 & 1.2 & 6 \\ 3 & 6 & 5 \\ 4 & 5 & 4 \\ 2 & 3 & 2 \end{bmatrix}$$

$$[T_{ij}] = \begin{matrix} \\ 1 \\ 2 \\ 3 \\ 4 \\ 5 \end{matrix} \begin{matrix} 1 & 2 & 3 \\ \begin{bmatrix} 3 & 8 & 7 \\ 1 & 1 & 6 \\ 4 & 5 & 4 \\ 5 & 6 & 3 \\ 1 & 2 & 3 \end{bmatrix} \end{matrix}$$

$$[b_j] = [21, 20, 42]^{\mathrm{T}}$$

Substituting the data above to model (7.1)–(7.4), the following program is created:

$$\min 4x_{11} + 7x_{12} + 7x_{13} + 1.5x_{21} + \cdots + 2x_{53}$$

$$x_{11} + x_{12} + x_{13} = 1 \qquad \text{for batch } i = 1$$

$$x_{21} + x_{22} + x_{23} = 1 \qquad \text{for batch } i = 2$$

$$x_{31} + x_{32} + x_{33} = 1 \qquad \text{for batch } i = 3$$

$$x_{41} + x_{42} + x_{43} = 1 \qquad \text{for batch } i = 4$$

$$x_{51} + x_{52} + x_{53} = 1 \qquad \text{for batch } i = 5$$

$$3x_{11} + 1x_{21} + 4x_{31} + 5x_{41} + 1x_{51} \le 21 \qquad \text{for machine } j = 1$$

$$8x_{12} + 1x_{22} + 5x_{32} + 6x_{42} + 2x_{52} \le 20 \qquad \text{for machine } j = 2$$

$$7x_{13} + 6x_{23} + 4x_{33} + 3x_{43} + 3x_{53} \le 42 \qquad \text{for machine } j = 3$$

$$x_{ij} = 0, 1 \qquad \text{for } i = 1, \ldots, 5, j = 1, \ldots, 4$$

The following solution is obtained:

$$x_{11} = 1 \qquad x_{22} = 1 \qquad x_{31} = 1 \qquad x_{43} = 1 \qquad x_{53} = 1$$

This solution is interpreted as follows: batches 1 and 3 are processed on machine 1, batch 2 is processed on machine 2, and batches 4 and 5 are processed on machine 3.

A frequently used constraint in manufacturing systems, besides the ones discussed previously, is a tool magazine capacity constraint. Model 7.1 is extended to Model 7.2, which incorporates the tool magazine capacity constraint.

Model 7.2.   In addition to the notation used in Model 7.1, the following are defined:

k_{ij} = space occupied in a tool magazine by tools required by batch i at machine j, $i \in I, j \in J$

f_j = capacity of the tool magazine on machine $j, j \in J$

q_j = penalty for using the tool magazine on machine $j, j \in J$

Z_j = upper limit on the number of tool magazines to be used on machine $j, j \in J$

z_j = number of tool magazines required on machine $j, j \in J$

The objective function (7.5) minimizes the total processing cost of all batches of parts (operations) and the penalty for tool magazine changeovers:

$$\min \sum_{i \in I} \sum_{j \in J} C_{ij} x_{ij} + q_j z_j \tag{7.5}$$

$$\sum_{j \in J} x_{ij} = 1 \qquad i \in I \tag{7.6}$$

$$\sum_{i \in I} T_{ij} x_{ij} \le b_j \qquad j \in J \tag{7.7}$$

$$\sum_{i \in I} k_{ij} x_{ij} \le f_j z_j \qquad \text{for each } j \in J \tag{7.8}$$

$$x_{ij} = 0, 1 \qquad \text{for each } i \in I, j \in J \tag{7.9}$$

$$z_j \le Z_j \qquad \text{integer for each } j \in J \tag{7.10}$$

Constraints (7.6) and (7.7) are the same as constraints (7.2) and (7.3) in Model 7.1. Constraint (7.8) limits the tool capacity at each machine j in $z_j - 1$ tool magazine changeovers. The last two constraints (7.9) and (7.10) ensure integrality.

In some cases of balancing the machine capacity, the splitting of batches is allowed, that is, a batch of parts can be assigned to more than one machine, which is considered in Model 7.3.

Model 7.3. In addition to the notation introduced in Model 7.1, the following are defined:

t_{ij} = processing time of each operation from batch i on machine j
c_{ij} = processing cost of an operation from batch i on machine j
a_i = required number of operations in batch i (size of batch i)
y_{ij} = number of operations of batch i to be processed on machine j

The objective function (7.11) minimizes the total processing cost of all operations:

$$\min \sum_{i \in I} \sum_{j \in J} c_{ij} y_{ij} \tag{7.11}$$

$$\sum_{j \in J} y_{ij} = a_i \qquad i \in I \tag{7.12}$$

$$\sum_{i \in I} t_{ij} y_{ij} \le b_j \qquad j \in J \tag{7.13}$$

$$y_{ij} \ge 0 \quad \text{integer } i \in I, j \in J \tag{7.14}$$

Constraint (7.12) ensures that all operations required are processed. Inequality (7.13) is the machine capacity constraint. The integrality of variable y_{ij} is imposed by constraint (7.14). The model (7.11)–(7.14) is illustrated in Example 7.2.

Example 7.2.   Given the following set of parameters:

1. Number of operation types $|I| = 10$
2. Number of machine types $|J| = 3$
3. Matrix of machining times

$$[t_{ij}] = \begin{array}{c} \\ 1 \\ 2 \\ 3 \\ 4 \\ 5 \\ 6 \\ 7 \\ 8 \\ 9 \\ 10 \end{array} \begin{array}{ccc} 1 & 2 & 3 \\ \left[\begin{array}{ccc} 29.1 & 24.5 & \infty \\ 18.4 & \infty & 20.0 \\ 31.2 & \infty & 28.0 \\ \infty & 14.5 & 16.5 \\ 24.5 & 22.0 & \infty \\ 16.5 & 14.5 & 17.4 \\ 8.5 & 6.4 & \infty \\ 35.4 & \infty & 39.1 \\ 19.4 & 18.1 & \infty \\ 24.1 & 26.8 & \infty \end{array}\right] \end{array}$$

4. Vector of batch sizes

$$[a_i] = [18, 17, 15, 14, 15, 20, 12, 18, 12, 16]$$

5. Vector of machine capacity

$$[b_j] = [1800, 1000, 1500]$$

Assign the 10 operations to machines so that the total machining time is minimized.

Solving the model (7.11)–(7.14) for the above set of data results in the following solution: $y_{12} = 18$, $y_{21} = 17$, $y_{33} = 15$, $y_{42} = 4$, $y_{43} = 10$, $y_{51} = 9, y_{52} = 6$, $y_{62} = 20$, $y_{72} = 12$, $y_{81} = 18$, $y_{91} = 12$, and $y_{10,1} = 16$. This integer solution can be interpreted as follows:

1. Quantity of operations {17 of type 2, 9 of type 5, 18 of type 8, 12 of type 9, 16 of type 10} to be performed on machine 1

2. Quantity of operations {18 of type 1, 4 of type 4, 6 of type 5, 20 of type 6, 12 of type 7} to be performed on machine 2

3. Quantity of operations {15 of type 3, 10 of type 4} to be performed on machine 3

This solution indicates that only batches with operations of types 4, 5, and 9 have been split. Type 4 operations: 4 are processed on machine 2 and 10 on machine 3; type 5 operations: 9 are processed on machine 1 and 6 on machine 2.

In the three models discussed, operations are being assigned to machines. As each part includes a number of different operations, the same batch of parts can be assigned to different machines. Restricting the visit of a particular batch of parts to different machines can be accomplished by assigning the value of ∞ to an appropriate entry in the cost or time matrix, (e.g., see the matrix $[t_{ij}]$ in Example 7.2), by defining the range of subscripts, or by aggregating a number of different operations into one operation.

7.3 ASSEMBLY LINE BALANCING

Assembly line balancing falls into the same category of problems as capacity balancing discussed in Section 7.2. In assembly an attempt is made to uniformly distribute tasks across all stations. The goal is to balance the line so as to minimize the idle time at each station and maximize system throughput. In the absence of a limit on the assembly system capacity, the throughput would only be limited by the duration of the longest task.

The minimum number of stations required to achieve a desired throughput is determined from the formula

$$\text{Minimum number of stations required} = \frac{\text{desired throughput} \times \text{total task time}}{\text{total time available per day}}$$

The cycle time is defined as the time between successive units leaving the assembly line:

$$\text{Cycle time CT} = \frac{\text{total time available per day}}{\text{throughput}}$$

The minimum possible cycle time is equal to the length of the longest task. The total of the task times gives the maximum cycle time.

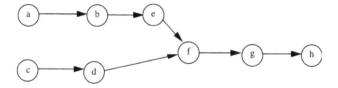

Figure 7.6. Precedence diagram for product P.

TABLE 7.1. Task Times for Product P

Task	Immediate Predecessor	Task Time (min)
a	—	0.2
b	a	0.2
c	—	0.8
d	c	0.6
e	b	0.3
f	d, e	1.0
g	f	0.4
h	g	0.3
	Total time	3.8

The line balancing problem is illustrated with the help of the following example (Stevenson, 1993).

Example 7.3. The precedence diagram for product P is shown in Figure 7.6 and the corresponding task times are given in Table 7.1. If the desired throughput for the above design is 400 units/day and the total time available per day is 480 minutes, then the minimum number of stations required is given by

$$\frac{400 \times 3.8}{480} = 3.17 \text{ (rounded up to 4)}$$

However, because of the precedence structure, one may not be able to achieve such a minimum. Assigning individual tasks to stations such that the precedence constraints are not violated gives a total idle time of 1 minute per part. One such assignment of individual tasks to stations is shown in Figure 7.7. Alternatively, if the number of stations available is 3, then the maximum throughput is

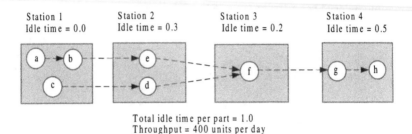

Total idle time per part = 1.0
Throughput = 400 units per day

Figure 7.7. Assignment of tasks of product P to four stations.

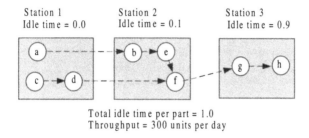

Station 1
Idle time = 0.0

Station 2
Idle time = 0.1

Station 3
Idle time = 0.9

Total idle time per part = 1.0
Throughput = 300 units per day

Figure 7.8. Assignment of tasks of product P to three stations.

$$\left\lceil \frac{480 \times 3}{3.8} \right\rceil = 379 \text{ units per day}$$

However, because of the precedence structure, we may not be able to achieve such a maximum. As shown in Figure 7.8, a throughput of only 300 units per day is possible for this design. Mathematical models similar to the ones discussed in Section 7.2 can be used to model the assembly line balancing problem. Balancing is a stepping stone to optimizing the line flow.

7.4 MANUFACTURING SCHEDULING

In the theory of deterministic machine scheduling, a set of parts (jobs) is to be processed on a set of machines (processors) in order to minimize (maximize) a certain performance measure. A part (job) may consist of a number of operations. All machining parameters are assumed to be known in advance. Each operation is to be processed by at most one machine at a time.

Depending on the way the parts visit machines, there are two modes of processing, flow shop and job shop (see Figure 7.9).

In the flow shop (Figure 7.9a), all parts flow in one direction, whereas in the job shop (Figure 7.9b), the parts may flow in different directions; also, a machine can be visited more than once by the same part. In both cases, a part does not have to visit all machines.

Operation o_i can be characterized by the following data:

t_{ij} = processing time of operation o_i on machine M_j

r_i = readiness of operation o_i for processing, that is, the time operation o_i is available for scheduling

d_i = due date, that is, the promised delivery time of operation o_i

w_i = weight (priority), which expresses the relative urgency of operation o_i

The mode of processing is called *preemptive* if an operation may be preempted at any time and restarted later at no cost, perhaps on another machine.

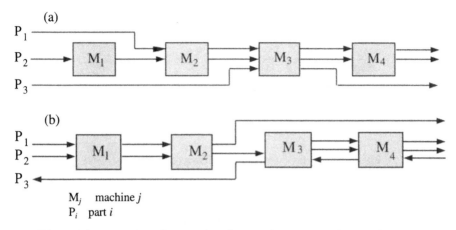

Figure 7.9. Illustration of two modes of processing: (*a*) flow shop; (*b*) job shop.

In a set of operations precedence constraints among them may be defined. The inequality $o_i < o_j$ means that the processing of o_i must be completed before o_j can be started. In other words, the set of operations is partially ordered by $<$. A set of operations ordered by the precedence relation is usually represented as a directed graph (a digraph) in which nodes correspond to operations and arcs correspond to precedence constraints.

For each operation o_i, $i = 1, \ldots, n$, in a given schedule, the following parameters are defined:

$C_i =$ completion time of operations o_i

$F_i =$ flow time (the difference between completion time and readiness), $F_i = C_i - r_i$

$L_i =$ lateness (the difference between completion time and due date), $L_i = C_i - d_i$

$T_i =$ tardiness, $= \max\{C_i - d_i, 0\}$

A schedule for which the value of a particular performance measure attains a minimum is called an optimal schedule. To evaluate schedules, the following basic performance measures are used:

- Schedule length (makespan): $C_{\max} = \max\{C_i\}$
- Mean flow time: $F = 1/n \sum_{i=1}^{n} F_i$
- Mean weighted flow time: $F_w = \sum_{i=1}^{n} w_i F_i / \sum_{i=1}^{n} w_i$
- Maximum lateness: $L_{\max} = \max\{L_i\}$
- Mean tardiness: $T = 1/n \sum_{i=1}^{n} T_i$
- Mean weighted tardiness: $T_w = \sum_{i=1}^{n} w_i T_i / \sum_{i=1}^{n} w_i$
- Number of tardy jobs: $N_T = \sum_{i=1}^{n} N_i$

where

$$N_i = \begin{cases} 1 & \text{if } C_i > d_i \\ 0 & \text{otherwise} \end{cases}$$

7.4.1 Scheduling *n* Operations on a Single Machine

One of the simplest scheduling problems is to schedule n operations on a single machine. Two cases of this problem are considered:

Case 1: No constraints are imposed.

Case 2: Due dates are imposed for all (some) operations.

Suppose that one attempts to minimize the mean flow time F in the one-machine scheduling problem without any constraints.

Theorem 7.1 (SPT Scheduling Rule).   For a one-machine scheduling problem, the mean flow time is minimized by the sequence

$$t_{(1)} \le t_{(2)} \le t_{(3)} \le \cdots \le t_{(i)} \le \cdots t_{(n)}$$

where $t_{(i)}$ is the processing time of the operation that is processed ith. The schedule obtained is optimal and is called the shortest processing time (SPT) schedule. The SPT rule is illustrated in Example 7.4.

Example 7.4.   Determine an optimal schedule for six operations to be processed on one machine:

Operation number 1 2 3 4 5 6

Processing time   4 7 1 6 2 3

The optimal SPT schedule is (3, 5, 6, 1, 4, 2), as illustrated in Figure 7.10. The flow for each operations in the schedule in Figure 7.10 can be easily computed: $F_3 = 1$, $F_5 = 3$, $F_6 = 6$, $F_1 = 10$, $F_4 = 16$, $F_2 = 23$. It can be also checked that the mean flow time $F = 14.03$ is the lowest of all schedules.

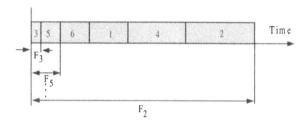

Figure 7.10. Gantt chart of single-machine schedule.

Consider a one-machine scheduling problem with due dates where the objective is to minimize the maximum lateness L_{max}.

Theorem 7.2 (EDD Scheduling Rule).   For the one-machine scheduling problem with due dates, the maximum lateness is minimized by sequencing such that

$$d_{(1)} \le d_{(2)} \le d_{(3)} \le \cdots \le d_{(i)} \le \cdots \le d_{(n)}$$

where $d_{(i)}$ is the due date of the operation that is processed ith. The schedule obtained is optimal and is called the earliest due date (EDD) schedule. Example 7.5 illustrates the EDD rule.

Example 7.5.   Determine the EDD schedule for a problem with six operations for the following data:

Operation number 1 2 3 4 5 6
Processing time   1 1 2 5 1 3
Due date   6 3 8 14 9 3

The values of lateness and tardiness for each operation in the sequence are shown in the following table:

Operation Number	Due Date	Completion Time	Lateness	Tardiness
6	3	3	0	0
2	3	4	1	1
1	6	5	−1	0
3	8	7	−1	0
5	9	8	−1	0
4	14	13	−1	0

From this table the optimal EDD schedule (6, 2, 1, 3, 5, 4) is determined. One can also note in the table that operation 2 is late ($L_2 = 1$).

7.4.2 Scheduling Flexible Forging Machine

In this section, a single-machine scheduling problem with changeover costs and precedence constraints is considered. The problem is discussed using the example of a flexible forging machine.

One of the most frequently emphasized aspects of manufacturing systems deals with changeover costs. It will be shown that changeover costs exist even in modern machine tools, and their reduction is a matter of a planning methodology. A manufacturing system, if not properly planned, may in fact result in unnecessarily large changeover costs.

Consider the flexible forging machine in Figure 7.11, which is computer controlled and has the following four automatic options:

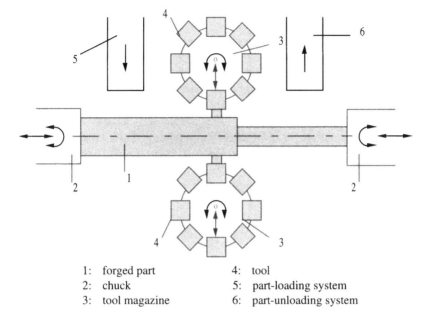

1: forged part   4: tool
2: chuck   5: part-loading system
3: tool magazine   6: part-unloading system

Figure 7.11. Schematic top-down view of a flexible forging machine (FFM).

- Part loading
- Chuck rotation and horizontal movement
- Tool magazine rotation and oscillation
- Part loading and unloading

Each of the two identical tool magazines visible in Figure 7.11 can handle eight tools. Tools are heavy, and it takes a considerable amount of time to change them. This will have some implications on the modeling and solution approach discussed in this section.

7.4.2.1 Features of Flexible Forging Machine Scheduling Model

Changeover Costs. Due to the progress in modern hardware and advancement of the planning methodologies, it has been possible to reduce the changeover cost but not to eliminate it entirely. It will be shown that in flexible forging machine discussed, changeover costs occur. Reduction of these costs is a matter of the flexible forging machine scheduling. The changeover costs in the flexible forging machine in Figure 7.11 are caused by:

1. Parts
2. NC programs (NC = numerical control)

3. Chucks holding parts
4. Tools

The changeover cost imposed by loading and unloading parts is dictated by the hardware design and its reduction is beyond operational control. The NC program is being changed during the loading of a new part and the unloading of the already-forged part. The NC program changeover time is only a small fraction of the part-loading and part-unloading time. The chucks are capable of holding many different parts, and they are very seldom changed. Tool changeover imposes significant costs, and this issue is of main concern for the optimization model presented later in this section. Moreover, these costs are sequence dependent. Before a new part (or a batch of parts) is forged, in general, it is necessary, to change some tools stored in the tool magazines.

Precedence Constraints. Classical scheduling problems might be formulated with two types of constraints:

1. Due dates
2. Precedences

The precedence constraints are more appropriate than the due date constraints in modeling the flexible forging machine scheduling problem. The flexible forging machine is to be integrated with the machining and assembling systems. Often final products are assembled in a sequence imposed by the market demand. In this case, in order to reduce a volume of in-process inventory, it is desirable to perform the machining and forging process in a sequence imposed by the assembly process. If this sequence is not preserved, a delay in product delivery may occur.

7.4.2.2 *Model without Precedence Constraints.* Based on the previous consideration, one can identify the underlying flexible forging machine scheduling problem as the single-machine scheduling problem with sequence-dependent changeover costs and precedence constraints. It is known from scheduling theory (see, e.g., Baker, 1974, p. 94) that the single-machine scheduling problem with sequence-dependent changeover costs is equivalent to the traveling salesman problem (TSP).

One of the most frequently encountered formulations of the traveling salesman problem is a mixed-integer programming formulation.

Denote:

$n =$ number of parts (cities in the TSP)

$c_{ij} =$ changeover cost from part i to part j

$u_i =$ nonnegative (intermediate) variable

$$x_{ij} = \begin{cases} 1 & \text{if part } i \text{ immediately precedes part } j \\ 0 & \text{otherwise} \end{cases}$$

The objective function (7.15) minimizes the total changeover cost:

$$\min \sum_{i=1}^{n} \sum_{j=1}^{n} c_{ij} x_{ij} \tag{7.15}$$

such that

$$\sum_{i=1}^{n} x_{ij} = 1 \qquad j = 1, \ldots, n \tag{7.16}$$

$$\sum_{j=1}^{n} x_{ij} = 1 \qquad i = 1, \ldots, n \tag{7.17}$$

$$u_i - u_j + n x_{ij} \leq n - 1 \quad i = 2, \ldots, n; \quad j = 2, \ldots, n; \quad i \neq j \tag{7.18}$$

$$x_{ij} = 0, 1 \qquad i, j = 1, \ldots, n \tag{7.19}$$

$$u_i \geq 0 \qquad i = 1, \ldots, n \tag{7.20}$$

Constraint (7.16) ensures that in a schedule only one part i immediately precedes part j. Constraint (7.17) imposes that part i is followed by exactly one part j. The formulation (7.15), (7.16), (7.17), and (7.19) is known in the operations research literature as the assignment problem. Constraint (7.18) eliminates subschedules generated by solving the assignment problem (7.15)–(7.17). The integrality of decision variable x_{ij} is imposed by constraint (7.19). Constraint (7.20) ensures nonnegativity of the intermediate variable u_i.

To illustrate the application of model (7.15)–(7.20) for scheduling the forging machine, consider Example 7.6.

Example 7.6. Find the optimal schedule for six parts. Each part is to visit a flexible forging machine only once. The matrix of sequence-dependent changeover costs is

		1	2	3	4	5	6
	1	∞	7	3	12	5	8
	2	4	∞	2	10	9	3
Part	3	6	7	∞	11	1	7
number	4	7	3	1	∞	8	8
	5	2	10	2	7	∞	3
	6	4	11	7	6	3	∞

Solving the formulation (7.15)–(7.20) by a standard mathematical programming computer code (e.g., LINDO; Schrage, 1984) or a specialized algorithm (see, e.g., Lawler et al., 1985), the following optimal solution is obtained:

$$x_{12} = x_{26} = x_{64} = x_{43} = x_{35} = x_{51} = 1$$

with the corresponding cost of 20.

The formulation (7.15)–(7.20) for the data from Example 7.6 is shown in the Appendix to this chapter.

This solution can also be expressed as (1, 2, 6, 4, 3, 5, 1), where 1's at the beginning and the end of the sequence indicate that the schedule starts with part 1 and ends with part 1. To make the interpretation of the solution of model (7.15)–(7.20) more natural, a dummy row and column are introduced in Example 7.7.

The flexible forging machine scheduling problem can be described as the TSP with precedence constraints. Before the scheduling model with precedence constraints is presented, consider the following example.

Example 7.7. For a given time horizon, assume that five parts are to be forged on the flexible forging machine in Figure 7.11. Each part requires a set of eight tools, as indicated in Table 7.2.

TABLE 7.2. Tools Required for Forging Five Parts

Tool Number	Part Number	1	2	3	4	5
1					1	
2		1		1	1	1
3			1			
4		1		1	1	1
5			1			
6		1		1		1
7					1	
8		1		1		1
9			1			1
10					1	
11		1		1		
12					1	
13		1				
14		1			1	
15			1			1
16				1	1	
17			1			
18				1		1
19		1	1			
20					1	
21			1			
22			1			1

For part $i = 1, 2, \ldots, 5$ and tool $k = 1, 2, \ldots, 22$, an entry a_{ki} in Table 7.2 is defined as follows:

$$a_{ki} = \begin{cases} 1 & \text{if tool } t_k \text{ is used for part } P_i \\ 0 & \text{otherwise} \end{cases}$$

For any two parts P_i and P_j, define the Hamming distance

$$d_{ij} = \sum_{k=1}^{22} \delta(a_{ki}, a_{kj})$$

where

$$\delta(a_{kik}, a_{kj}) = \begin{cases} 1 & \text{if } a_{ki} \neq a_{kj} \\ 0 & \text{otherwise} \end{cases}$$

The matrix $[d_{ij}]$ of Hamming distances is obtained for the data in Table 7.2:

$$[d_{ij}] = \begin{array}{c c} & \begin{array}{c c c c c} 1 & 2 & 3 & 4 & 5 \end{array} \\ \begin{array}{c} 1 \\ 2 \\ 3 \\ 4 \\ 5 \end{array} & \begin{bmatrix} \infty & 14 & 6 & 10 & 8 \\ 14 & \infty & 16 & 16 & 10 \\ 6 & 16 & \infty & 10 & 6 \\ 10 & 16 & 10 & \infty & 12 \\ 8 & 10 & 6 & 12 & \infty \end{bmatrix} \end{array} \qquad (7.21)$$

The costs c_{ij} in the objective function (7.15) are set equal to the Hamming distances d_{ij}.

For the convenience of interpreting solutions, we introduce a dummy batch (batch number $s = 0$) and a row and column vector of zero changeover costs. The dummy batch indicates the initial and final states of the tool magazine:

$$[d_{ij}] = \begin{array}{c} 0 \\ 1 \\ 2 \\ 3 \\ 4 \\ 5 \end{array} \begin{bmatrix} \infty & 0 & 0 & 0 & 0 & 0 \\ 0 & \infty & 14 & 6 & 10 & 8 \\ 0 & 14 & \infty & 16 & 16 & 10 \\ 0 & 6 & 16 & \infty & 10 & 6 \\ 0 & 10 & 16 & 10 & \infty & 12 \\ 0 & 8 & 10 & 6 & 12 & \infty \end{bmatrix} \qquad (7.22)$$

Impose the precedence constraints in Figure 7.12.

The following solution has been generated with the solution approach discussed in the next section:

$$x_{02} = 1, \quad x_{25} = 1, \quad x_{51} = 1, \quad x_{13} = 1, \quad x_{34} = 1, \quad x_{40} = 1$$

This solution can be expressed as the schedule beginning and ending with state 0 of the tool magazine: $(0, 2, 5, 1, 3, 4, 0)$.

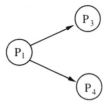

Figure 7.12. Precedence constraints.

7.4.2.3 Model with Precedence Constraints. In this section, the single machine scheduling problem with sequence-dependent setup times and precedence constraints is considered. The problem involves sequencing n parts on a single machine, where part i requires processing time t_i. Setup time may occur when part j is processed immediately after part i. Precedence constraints are represented by a directed acyclic graph G. A directed path from vertex i to vertex j indicates that part i precedes part j. Given a feasible sequence of parts, the total setup time can be determined. The objective is to find a feasible sequence of parts that minimizes the total setup time, or setup cost.

Most research papers on the single-machine scheduling problem do not consider sequence-dependent setup times. Lawler (1978) developed an algorithm for solving the single-machine problem with serial and parallel precedence constraints. For general precedence constraints, Lawler (1978) and Lenstra and Rinnooy Kan (1978) showed that the problem is NP hard. Potts (1985) developed a Lagrangian-based branch-and-bound algorithm that solved the TSP with up to 100 parts.

The problem of scheduling parts with sequence-dependent setup times on a single machine is equivalent to the TSP.

Kusiak and Finke (1987) developed a branch-and-bound algorithm for solving the single-machine scheduling problem with sequence-dependent setup times and precedence constraints. A lower bound was determined by solving a network formulation. The heuristic algorithm developed by Martello (1983) was used to detect a feasible tour in the flow graph. The network formulation was solved by a standard linear programming code.

In the next section, a mixed-integer model of the single-machine scheduling problem with sequence-dependent setup times and precedence constraints is considered (He and Kusiak, 1992).

Model Formulation. The single-machine scheduling problem with sequence-dependent setup times is equivalent to the TSP. The standard integer formulation of the TSP becomes cumbersome when precedence constraints are considered. To avoid the difficulty with modifying the standard TSP formulation, consider the model presented next (He and Kusiak, 1992), with the notation defined as follows:

$i, j =$ part index, $i, j = 1, \ldots, n$

$n =$ number of parts to processed

t_i = processing time of part i
t_{ij} = setup time from part i to j
τ_{ij} = $t_{ij} + t_j$ adjusted setup time
T = makespan
G = set of parts (i, j), where part i precedes part j
Q = set of parts (i, j), where part i and j can be processed in any order
M = an arbitrary large positive number
x_i = completion time of part i
$x_{ij} = \begin{cases} 1 & \text{if part } i \text{ precedes part } j \\ 0 & \text{otherwise} \end{cases}$

The objective function (7.23) minimizes the schedule makespan

$$\min T \tag{7.23}$$

such that

$$x_i \leq T \qquad \text{for all } i \tag{7.24}$$

$$x_j - x_i \geq \tau_{ij} \qquad [i, j] \in G, i \neq j \tag{7.25}$$

$$x_i - x_j + M x_{ij} \geq \tau_{ij} x_{ij} + \tau_{ji} x_{ji} \quad [i, j] \in Q, i \neq j \tag{7.26}$$

$$x_{ij} + x_{ji} = 1 \qquad [i, j] \in Q, i \neq j \tag{7.27}$$

$$x_i \geq t_i \qquad \text{for all } i \tag{7.28}$$

$$x_{ij} = 0, 1 \qquad [i, j] \in Q, i \neq j \tag{7.29}$$

Constraint (7.24) imposes that the completion time of each part is not greater than the minimum makespan. Constraints (7.25) and (7.26) impose that two parts cannot be processed at the same time. Constraint (7.27) implies that any two parts are to be processed in only one sequence. Constraint (7.28) ensures that the completion time of each part is not less than its processing time. Constraint (7.29) imposes integrality.

The formulation (7.23)–(7.29) has the following features:

1. The minimization of the makespan is equivalent to the minimization of the total sequence-dependent setup time as the total processing time is constant.

2. In order to minimize the total setup time, the processing time does not have to be considered in the formulation. A nonzero positive number can be assigned to all parts. For the sake of simplicity, the processing time is set equal to 1 for all parts, that is, $t_i = 1$, $i = 1, \ldots, n$, and $\tau_{ij} = t_{ij} + 1$. Constraint (7.28) is replaced with constraint (7.30):

$$x_i \geq 1 \quad \text{for all } i \qquad (7.30)$$

When matrix $[\tau_{ij}]$ is symmetric, constraint (26) takes the form of constraint (7.31):

$$x_i - x_j + Mx_{ij} \geq \tau_{ji} \quad [i, j] \in Q \quad \text{for all } i \neq j \qquad (7.31)$$

Algorithm 7.1.   It is known that the single-machine scheduling model with precedence constraint is an NP-complete problem (Lawler, 1978). Here, a heuristic algorithm (He and Kusiak, 1992) is presented to solve the scheduling model with sequence-dependent setup times and precedence constraints.

Step 0. Set $K = \emptyset$.

Step 1. From the precedence graph G, select vertices (parts) that do not have predecessors and place them in K.

Step 2. From the setup time matrix, select rows corresponding to the vertices in K.

Step 3. Compute $t_{i^*j^*} = \min\{t_{ij}: i \in K, j = 1, \ldots, n\}$ until such i^* and j^* are found that either j^* immediately follows i^* or j^* has no predecessors in the precedence graph.

Step 4. Output i^*, replace K with j^*, remove the row and column corresponding to i^* from the matrix $[t_{ij}]$, and delete vertex i^* and all edges leading out of vertex i^* from the precedence graph G.

Step 5. Go back to Step 3 until the final solution is obtained.

Algorithm 7.1 is illustrated in Example 7.8.

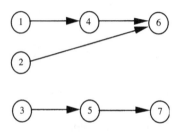

Figure 7.13. Graph G of precedences among parts.

TABLE 7.3. Matrix of Setup Times

$$
[t_{ij}] = \begin{array}{c} \\ 1 \\ 2 \\ 3 \\ 4 \\ 5 \\ 6 \\ 7 \end{array}
\begin{array}{c}
\begin{array}{ccccccc} 1 & 2 & 3 & 4 & 5 & 6 & 7 \end{array} \\
\left[\begin{array}{ccccccc}
- & 8 & 4 & 6 & 2 & 5 & 4 \\
10 & - & 9 & 5 & 2 & 6 & 13 \\
5 & 13 & - & 11 & 4 & 9 & 10 \\
5 & 2 & 7 & - & 5 & 8 & 4 \\
8 & 5 & 4 & 6 & - & 3 & 6 \\
13 & 5 & 4 & 8 & 4 & - & 10 \\
2 & 11 & 5 & 10 & 8 & 9 & -
\end{array} \right]
\end{array}
$$

Example 7.8.   Given the precedence constraints in Figure 7.13 and the time matrix in Table 7.3, solve the single-machine scheduling problem.

For the above data, the mixed-integer formulation (7.23)–(7.29) is solved by LINDO (Schrage, 1986). The optimal sequence of operations is {3, 5, 7, 1, 4, 2, 6} with the corresponding minimum total setup time of 26.

An iteration of Algorithm 7.1 applied to solve the same model is presented next (He and Kusiak, 1992).

Step 0. $K = \emptyset$.

Step 1. $K = \{1, 2, 3\}$ as the vertices (parts) 1, 2, and 3 have no predecessors in the precedence graph G.

Step 2. Select row 1, row 2, and row 3 in time matrix $[t_{ij}]$.

Step 3. Compute $t_{i^*j^*} = \min\{t_{12}, t_{13}, t_{14}, t_{21}, t_{23}, t_{26}, t_{31}, t_{32}, t_{35}\} = \min\{8, 4, 6, 10, 9, 6, 5, 13, 4\} = 4$, $i^* = 1$, $j^* = 3$.

Step 4. Output 1, {1}. Replace K with $K = \{3\}$. Remove row 1 and column 1 from matrix $[t_{ij}]$ and delete vertex 1 and all edges leading out of vertex 1 from precedence graph G.

Step 3. Compute $t_{i^*j^*} = \min\{t_{32}, t_{34}, t_{35}\} = \min\{13, 11, 4\} = 4$, $i^* = 3$, $j^* = 5$.

Step 4. Output 3, {1, 3}. Replace K with $K = \{5\}$. Remove row 3 and column 3 in matrix $[t_{ij}]$ and delete vertex 3 and all edges leading out of vertex 3 from the precedence graph G.

Step 3. Compute $t_{i^*j^*} = \min\{t_{52}, t_{54}, t_{57}\} = \min\{5, 6, 6\} = 5$, $i^* = 5$, $j^* = 2$.

Step 4. Output 5, {1, 3, 5}. Replace K with $K = \{2\}$. Remove row 5 and column 5 from matrix $[t_{ij}]$ and delete vertex 5 and all edges leading out of vertex 5 from the precedence graph G.

After 17 iterations the final solution {1, 3, 5, 2, 4, 7, 6} is obtained. The total sequence-dependent setup time is 31, as opposed to 26 in the optimal solution.

7.4.3 Two-Machine Flowshop Model

Consider scheduling n parts on two machines, where each part is processed in the order (machine M_1, machine M_2), so that the maximum flow time F_{max} is minimized.

To solve this problem optimally, the Johnson (1954) algorithm can be used. Kusiak (1986) developed an efficient implementation of the Johnson algorithm, which is presented next.

Algorithm 7.2

Step 1. Set $k = 1$, $l = n$.

Step 2. For each part, store the shortest processing time and corresponding machine number.

Step 3. Sort the resulting list of the triplets "part number/time/machine number" in increasing value of processing time.

Step 4. For each entry in the sorted list:

IF machine number is 1, then

    (i)   set the corresponding part number in position k,

    (ii) set $k = k + 1$.

ELSE

    (i)   set the corresponding part number in position l,

    (ii) set $l = l - 1$.

END-IF.

Step 5. Stop if the entire list of parts has been exhausted.

Algorithm 7.2 is illustrated in Example 7.9.

Example 7.9. Solve the two-machine flow shop scheduling model for the following set of data:

Part Number	Processing Time t_{ij} of Part i on Machine j	
i	$j = 1$	$j = 2$
1	6	3
2	2	9
3	4	3
4	1	8
5	7	1
6	4	5
7	7	6

The result of Step 2 is the following set of triplets:

Part Number	$\min\{t_{i1}, t_{i2}\}$	Machine Number
1	3	2
2	2	1
3	3	2
4	1	1
5	1	2
6	4	1
7	6	2

Sorting in Step 3 the triplets in the increasing value of the processing time results in the following set of triplets:

$$(4, 1, 1), (5, 1, 2), (2, 2, 1), (3, 3, 2), (1, 3, 2), (6, 4, 1), (7, 6, 2)$$

Step 4 produces the optimal schedule (4, 2, 6, 7, 1, 3, 5) illustrated in Figure 7.14.

Algorithm 7.2 can also be applied to solve the special case of the two-machine job shop problem and three machine flow shop problem discussed next.

7.4.4   Two-Machine Job Shop Model

Partition a set of n parts into the following four types:

Type A: parts to be processed only on machine M_1.
Type B: parts to be processed only on machine M_2.
Type C: parts to be processed on both machines in the order M_1, then M_2.
Type D: parts to be processed on both machines in the order M_2, then M_1.

Based on the above partition, an optimal schedule is constructed using Algorithm 7.3.

Algorithm 7.3

Step 1.   Schedule the parts of type A in any order to obtain the sequence S_A.

Step 2.   Schedule the parts of type B in any order to obtain the sequence S_B.

Step 3.   Scheduling the parts of type C according to Algorithm 7.2 produces the sequence S_C.

Step 4.   Scheduling the parts of type D according to Algorithm 7.2 produces the sequence S_D. (Note that M_2 is the first machine, whereas M_1 is the second one.)

Step 5.   Construct an optimal schedule as follows:

Machine	Processing Order
M_1	(S_C, S_A, S_D)
M_2	(S_D, S_B, S_C)

To see that the schedule generated in Step 5 is optimal, remember that no time is wasted, and hence F_{max} increases if either M_2 is kept idle waiting for part of type C to

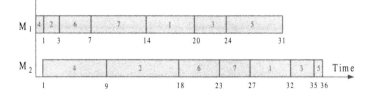

Figure 7.14. Optimal schedule.

be completed on M_1 or M_1 is kept idle waiting for parts of type D to be completed on M_2. This schedule clearly minimizes such idle time. Algorithm 7.3 is illustrated in Example 7.10.

Example 7.10.   Consider the problem of scheduling nine parts on two machines for the following data:

Processing Order and Time

Part Number	First Machine		Second Machine	
	Order	Time	Order	Time
1	M_1	7	M_2	1
2	M_1	6	M_2	5
3	M_1	9	M_2	7
4	M_1	4	M_2	6
5	M_2	6	M_1	6
6	M_2	5	M_1	5
7	M_1	4		
8	M_1	5		
9	M_2	1		
10	M_2	5		

To determine an optimal schedule, separate the parts into four types:

Type A Parts: Parts 7 and 8 are to be processed on machine M_1 alone. An arbitrary order $S_A = (7, 8)$ is selected.

Type B Parts: Parts 9 and 10 require machine M_2 alone. Select an arbitrary order $S_B = (9, 10)$.

Type C Parts: Parts 1, 2, 3, and 4 require machine M_1 first and then machine M_2. Algorithm 7.2 for the four-part problem produces the sequence $S_C = (4, 3, 2, 1)$.

Type D Parts: Parts 5 and 6 require machine M_2 first and then machine M_1. Algorithm 7.2 for the two-part problem produces the sequence $S_D = (5, 6)$. (Note that machine M_1 is now the second machine.)

The optimal solution to this problem is as follows:

Processing Sequence
Machine M_1 (4, 3, 2, 1, 7, 8, 5, 6)
Machine M_2 (5, 6, 9, 10, 4, 3, 2, 1)

The resulting Gantt diagram is given in Figure 7.15, where one can see that $F_{max} = 46$ for an optimal schedule.

7.4.5   Special Case of Three-Machine Flow Shop Model

Algorithm 7.2 for the two-machine flow shop problem can be extended to a special case of a three-machine flow shop problem.

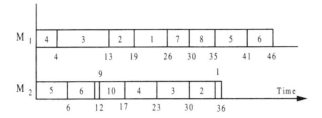

Figure 7.15. Gantt chart for the optimal solution to the problem in Example 7.10.

A condition is needed (French, 1982):

$$\text{Either} \qquad \min_{i=1}^{n}\{t_{i1}\} \geq \max_{i=1}^{n}\{t_{i2}\}$$

(7.32)

$$\text{or} \qquad \min_{i=1}^{n}\{t_{i3}\} \geq \max_{i=1}^{n}\{t_{i2}\}$$

that is, the maximum processing time on the second machine is no greater than the minimum time on either the first or the third machine. If (7.32) holds, an optimal schedule for the problem may be found by letting

$$a_i = t_{i1} + t_{i2} \qquad b_i = t_{i2} + t_{i3}$$

and scheduling the parts as if they were to be processed on two machines only but with the processing time of each part being a_i and b_i on the first and second machines, respectively. The special case of three-machine flow shop model is illustrated in Example 7.11.

Example 7.11. Solve the three-machine flow shop scheduling problem with the data in Table 7.4. First we check whether (7.32) holds for this problem. Here we have

$$\min_{i=1}^{6}\{t_{i1}\} = 3 \qquad \max_{i=1}^{6}\{t_{i2}\} = 3 \qquad \min_{i=1}^{6}\{t_{i3}\} = 1$$

It is easy to note that $\min_{i=1}^{6}\{t_{i1}\} = 3 \geq 3 = \max_{i=1}^{6}\{t_{i2}\}$ and therefore (7.32) holds (only one of the inequalities needs to hold).

The constructed times a_i and b_i are given in Table 7.4. Algorithm 7.2 produces the sequence (2, 4, 5, 1, 3, 6) illustrated with the Gantt chart in Figure 7.16.

7.4.6 Scheduling Model for *m* Machines and *n* Operations

One of the approaches to scheduling is to model the scheduling problem and then solve it using a commercial computer code or specialized algorithm. In this section a

TABLE 7.4. Scheduling Data

Part Number	Actual Processing Times[a]			Constructed Processing Times	
	t_{i1}	t_{i2}	t_{i3}	First Machine, a_i	Second Machine, b_i
1	4	1	2	5	3
2	6	2	10	8	12
3	3	1	2	4	3
4	5	3	6	8	9
5	7	2	6	9	8
6	4	1	1	5	2

[a]t_{i1}, machine M_1; t_{i2}, machine M_2; t_{i3}, machine M_3.

new mixed-integer programming model for the problem of scheduling n operations with precedence constraints on m machines is presented. The scheduling model considered is a generalization of the scheduling problem presented in Baker (1974).

To present the scheduling model, the following notation is introduced:

n = number of parts

m = number of machines

R_i = set of pairs of operations $[k, l]$ for part P_i, where operation k precedes operation l, $i = 1, \ldots, n$

Q_i = set of pairs of operations $[k, l]$ for part P_i, where k and l can be performed in any order, $i = 1, \ldots, n$

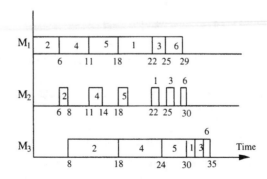

Figure 7.16. Gantt chart for the optimal solution to the problem in Example 7.11.

I_i = set of operations without precedence constraints, $i = 1, \ldots, n$

N_p = set of operations to be performed on machine p, $p = 1, \ldots, m$

n_i = number of operations in part P_i, $i = 1, \ldots, n$

t_{il} = processing time of operation l of part P_i, $i = 1, \ldots, n$, $i = 1, \ldots, n$

M = an arbitrary large positive number

x_{ik} = completion time of operation k of part P_i, $k = 1, \ldots, n$, $i = 1, \ldots, n$

$x_{[i]}$ = completion time of the root operation (the last in the sequence) of part P_i, $i = 1, \ldots, n$

$y_{kl} = \begin{cases} 1 & \text{if operation } k \text{ precedes operation } l \\ 0 & \text{otherwise} \end{cases}$

for all $[k, l] \in Q_i$, $i = 1, \ldots, n$

$z_{kl} = \begin{cases} 1 & \text{if operation } k \text{ precedes operation } l \\ 0 & \text{otherwise} \end{cases}$

for all $[k, l] \in N_p$, $p = 1, \ldots, m$

The objective of the scheduling model is to minimize the total completion time of all parts:

$$\min \sum_{i=1}^{n} x_{[i]} \tag{7.33}$$

subject to

$$x_{il} - x_{ik} \geq t_{il} \qquad \text{for all } [k, l] \in R_i, \text{ for all } i \tag{7.34}$$

$$x_{il} - x_{ik} + M(1 - y_{kl}) \geq t_{il} \qquad \begin{array}{l} \text{for all } [k, l] \in Q_i \\ \text{for all } i \end{array} \tag{7.35}$$

$$x_{ik} - x_{il} + M y_{kl} \geq t_{ik} \tag{7.36}$$

$$x_{jl} - x_{ik} + M(1 - z_{kl}) \geq t_{jl} \qquad \text{for all } [k, l] \in N_p, \text{ for all } p \tag{7.37}$$

$$x_{ik} - x_{jl} + M z_{kl} \geq t_{ik} \qquad i \neq j \tag{7.38}$$

$$x_{ik} \geq \begin{cases} t_{ik} & \text{for } [i, k] \in I_i, i = 1, \ldots, n \\ 0 & \text{for all other } i, k \end{cases} \tag{7.39}$$

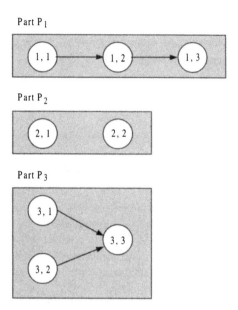

Figure 7.17. Structure of three parts.

$$y_{kl} = 0, 1 \qquad \text{for all } [k, l] \in Q_i, \text{ for all } i \qquad (7.40)$$

$$z_{kl} = 0, 1 \qquad \text{for all } [k, l] \in N_p, \text{ for all } p \qquad (7.41)$$

Constraint (7.34) imposes that the operations of each part are processed according to the precedences required. Constraints (7.35) and (7.36) ensure that any two operations belonging to the same part cannot be processed at the same time. Constraints (7.37) and (7.38) ensure that a machine cannot process more than one part at the same time. Constraints (7.39)–(7.41) imply nonnegativity and integrality of the corresponding variables.

The application of the scheduling model is illustrated in the following example.

Example 7.12. Schedule eight operations belonging to the three parts shown in Figure 7.17 on two machines. The required data are given in Table 7.5. The LINDO

TABLE 7.5. Scheduling Data

Part number	1			2		3		
Operation number	1	2	3	1	2	1	2	3
Machine number	1	1	2	1	2	2	1	2
Processing time	3	5	6	8	4	9	2	7

input file for the data in Table 7.5 and the precedence constraints in Figure 7.17 is presented next:

```
MIN
X13+X21+X22+X33
SUBJECT TO
```

Four constraints in the form (7.34) for each pair of operations are linked by a precedence constraint (an arc in Figure 7.17) within the same part:

```
X12-X11>=5
X13-X12>=6
X33-X31>=7
X33-X32>=7
```

Two pairs of constraints in the form (7.35) and (7.36) for the two operations in part P_2 are not linked by a precedence constraint:

```
X21-X22+999Y1>=8
X22-X21-999Y1>=-995   (derived from X22 – X21 + 999 (1 – Y1) ≥ 4)
X31-X32+999Y2>=9
X32-X31-999Y2>=-997
```

Ten pairs of constraints (7.37) and (7.38) for each pair of operations that belong to different parts and are to be performed on the same machine:

```
X11-X21+999Z1>=3
X21-X11-999Z1>=-991
X11-X32+999Z2>=3
X32-X11-999Z2>=-997
X12-X21+999Z3>=5
X21-X12-999Z3>=-991
X12-X32+999Z4>=5
X32-X12-999Z4>=-997
X21-X32+999Z5>=8
X32-X21-999Z5>=-997
X13-X22+999Z6>=6
X22-X13-999Z6>=-995
X13-X31+999Z7>=6
X31-X13-999Z7>=-990
X13-X33+999Z8>=6
X33-X13-999Z8>=-992
X22-X31+999Z9>=4
X31-X22-999Z9>=-990
X22-X33+999Z10>=4
X33-X22-999Z10>=-992
```

Five constraints (7.39) for operations without precedences:

```
X11>=3
X21>=8
X22>=4
X31>=9
X32>=2
END
```

Twelve integrality constraints (7.40) and (7.41) are for integer variables y and z:

```
INTEGER Y1
INTEGER Y2
INTEGER Z1
INTEGER Z2
INTEGER Z3
INTEGER Z4
INTEGER Z5
INTEGER Z6
INTEGER Z7
INTEGER Z8
INTEGER Z9
INTEGER Z10
```

Using scheduling model (7.33)–(7.41) to solve this problem, the optimal schedule $x_{32} = 4$, $x_{21} = 12$, $x_{11} = 15$, $x_{12} = 20$, $x_{22} = 4$, $x_{31} = 13$, $x_{33} = 20$, and $x_{13} = 26$ is generated (see Figure 7.18). The completion time for parts P_1, P_2, and P_3 in the schedule in Figure 7.18 is $C_1 = 26$, $C_2 = 12$, and $C_3 = 20$, respectively.

The model (7.33)–(7.41) involves machines and operations only. In order to incorporate other resources such as tools and fixtures as well as due dates, the formulation (7.33)–(7.41) needs to be extended.

Define:

Figure 7.18. Optimal schedule.

m = number of resource types

r_s = number of resources of type s, $s = 1, \ldots, m$

$d_{[i]}$ = due date of part number i

N_q = set of operations using resource q, $q = 1, \ldots, r_s$, $s = 1, \ldots, m$

The extended formulation of the n-operation, m-machine scheduling problem with precedence constraints, limited resources, and due dates is presented next.

$$\min \sum_{i=1}^{n} x_{[i]} \tag{7.42}$$

subject to

$$x_{il} - x_{ik} \geq t_{il} \qquad \text{for all } [k, l] \in R_i, \text{ for all } i \tag{7.43}$$

$$x_{il} - x_{ik} + M(1 - y_{kl}) \geq t_{il} \qquad \text{for all } [k, l] \in Q_i \tag{7.44}$$

$$x_{ik} - x_{il} + M y_{kl} \geq t_{ik} \qquad \text{for all } i \tag{7.45}$$

$$x_{jl} - x_{ik} + M(1 - z_{kl}) \geq t_{jl} \qquad \text{for all } [k, l] \in N_q, \text{ for all } q \tag{7.46}$$

$$x_{ik} - x_{jl} + M z_{kl} \geq t_{ik} \qquad i \neq j \tag{7.47}$$

$$x_{[i]} \leq d_{[i]} \qquad \text{for all } i \tag{7.48}$$

$$x_{ik} \geq \begin{cases} t_{ik} & \text{for } [i, k] \in I_i, i = 1, \ldots, n \\ 0 & \text{for all other } i, k \end{cases} \tag{7.49}$$

$$y_{kl} = 0, 1 \qquad \text{for all } [k, l] \in Q_i, \text{ for all } i \tag{7.50}$$

$$z_{kl} = 0, 1 \qquad \text{for all } [k, l] \in N_q, \text{ for all } q \tag{7.51}$$

Constraint (7.43) imposes that the operations of each part are processed according to the precedences required. Constraints (7.44) and (7.45) ensure that any two operations belonging to the same part cannot be processed at the same time. Constraints (7.46) and (7.47) ensure that any resource cannot process more than one

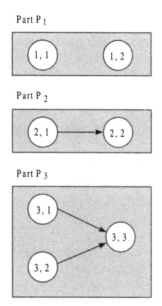

Figure 7.19. Structure of three parts.

part at the same time. Constraint (7.48) imposes due dates. Constraints (7.49)–(7.51) impose nonnegativity and integrality.

The application of the scheduling model (7.42)–(7.51) is illustrated in the Example 7.13.

Example 7.13.   Schedule seven operations belonging to the three parts in Figure 7.19 on two machines. The data required are given in Table 7.6.

Solving the scheduling model for the data in Table 7.6, the optimal schedule is generated (see Figure 7.20).

The schedule in Figure 7.20 was generated under the assumption that only one copy of each tool was available. Relaxing this assumption results in the schedule $x_{11} = 2$, $x_{12} = 6$, $x_{22} = 4$, and $x_{33} = 8$ presented in Figure 7.21. The completion time of part P_3

TABLE 7.6. Scheduling Data

Part number	1		2		3		
Operation number	1	2	1	2	1	2	3
Machine number	1	2	2	1	1	2	1
Processing time	2	3	1	2	3	2	1
Tool number	1	1	2	2	2	1	2

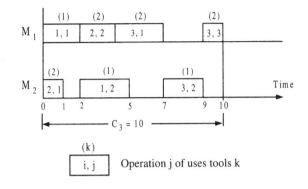

Figure 7.20. Optimal schedule with limited number of tools.

is $C_3' = 8$, rather than $C_3 = 10$ for the solution in Figure 7.20. As one can see in Figure 7.21, two copies of tool 1 are used to perform operations (1, 2) and (3, 2), and two copies of tool 2 for operations (2, 1) and (3, 1).

7.4.7 Heuristic Scheduling of Multiple Resources

In this section, a general heuristic for solving a model with multiple resources and precedence constraints is discussed. An application of this heuristic is in scheduling manufacturing systems.

Consider a part with a number of operations, where each operation is to be processed on one machine and may require other resources such as tools, pallets, or fixtures. Precedence constraints among operations may also exist. A definition of schedulable operations is introduced.

An operation is *schedulable* at time t if all the following three conditions are satisfied:

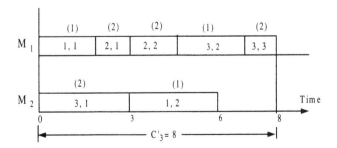

Figure 7.21. Optimal schedule with unlimited number of tools.

- No other operation that belongs to the same part is being processed at time t.
- All operations preceding the operation considered have been completed before time t.
- All resources (e.g., machines, tools, and fixtures) required for performing the operation are available at time t.

Before Algorithm 7.4 is presented, some terms that are used in Step 2 of the algorithm are defined. Consider part P, which has the structure shown in Figure 7.22.

All operations but 1 and 6 have precedence constraints. Assume that part P is being scheduled and operation 1 has been completed. In this case the number of *unprocessed operations* is then five (operations 2, 3, 4, 5, and 6). One notes in Figure 7.22 that operation 2 is followed by four *successive operations* 3, 4, 5, and 6, whereas operation 6 has no successive operations. The operations 3 and 4 are *immediate successive* operations of operation 2.

A *slack time* of an operation is the difference between its due date and the current time of this operation.

Algorithm 7.4

Step 1. Initialize:
- Current time
- Set of schedulable operations
- Set of completed operations

Step 2. From the set of schedulable operations, select an operation using the following priority rules:
- P1: with the largest number of successive operations
- P2: belonging to a part with the minimum number of schedulable operations
- P3: with the largest number of immediate successive operations
- P4: belonging to a part with the largest number of unprocessed operations
- P5: with the shortest processing time
- P6: belonging to a part with the corresponding shortest slack time

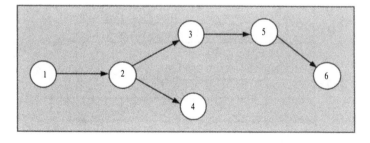

Figure 7.22. Example part P with six operations.

- P7: arbitrary selection

Step 3. Schedule the operation selected in Step 2. Update:
- The resource status
- The set of schedulable operations

If the set of schedulable operations is empty, go to Step 4; otherwise, go to Step 2.

Step 4. Calculate the completion time of each operation scheduled but not completed at the current time. Set the current time equal to the completion time of the operation with the least remaining processing time. Add this operation (or operations in case of a tie) to the set of completed operations. Update:
- The resource status
- The set of schedulable operations

If there is no unprocessed operations, stop; otherwise, go to Step 5.

Step 5. If the set of schedulable operations is empty, go to Step 4; otherwise go to Step 2.

The sequence of the priority rules used in Step 2 depends on the characteristics of the problem considered. The general heuristic algorithm is illustrated with the following example.

Example 7.14. Schedule eight operations that belong to three parts shown in Figure 7.23 on two machines. The data for the problem are presented in Table 7.7.

The first iteration of Algorithm 7.4 is shown next:

Step 1. Initialize:
- Current time $t = 0$
- The set of schedulable operations $S_1 = \{1, 3, 4, 6, 7\}$
- The set of completed operations $F = \varnothing$

Step 2. Using rule P1, operations 1, 3, 4, 7 have been selected. Priority rule P2 selects operation 1 as shown in Table 7.7.

Step 3. Operation 1 is scheduled on machine 1 as indicated next.

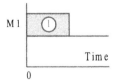

Operation 1 and all operations to be processed on machine 1 have been crossed out in the graph representing the parts and the table, as shown next.

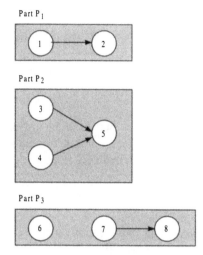

Figure 7.23. Structure of three parts.

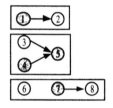

Part No.	1		2			3		
Operation No.	1	2	3	4	5	6	7	8
Machine No.	1	2	2	1	1	2	1	2
Machining time	4	2	3	2	2	3	2	1

The set of schedulable operations is updated to $S_1 = \{3, 6\}$. Note that operation 2 is not schedulable as the current time has not been updated and part P_1 is occupying machine M_1. Operation 8 cannot be considered for scheduling due to the precedence constraint. Go to step 2.

Step 2. Rule P1 selects operation 3.

Step 3. Operation 3 is scheduled on machine 2.

M 1 [①]

M 2 [③]

The graph and table have been updated as follows:

Part No.	1		2			3		
Operation No.	1	2	3	4	5	6	7	8
Machine No.	1	2	2	1	1	2	1	2
Machining time	4	2	3	2	2	3	2	1

TABLE 7.7. Scheduling Data

Part number	1		2			3		
Operation number	1	2	3	4	5	6	7	8
Machine number	1	2	2	1	1	2	1	2
Machining time	4	2	3	2	2	3	2	1

The set of schedulable operations is updated to $S_1 = \varnothing$. Go to Step 4.

Step 4. The completion time of operations 1 and 3 is 4 and 3, respectively, and current time is set to 3 (see Figure 7.24). Operation 3 is added to the set of completed operations F, $F = \{3\}$. Machine 2 becomes available and its status has been updated. The set of schedulable operations is updated to $S_1 = \{6\}$. Go to Step 5.

Step 5. Since the set of schedulable operations $S_1 = \{6\} \neq \varnothing$, go to Step 2.

Step 2. Operation 6 is selected.

Step 3. Operation 6 is scheduled on machine 2.

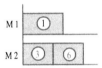

The set of schedulable operations $S_1 = \{4\} \neq \varnothing$.

Step 2. Operation 4 is selected.

Step 3. Operation 4 is scheduled on machine 1.

Part No.	1		2			3		
Operation No.	1	2	3	4	5	6	7	8
Machine No.	1	2	2	1	1	2	1	2
Machining time	4	2	3	2	2	3	2	1

After six iterations, the final schedule with the makespan of 10 is shown in Figure 7.24.

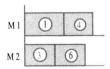

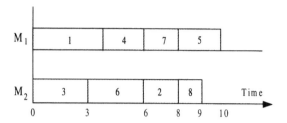

Figure 7.24. Final schedule.

7.4.8 Resource-Based Scheduling Rule

Priority rules P1–P6 used in Step 2 of Algorithm 7.4 as well as most scheduling rules existing in the literature do not consider the multiple-resource aspect of manufacturing systems. The project planning systems, (e.g., PERT) and other approaches published in the project scheduling literature consider resources; however, the nature of models considered there is different from manufacturing systems. In an automated manufacturing system, a part cannot be scheduled if all the manufacturing resources required are not available (e.g., machines, tools, fixtures). In this section, the most dissimilar resources (MDR) dispatching rule (Ahn et al., 1993) is presented for effective scheduling of operations in an automated machining system where the maximization of the utilization rate of manufacturing resources is a major concern because of the following:

- Considerable capital investment is needed to install the system.
- Production is lost when manufacturing resources are idle.
- Scheduling with the minimum makespan and minimum tardiness criteria tends implicitly to maximize resource utilization over the scheduling horizon

The MDR dispatching rule selects the largest set of schedulable operations, which satisfies the following conditions:
- Each operation belongs to a distinct part, that is, more than one operation of a part cannot be processed at the same time.
- All the operations in the set can be processed at the same time, that is, no resource conflict exists.

The idea of the MDR dispatching rule can be explained with an operation–resource graph. For *n* parts, the operation–resource graph becomes an *n*-partite graph and is not necessarily complete. In the operation–resource graph, each edge indicates the nodes (operations) that can be processed simultaneously, that is, the nodes that do not share the same resources. Note that the MDR dispatching rule attempts to select the largest set of nodes such that each node in the set belongs to a distinct part and there is no resource conflict between the nodes in the set.

A clique in a simple graph, G, is defined as a subgraph $G[S]$ such that $G[S]$ is complete (Bondy and Murty, 1976). Thus, a clique in the operation–resource graph satisfies the conditions under which the MDR dispatching rule selects a set of schedulable operations from the operation–resource incidence matrix, and the maximum clique in the graph corresponds to the largest set of schedulable operations obtained from the incidence matrix. Therefore, selecting the largest set of schedulable operations is equivalent to the problem of finding the maximum clique in the operation–resource graph G.

In real time scheduling, it is important that the maximum clique problem (NP complete) is solved almost instantaneously. A standard integer programming code (e.g., LINDO) is not likely to meet the latter requirement for the problem sizes encountered in industry. A more efficient way to solve the maximum clique problem is to use a specialized algorithm. A simple pairwise counting heuristic of computational time complexity $O(n^2)$ is presented next.

Before the heuristic is presented, the following notation is introduced:

$P =$ set of operations to be dispatched

$O_i =$ set of operations of part i

$c =$ resource type index

$r =$ resource index

$c^r =$ rth resource of type c

$[m_{kc^r}] =$ operation–resource incidence matrix, where

$$m_{kc^r} = \begin{cases} 1 & \text{if operation } k \text{ requires resource } c^r \\ 0 & \text{otherwise} \end{cases}$$

$[d_{kl}] =$ distance matrix, where

$$d_{kl} = \sum_{c^r} d(m_{kc^r}, m_{lc^r}) \text{ is the distance between operations } k \text{ and } l \qquad (7.52)$$

for

$$d(m_{kc^r}, m_{lc^r}) = \begin{cases} 1 & \text{if } m_{kc^r} m_{lc^r} = 1 \text{ or } (k, l) \in O_i \\ 0 & \text{otherwise} \end{cases}$$

$D(k) = distance\ index$ of operation k, defined as the number of parts such that at least one of their operations has a value 0 in the kth column of the distance matrix$[d_{kl}]$

The resource type defined above refers to different types of machines, tools, and fixtures used in manufacturing systems. For example, in the case of four drilling machines and five milling machines two types of resources are considered ($c = 1, 2$). For the milling machine the number of resources is 4 ($r = 1, \ldots, 4$) and for the drilling machine the number of resources is 5 ($r = 1, \ldots, 5$).

The above-defined distance matrix and distance index are illustrated in Example 7.15.

Example 7.15. Schedule three parts in a manufacturing system with three machines, two tools, and three fixtures. The operation–resource incidence matrix $[m_{kc'}]$ showing the resources required to manufacture the three parts is as follows:

		Machine				Tool				Fixture			
		M_1	M_2	M_3	M_4	t_1	t_2	t_3	t_4	f_1	f_2	f_3	f_4
Part 1	o_1	1	0	0	0	0	1	0	0	0	0	1	0
	o_2	0	0	0	1	0	0	0	1	0	0	0	1
Part 2	o_3	0	1	0	0	1	0	0	0	0	1	0	0
	o_4	1	0	0	0	0	0	1	0	0	1	0	0
	o_5	0	0	1	0	0	1	0	0	1	0	0	0
Part 3	o_6	0	0	1	0	0	0	1	0	0	0	0	1

$[m_{kc'}] =$ (the above matrix)

The distance matrix $[d_{kl}]$ is computed from (7.52) based on the above matrix $[m_{kc'}]$:

The distance index $D(k)$ is calculated as follows:

		Operation					
		o_1	o_2	o_3	o_4	o_5	o_6
Part 1	o_1	1	1	0	1	1	0
	o_2	1	1	0	0	0	1
Part 2	o_3	0	0	1	1	1	0
	o_4	1	0	1	1	1	1
	o_5	1	0	1	1	1	1
Part 3	o_6	0	1	0	1	1	1

$[d_{kl}] =$ (the above matrix)

$D(1) = 1$(operation 3 of part 2) + 1(operation 6 of part 3) = 2

$D(2) = 1$(operation 3, 4, and 5 of part 2)

$D(3) = 1$(operation 1 and 2 of part 1) + 1(operation 6 of part 3) = 2, and so on

The algorithm constructs a set of operations to be dispatched, P, from a set of schedulable operations, S.

Algorithm 7.5

Step 1. From the set of schedulable operations S, construct an operation–resource incidence matrix $[m_{kc'}]$.

Step 2. From the operation–resource incidence matrix $[m_{kc'}]$, construct the distance matrix $[d_{kl}]$.

Step 3. For each operation $k \in S$, compute the distance index $D(k)$ from the operation–resource incidence matrix $[m_{kc'}]$.

Step 4. Find the maximum $D(k^*)$. If a tie occurs, break it—arbitrary.

Step 5. Move operation k^* to P.

7.5 RESCHEDULING

Rescheduling is an area that has begun gaining recognition in the literature. There are a few research articles published on this issue. Peng and Chen (1998) developed a simulation and optimization based framework for manufacturing scheduling enabling the adoption of various short-term scheduling policies when responding to dynamic shop floor changes. Laursen et al. (1998) presented an architecture for real-time control of one-of-a-kind production. The framework includes off-line and real-time components. The off-line module performs scheduling and job decomposition while dispatching, shop floor, observer, simulator, and monitoring are considered by the real-time part.

Fang and Xi (1997) studied job shop scheduling where jobs arrive continuously, machines break down, and due dates of jobs may change during processing. They presented an event-driven strategy adapted to continuous processing in changing the environment by using a rolling horizon optimization method from predictive control technology. Their scheduling algorithm is a hybrid including a genetic algorithm and dispatching rules for solving the job shop scheduling problem with sequence-dependent set-up time and due date constraints. Li et al. (1996) investigated the flow time performance of a job shop by constructing a modular simulation model using SLAM, a discrete-event simulation language. They considered the rescheduling procedure as a two-phase iterative process: (a) rescheduling based on changes demands, conditions, or constraints; (b) if the result of the new schedule is reasonable, then stop; else, determine an improved solution. The rescheduling method assigned a priority level to each job, and the job with highest priority is assigned first. In a dynamic process, the priority of the job changes depending on the different states of the system of the workstation queues.

Li et al. (1993) defined several rescheduling factors and termed any operation affected by these as an "independent affected operation" and any other operation not directly affected by the rescheduling factor as a "dependent affected operation." Furthermore, the occurrence of any such factor has a time effect on an independent affected operation. They proposed an algorithm that constructs a scheduling binary tree for the affected dependent and independent operations. The algorithm can be embedded in a simulation-based scheduling system or an electronic Gantt chart.

7.6 SUMMARY

In this chapter, a three-level hierarchical approach planning and scheduling manufacturing system was presented. At the top level, an MRP is fed by the master production plan to generate the requirements for parts and schedule them accordingly. The MRP output is in turn used for capacity balancing and scheduling at the next two levels. The goal of capacity balancing is to assign operations to machines so that the existing manufacturing capacity and resource requirements are met. In practice, capacity balancing is the final formal production planning activity. Once the parts have been assigned to machines, they are released for production according to less formal priority rules, for example, first-in first-out, degree of urgency, or even at random. Numerous scheduling models and algorithms were discussed.

APPENDIX: INTEGER PROGRAMMING FORMULATION OF THE PROBLEM IN EXAMPLE 7.6

```
MIN
999X11+7X12+3X13+12X14+5X15+8X16+4X21+999X22+2X23+
10X24+9X25+3X26+6X31+7X32+999X33+11X34+X35+7X36+
7X41+3X42+X43+999X44+8X45+8X46+2X51+10X52+2X53+7X54+
999X55+3X56+4X61+11X62+7X63+6X64+3X65+999X66
SUBJECT TO
X11+X21+X31+X41+X51+X61=1
X12+X22+X32+X42+X52+X62=1
X13+X23+X33+X43+X53+X63=1
X14+X24+X34+X44+X54+X64=1
X15+X25+X35+X45+X55+X65=1
X16+X26+X36+X46+X56+X66=1
X11+X12+X13+X14+X15+X16=1
X21+X22+X23+X24+X25+X26=1
X31+X32+X33+X34+X35+X36=1
X41+X42+X43+X44+X45+X46=1
X51+X52+X53+X54+X55+X56=1
X61+X62+X63+X64+X65+X66=1
6X23+U2-U3<=5
6X24+U2-U4<=5
6X25+U2-U5<=5
6X26+U2-U6<=5
6X32+U3-U2<=5
6X34+U3-U4<=5
6X35+U3-U5<=5
6X36+U3-U6<=5
6X42+U4-U2<=5
6X43+U4-U3<=5
6X45+U4-U5<=5
6X46+U4-U6<=5
6X52+U5-U2<=5
6X53+U5-U3<=5
6X54+U5-U4<=5
6X56+U5-U6<=5
```

```
6X62+U6-U2<=5
6X63+U6-U3<=5
6X64+U6-U4<=5
6X65+U6-U5<=5
END
INTEGER 36
```

REFERENCES

Ahn, J., W. He, and A. Kusiak (1993), Scheduling with alternative operations, *IEEE Transactions on Robotics and Automation*, Vol. 9, No. 3, pp. 297–303.

Baker, K. R. (1974), *Introduction to Sequencing and Scheduling*, John Wiley, New York.

Belhe, U., and A. Kusiak (1997), Production management, in J. D. Irwin (Ed.), *Industrial Electronics Handbook*, CRC Press, Boca Raton, FL.

Berry, W. L., T. E. Vollmann, and D. C. Whybark (1979), *Master Production Scheduling: Principles and Practice*: American Production and Inventory Control Society, Fall Church, VA.

Blackburn, J. D., D. H. Kropp, and R. A. Millen (1986), A comparison of strategies to dampen nervousness in MRP systems, *Management Science*, Vol. 32, No. 4, pp. 413–429.

Bondy, J. A., and U. S. R. Murty (1976), *Graph Theory with Applications*, Elsevier, New York.

Chase, R. B., N. J. Aquilano, and F. R. Jacobs (1998), *Production and Operations Management: Manufacturing and Services*, Irwin/McGraw-Hill, Boston, MA.

Chung, C. H., and L. J. Krajewski (1986), Replanning frequencies for master production schedules, *Decision Sciences*, Vol. 17, No. 2, pp. 263–273.

Durmusoglu, S., H. Sumen, and V. Z. Yenen (1996), The state-of-the-art of MRP/MRP II implementation in Turkey, *Production Planning and Control*, Vol. 7, No. 1, pp. 2–10.

Fang, J., and Y. G. Xi (1997), A rolling horizon job shop rescheduling strategy in the dynamic environment, *International Journal of Advanced Manufacturing Technology*, Vol. 13, No. 3, pp. 227–232.

French, S. (1982), *Sequencing and Scheduling: An Introduction to the Mathematics of the Job Shop*, John Wiley, New York.

He, W., and A. Kusiak (1992), Scheduling manufacturing Systems, *Computers in Industry*, Vol. 20, pp. 163–175.

Japan Management Association (Ed.) (1986), *Kanban: Just-in-Time at Toyota*, Productivity Press, Cambridge, MA, pp. 87–92.

Johnson, S. M. (1954), Optimal two- and three-stage production schedules with set-up times included, *Naval Research Logistics Quarterly*, Vol. 1, pp. 61–68.

Kropp, D. H., R. C. Carlson, and J. V. Jucker (1983), Heuristic lot-sizing approaches for dealing with MRP system nervousness, *Decision Sciences*, Vol. 14, No. 2, pp. 156–168.

Kurisu, T. (1979), Single machine sequencing with precedence constraints and deferral costs, *Journal of Operational Research Society of Japan*, Vol. 22, pp. 1–14.

Kusiak, A. (1986), Efficient implementation of Johnson's scheduling algorithm, *IIE Transactions*, Vol. 18, No. 2, pp. 215–216.

Kusiak, A., Ed. (1992), *Intelligent Design and Manufacturing*, John Wiley, New York.

Kusiak, A., and G. Finke (1987), Modeling and solving the flexible forging module scheduling problem, *Engineering Optimization*, Vol. 12, No. 1, pp. 1–12.

Kusiak, A., and H. H. Yang (1995), A knowledge-based approach for production scheduling in an MRP environment, in P. Brandimarte and A. Villa (Eds.), *Optimization Models and Concepts in Production Management*, Gordon and Breach, Basel, Switzerland, pp. 257–275.

Kusiak, A., J. Ahn, and K. Park (1993), Artificial intelligence in planning, scheduling, and control of manufacturing systems, in S. G. Tzafestas (Ed.), *Applied Control: Current Trends and Modern Methodologies*, Dekker, New York, pp. 937–976.

Laursen, R. P., C. Orum-Hansen, and E. Trostmann (1998), The concept of state within one-of-a-kind real-time production control systems, *International Journal of Production Planning and Control*, Vol. 9, No. 6, pp. 542–552.

Lawler, E. L. (1978), Sequencing jobs to minimize total weighted completion time subject to precedence constraints, *Annals of Discrete Mathematics*, Vol. 2, pp. 75–90.

Lawler, E. L., J. K. Lenstra, A. H. G. Rinnooy Kan, and D. B. Shmoys, Eds. (1985), *The Traveling Salesman Problem: A Guided Tour of Combinatorial Optimization*, John Wiley, New York.

Lenstra, J. K., and A. H. G. Rinnooy Kan (1978), Complexity of scheduling under precedence constraints, *Operations Research*, Vol. 26, pp. 22–35.

Li, R. K., Y. T. Shyu, and S. Adiga (1993), A heuristic rescheduling algorithm for computer-based production scheduling systems, *International Journal of Production Research*, Vol. 31, No. 8, pp. 1815–1826.

Li, Y. C. E., W. H. Shaw, and L. A. Martin-Vega (1996), Flow-time performance of modified scheduling heuristics in a dynamic rescheduling environment, *Computers and Industrial Engineering*, Vol. 31, Nos. 1/2, pp. 213–216.

Lubben, R. (1988), in *Just-in-Time Manufacturing: An Aggressive Manufacturing Strategy*, McGraw-Hill, New York, pp. 13–14.

Martello, S. (1983), An enumerative algorithm for finding Hamiltonian circuits in a directed graph, *ACM Transactions on Mathematical Software*, Vol. 14, pp. 256–268.

Minifie, J. R., and R. A. Davis (1990), Interaction effects on MRP nervousness, *International Journal of Production Research*, Vol. 28, No. 1, pp. 173–183.

Peng, C., and F. F. Chen (1998), Real-time control and scheduling of flexible manufacturing systems: An ordinal optimization based approach, *International Journal of Advanced Manufacturing Technology*, Vol. 14, No. 10, pp. 775–786.

Potts, C. N. (1980), An algorithm for the single machine sequencing problem with precedence constraints, *Mathematical Programming Study*, Vol. 13, pp. 78–87.

Potts, C. N. (1985), An Lagrangian based branch and bound algorithm for the single machine sequencing with precedence constraints to minimize total weighted completion time, *Management Science*, Vol. 31, No. 10, pp. 1300–1311.

Schrage, L. (1984), *Linear, Integer, and Quadratic Programming with LINDO*, Scientific Press, Palo Alto, CA.

Steele, D. C. (1975), The nervous MRP system: How to do battle, *Production and Inventory Management*, Vol. 16, No. 4, pp. 83–89.

Stevenson, W.J. (1993), *Production Operations Management*, 4th ed., Irwin, Boston.

Sridharan, V., and W. L. Berry (1990), Master production scheduling make-to-stock products: A framework for analysis, *International Journal of Production Research*, Vol. 28, No. 3, pp. 541–558.

Sridharan, V., W. L. Berry, and V. Udayabhanu (1988), Measuring master production schedule stability under rolling planning horizons, *Decision Sciences*, Vol. 19, No. 1, pp. 147–166.

Vollmann, T. E., W. L. Berry, and D. C. Whybark (1988), *Manufacturing Planning and Control Systems*, Irwin, Homewood, IL.

Winston, P. H., and B. K. P. Horn (1981), *LISP*, Addison-Wesley, Reading, MA.

Yano, C. A., and R. C. Carlson (1987), Interaction between frequency of rescheduling and the role of safety stock in material requirements planning systems, *International Journal of Production Research*, Vol. 25, No. 2, pp. 221–232.

QUESTIONS

7.1. What are the differences between MRP, MRP II, and ERP systems?

7.2. What are the basic functions of an MRP system?

7.3. What is MRP system nervousness?

7.4. What is OPT?

7.5. What is the basic goal of a just-in-time production system?

7.6. What data are needed to estimate the number of kanbans?

7.7. What is the goal of capacity balancing?

7.8. What is the relationship between capacity balancing and manufacturing scheduling?

7.9. What is the relationship between machine loading models and assembly line balancing?

7.10. What are the two simplest single-machine scheduling algorithms?

7.11. What is unique about scheduling models?

7.12. What is the basic model for solving scheduling problems with sequence-dependent set-up times?

7.13. What type of three-machine scheduling problem can be solved with Johnson's algorithm?

7.14. Is it possible to develop mathematical programming models of scheduling problems with a large number of machines, operations, and precedence constraints?

7.15. What algorithm would you use to solve a large-scale manufacturing scheduling problem?

PROBLEMS

7.1. Given the MRP record in Table 7.8 calculate the number of items on hand:

TABLE 7.8. MRP Record

Period		1	2	3	4	5
Gross requirements			20		35	16
Scheduled receipts		30				
Balance on hand	15	45	25	25	20	20
Planned order release			30	16		
Lead time = 2 periods						
Safety stock = 20						

(a) The lead time decreases from 2 units to 1 unit and all other values remain the same.

(b) The lot size is constant and equal to 50, and all other values remain the same.

(c) Determine the minimum lead time and lot size so that the level of inventory does not drop below a safety stock of 24 items, assuming an additional gross requirement of 48 items in period 3 has been generated.

7.2. Determine whether the MRP record in Table 7.9 is correct. If not, correct the error(s).

7.3. Definitions:

- Item: The name or number that identifies the item being scheduled.
- *LLC:*   Low-level code; the lowest level at which the item appears in a product structure.
- *Lot size:*   Normally, an order will be placed in *multiples* of this quantity, but it can also represent a *minimum* or *maximum* order quantity or the type of lot-sizing technique.
- *LT:*   Lead time; the time from when an order is placed until it is received.

TABLE 7.9. MRP Record for Weeks 1–9

Parameter	1	2	3	4	5	6	7	8	9
Gross requirements	0	0	200	0	100	0	175	0	0
Order receipts	0	0	300	0	0	0	175	0	0
On hand	0	0	0	100	100	0	0	125	0
Order releases	0	300	0	0	0	300	0	0	0

- *PD:* Past-due time bucket. If an order appears in the PD time bucket, the schedule is infeasible and an error massage will be generated. Projected on-hand entries in the PD column represent beginning inventory.

- *Gross requirement:* The demand for an item by time period. For an end item, this quantity is obtained from the master production schedule. For a lower level item, it is derived from the *planned order releases* of its parents.

- *Scheduled receipts:* The quantity of material that is already ordered (from work orders or purchase orders) and when it is expected to arrive. Once a planned order is released, it becomes a scheduled receipts.

- *Projected on hand:* The expected quantity in inventory at the end of a period that will be available for demand in subsequent periods. It is calculated by subtracting the *gross requirements* in the period from the sum of *scheduled receipts* for the same period, *projected on hand* from the previous period, and *planned order releases* from the $t - l$ period (where t represents the current period and l is the lead time).

- *Net requirement:* The net number of items that must be provided and when they are needed. It is calculated by subtracting the *scheduled receipts* in the period plus the *projected on hand* in the previous period from *gross requirements*. It appears in the same time period as gross requirements.

- *Planned order receipts:* The same as net requirements adjusted for lot sizing. If lot sizing is seldom used or lead time is negligible, this row can be deleted from the matrix.

- *Planned order releases:* Planned order receipts offset for lead times. It shows when an order should be placed (i.e., released) so that items are available when needed. *Planned order releases* at one level generate *gross requirements* at the next lower level.

Complete the MRP record in Table 7.10 and provide answers to the following three questions:

TABLE 7.10. Incomplete MRP Record

Item: X LLC: 1					Period				
LT: 2	PD	1	2	3	4	5	6	7	8
Lot Size: Min 50									
Gross requirement		20	30	50	50	60	90	40	60
Scheduled receipts			50						
Projected on hand	40								
Net requirement									
Planned order receipts									
Planned order releases									

(a) In what periods should orders be released and what should be the size of those orders? *Hint:* The order size is identical for all periods.

(b) How would the planned order releases change with no lot sizing? *Hint:* The order size may vary from period to period.

(c) How would the planned order releases change with no lot sizing and a safety stock of 20?

7.4. Five batches of parts are to be processed in a manufacturing system with three machines. Given the matrix of batch processing costs $[C_{ij}]$, the matrix $[T_{ij}]$ of processing of each batch on the three machines, the machine capacity vector $[b_j]$, the matrix $[k_{ij}]$ of space occupied by tools in tool magazines, the tool magazine capacity vector $[f_j]$, the penalty vector $[q_j]$, and the vector $[Z_j]$ of upper limits on the number of tool magazines to be used on machine j:

$$[C_{ij}] = \begin{array}{c} \\ 1 \\ 2 \\ 3 \\ 4 \\ 5 \end{array} \begin{array}{ccc} 1 & 2 & 3 \\ \left[\begin{array}{ccc} 4 & 6 & 5 \\ 2 & 3 & 6 \\ 3 & 8 & 5 \\ 4 & 5 & 4 \\ 7 & 3 & 9 \end{array}\right] \end{array} \qquad [T_{ij}] = \begin{array}{c} \\ 1 \\ 2 \\ 3 \\ 4 \\ 5 \end{array} \begin{array}{ccc} 1 & 2 & 3 \\ \left[\begin{array}{ccc} 6 & 9 & 7 \\ 2 & 1 & 5 \\ 4 & 3 & 2 \\ 4 & 7 & 5 \\ 2 & 3 & 1 \end{array}\right] \end{array} \qquad [b_j] = [15, 15, 15]^T$$

$$[k_{ij}] = \begin{array}{c} \\ 1 \\ 2 \\ 3 \\ 4 \\ 5 \end{array} \begin{array}{ccc} 1 & 2 & 3 \\ \left[\begin{array}{ccc} 2 & 2 & 2 \\ 4 & 4 & 4 \\ 3 & 3 & 3 \\ 5 & 5 & 5 \\ 5 & 5 & 5 \end{array}\right] \end{array} \qquad [f_j] = [8, 8, 10]^T$$

$$[q_j] = [1000, 1000, 1000]^T \qquad [Z_j] = [3, 2, 3]^T$$

Assign optimally the parts and tools to the machines. What is the number of tool magazines used on each of the three machines?

7.5. Five batches of parts are to be processed in a manufacturing system consisting of three machines. Given the matrix of batch processing costs $[C_{ij}]$, the matrix $[T_{ij}]$ of processing of each batch on the three machines, the machine capacity vector $[b_j]$, the matrix $[k_{ij}]$ of space occupied by tools in tool magazines, the tool magazine capacity vector $[f_j]$, and the penalty vector $[q_j]$:

$$[C_{ij}] = \begin{array}{c} \\ 1 \\ 2 \\ 3 \\ 4 \\ 5 \end{array} \begin{array}{ccc} 1 & 2 & 3 \\ \left[\begin{array}{ccc} 4 & 6 & \infty \\ \infty & 3 & 6 \\ 3 & 8 & 5 \\ 4 & \infty & 4 \\ 7 & 3 & 9 \end{array}\right] \end{array} \qquad [T_{ij}] = \begin{array}{c} \\ 1 \\ 2 \\ 3 \\ 4 \\ 5 \end{array} \begin{array}{ccc} 1 & 2 & 3 \\ \left[\begin{array}{ccc} 6 & 9 & \infty \\ \infty & 1 & 5 \\ 4 & 3 & 2 \\ 4 & \infty & 5 \\ 2 & 3 & 1 \end{array}\right] \end{array} \qquad [b_j] = [11, 11, 11]^T$$

$$[k_{ij}] = \begin{array}{c} \\ 1 \\ 2 \\ 3 \\ 4 \\ 5 \end{array} \begin{array}{ccc} 1 & 2 & 3 \\ \begin{bmatrix} 2 & 2 & 2 \\ 4 & 4 & 4 \\ 3 & 3 & 3 \\ 5 & 5 & 5 \\ 5 & 5 & 5 \end{bmatrix} \end{array} \qquad [f_j] = [9, 9, 10]^T \qquad [q_j] = [900, 900, 900]^T$$

(a) Assign optimally the parts and tools to the machines.

(b) What is the number of tool magazines used on each of the three machines?

(c) How does the value of $q_j, i = 1, 2, 3,$ impact the solution of the capacity-balancing problem?

7.6. Consider the capacity-balancing model (7.5)–(7.10) discussed in this chapter.

(a) Assume your own data (do not make the matrices too large) and solve the model with an integer programming software (e.g., LINDO). Did you get a feasible solution the first time? Illustrate the solution graphically.

(b) Multiply the penalty cost q_j in the objective function by 10 and see how the solution was affected.

(c) Divide the penalty cost q_j in the objective function by 10 and see how the solution was affected.

7.7. Consider the two schedules in Figure 7.25.

(a) What is the major difference between schedules (a) and (b) in Figure 7.25?

(b) Which of the two schedules, if any, is optimal? Justify the answer.

7.8. The matrix $[t_{ij}]$ represents the sequence-dependent changeover times of a printing press where six different production runs are planned:

$$[t_{ij}] = \begin{array}{c} 1 \\ 2 \\ 3 \\ 4 \\ 5 \\ 6 \end{array} \begin{bmatrix} \infty & 9 & 5 & 8 & 2 \\ 4 & \infty & 9 & 10 & 8 \\ 3 & 14 & \infty & 8 & 10 \\ 6 & 6 & 8 & 10 & 6 \\ 2 & 10 & 12 & \infty & 12 \\ 9 & 8 & 10 & 4 & \infty \end{bmatrix}$$

The processing times are given in matrix $[t_i]$:

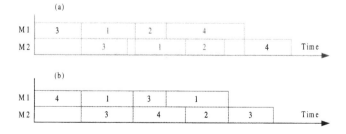

Figure 7.25. Two two-machine schedules.

Figure 7.26. Precedence constraints.

$$[t_i] = [4, 7, 2, 9, 4, 3]$$

(a) Find the sequence of production runs with the minimum total changeover and processing time.

(b) Draw the schedule Gantt chart.

The same printing press was used to process another six production runs. Though the changeover times remained the same, the precedence constraints in Figure 7.26 must be obeyed.

(c) Find the sequence of production runs with the minimum total changeover and processing time. (Find two solutions—heuristic and optimal.)

(d) Draw the schedule Gantt chart.

7.9. Given the following matrix of sequence-dependent setup times:

$$[t_{ij}] = \begin{array}{c} \\ 1 \\ 2 \\ 3 \\ 4 \\ 5 \\ 6 \end{array} \begin{array}{cccccc} 1 & 2 & 3 & 4 & 5 & 6 \\ \left[\begin{array}{cccccc} - & 9 & 4 & 7 & 2 & 10 \\ 10 & - & 9 & 5 & 2 & 6 \\ 5 & 13 & - & 16 & 4 & 9 \\ 5 & 2 & 7 & - & 5 & 9 \\ 8 & 6 & 4 & 6 & - & 3 \\ 19 & 5 & 4 & 11 & 8 & - \end{array}\right] \end{array}$$

(a) Formulate a scheduling model minimizing the total changeover time.

(b) Find the optimal schedule for manufacturing the six batches of parts.

7.10. The products in Figure 7.27 are to be assembled on a two-station assembly system. The assembly time for each part is provided in Table 7.11.

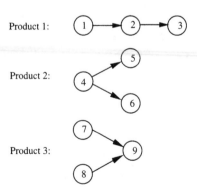

Figure 7.27. Structure of three products.

TABLE 7.11. Assembly Time

Part Number	Station 1	Station 2
1	6	4
2	2	2
3	5	3
4	5	4
5	1	3
6	3	2
7	3	2
8	3	5
9	2	1

(a) Find a schedule with the shortest makespan.

(b) Is this solution optimal?

(c) Are the two stations balanced?

(d) Could the makespan be reduced? If yes, explain how.

7.11. Six different parts are to be scheduled in a three-machine manufacturing. The processing times are provided in Table 7.12. Draw a Gantt chart of the schedule with min F_{max}.

7.12. Solve optimally (min F_{max}) the job shop scheduling problem with 12 operations and the processing data in Table 7.13. Illustrate the solution with a Gantt chart.

7.13. Determine a schedule with the shortest maximum flow time for 13 parts in Table 7.14. Draw the schedule Gantt chart.

7.14. Each of the three products P_1–P_3 in Figure 7.28 involves three to five assembly operations. Find the minimum makespan schedule for the three products to be assembled on a two-station line:

Operation number	1	2	3	4	5
Assembly time on station 1	3	4	1	3	2
Assembly time on station 2	1	5	2	2	4

TABLE 7.12. Scheduling Data

Part Number	Processing Time		
	M_1	M_2	M_3
1	10	9	10
2	2	4	13
3	3	1	10
4	6	9	12
5	9	10	10
6	4	10	11

TABLE 7.13. Scheduling Data

	First Machine		Second Machine	
Operation Number	Processing Order	Time	Processing Order	Time
1	M_1	7	M_2	4
2	M_1	9	M_2	11
3	M_2	12	M_1	7
4	M_2	7		
5	M_1	10	M_2	2
6	M_2	5	M_1	4
7	M_1	13		
8	M_1	4	M_2	10
9	M_1	9	M_2	3
10	M_2	13	M_1	8
11	M_2	9		
12	M_2	6		

TABLE 7.14. Scheduling Data

	First Machine		Second Machine	
Part Number	Processing Order	Time	Processing Order	Time
1	M_2	10	M_1	6
2	M_2	7	M_1	9
3	M_2	12	M_1	12
4	M_1	7		
5	M_1	12	M_2	4
6	M_2	5	M_1	4
7	M_2	13		
8	M_2	6	M_1	10
9	M_1	12	M_2	6
10	M_2	15	M_1	11
11	M_2	6		
12	M_1	8		
13	M_1	5	M_2	10

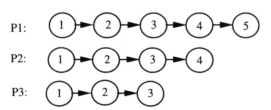

Figure 7.28. Assembly structure of three products and data.

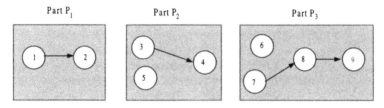

Figure 7.29. Structure of three parts.

TABLE 7.15. Scheduling Data

Part number	1		2			3			
Operation number	1	2	3	4	5	6	7	8	9
Machine number	1	2	2	1	2	1	2	1	1
Tool number	5	6	5	7	6	7	7	6	5
Processing time	2	2	4	3	2	4	2	1	2

Find a heuristic schedule of the two-machine problem with three parts in Figure 7.29 and data in Table 7.15. How would the schedule makespan change if there were no limit on the number of tools?

7.15. For the parts in Figure 7.30 and data in Table 7.16:

    (a) Schedule all operations belonging to the three parts on two machines with a suitable algorithm.

    (b) Compute the following performance measures of your schedule:

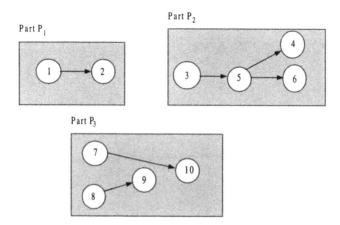

Figure 7.30. Structure of three parts.

TABLE 7.16. Scheduling Data

Part number	1			2			3			
Operation number	1	2	3	4	5	6	7	8	9	10
Machining time	1	4	5	3	2	1	1	3	2	4
Machine number	2	1	2	1	2	1	2	1	2	2
Tool number	4	5	4	5	5	6	4	5	7	4

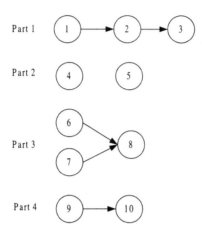

Figure 7.31. Structure of four parts.

    (i)   Makespan

    (ii)   Mean flow time

  (c) Is your schedule optimal?

7.16. Determine a suboptimal schedule of the four parts in Figure 7.31. The manufacturing time of each operation and the resources (tool set and operator) required are listed in Table 7.17. Draw a Gantt chart and determine the schedule makespan.

TABLE 7.17. Scheduling Data

Operation number	1	2	3	4	5	6	7	8	9	10
Manufacturing time	4	6	1	5	2	2	4	1	5	6
Machine number	1	2	2	1	3	2	1	2	3	1
Tool set	T1	T2	T1	T5	T4	T1	T4	T6	T1	T4
Operator	O2	O2	O1	O3	O4	O2	O5	O5	O2	O4

CHAPTER 8

KANBAN SYSTEMS

8.1 INTRODUCTION

Kanban is a Japanese word for a "visual record." The idea of a kanban system supporting the just-in-time production concept originated from supermarkets, where customers obtain (1) what is needed, (2) at the time it is needed, and (3) the amount needed. A supermarket manager maintains a certain amount of inventory on the shelves.

Sometimes the tem *kanban* is confused with another Japanese term, *kaizen*, describing efforts in continuous improvements. Teaming and problem solving are the focus of *kaizen*. *Kanban* and *kaizen* can in fact complement each other. The original kanban system relies on cards containing information, for example, job type and quantity of parts to be transferred (Huang and Kusiak, 1996).

The concept of "push systems" has been used in industry for a long time. In a push system, jobs are released to the first stage of manufacturing, and in turn this stage pushes the work in process to the succeeding stage, and so on, until the final products are obtained. The kanban system is known as a "pull system" in the sense that the production of the current stage depends on the demand of the subsequent stages, that is, the preceding stage must produce only the exact quantity withdrawn by the subsequent manufacturing stage. In this way, the kanban system was created to indicate what is needed at which production stage and to allow various stages to efficiently communicate with each other. The company's production plan is given only to the final assembly line. When parts or materials are withdrawn from the preceding stage, a chain of communication is established with each of the relevant preceding stages, and every stage automatically knows how much and when to produce the parts required. At each station, the information about the product name, code, volume, and so on, can be easily obtained from the kanbans. Figure 8.1 illustrates the general kanban system.

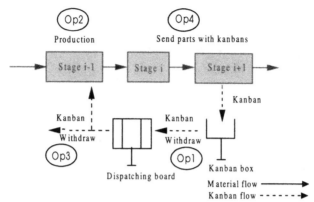

Op1: When demand from stage i+1 occurs, withdraw kanbans and place them on the dispatching board.
Op2: Production activity initiates when a kanban is placed on the dispatching board.
Op3: Simultaneously, demand is sent to stage i − 1 if the demand occurs at stage i.
Op4: Completed parts with kanbans are sent to stage i + 1.

Figure 8.1. General kanban system.

8.1.1 Operations Principles

The main operations principles of kanban systems are as follows (Hall, 1983):

1. Level production (balance the schedule) in order to achieve low variability of the number of parts from one time period to the next.
2. Avoid complex information and hierarchical control systems on a factory floor.
3. Do not withdraw parts without a kanban.
4. Withdraw only the parts needed at each stage.
5. Do not send defective parts to the succeeding stages.
6. Produce the exact quantity of parts withdrawn.

8.1.2 Kanban Functions

The key objective of a kanban system is to deliver the material just-in-time to the manufacturing workstations and to pass information to the preceding stage as to what and how much to produce.

A kanban fulfills the following functions:

1. *Visibility Function.* The information and material flow are combined together as kanbans move with their parts (work-in-progress, WIP).
2. *Production Function.* The kanban detached from the succeeding stage fulfills a production control function, which indicates the time, quantity, and part types to be produced.

3. Inventory Function.   The number of kanbans actually measures the amount of inventory. Hence, controlling the number of kanbans is equivalent to controlling the amount of inventory; that is, increasing (decreasing) the number of kanbans corresponds to increasing (decreasing) the amount of inventory. Controlling the number of kanbans is much simpler than controlling the amount of inventory itself.

8.1.3   Kanban Types

According to their functions, kanbans are classified as follows:

1. Primary Kanban: Travels from one stage to another among main manufacturing cells or production preparation areas. The primary kanbans are of two kinds: the "withdrawal kanban" (conveyor kanban), which is carried when going from one stage to the preceding stage, and the "production kanban" (see Figure 8.2), which is used to order production of the portion withdrawn by the succeeding stage. These two kanbans are always attached to the containers holding parts.

2. Supply Kanban: Travels from a warehouse or storage facility to a manufacturing facility (see Figure 8.3).

3. Procurement Kanban: Travels from outside of a company to the receiving area (see Figure 8.4).

4. Subcontract Kanban: Travels between subcontracting units.

5. Auxiliary Kanban: May take the form of an express kanban, emergency kanban, or a kanban for a specific application (Singh and Falkenburg, 1994).

8.1.4   Auxiliary Equipment

1. Kanban Box: To collect kanbans after they are withdrawn.

2. Dispatching Board: In which kanbans from the succeeding stage is placed in order to display the production schedule.

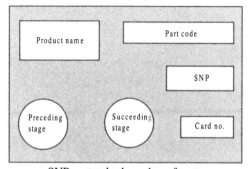

SNP: standard number of parts

Figure 8.2. Production kanban.

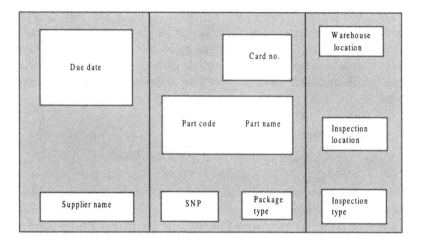

Figure 8.3. Supply kanban.

3. Kanban Management Account: An account to manage kanbans.
4. Supply Management Account: An account to manage the supply of raw materials.

8.1.5 Kanban Operations

For production stage i, when parts are processed and demand from its receiving stage $i + 1$ occurs, the production kanban is removed from a container and is placed in the dispatching board at stage i. The withdrawal kanban from stage $i + 1$ then replaces the production kanban and the container. This container along with the withdrawal kanban is then sent to stage $i + 1$ for processing.

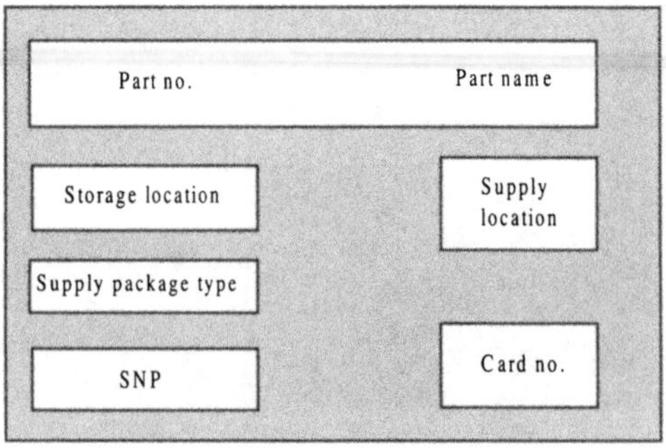

Figure 8.4. Procurement kanban.

Meanwhile at stage i, production activity takes place when a production kanban and a container with the withdrawal kanban are available. The withdrawal kanban is then replaced by the production kanban and sent back to stage $i-1$ to initiate production activity at stage $i-1$. This forms a cyclic production chain.

The kanban pulls (withdraws) parts instead of pushing parts from one stage to another to meet the demand at each stage. The kanban controls the move of products, and the number of kanbans limits the flow of products (Shingo, 1987). If no withdrawal is requested by the succeeding stage, the preceding stage will not produce at all, and hence no excess items are manufactured. Therefore, by the number of kanbans (containers) circulating in a just-in-time (JIT) system, non-stock production (NSP) may be achieved.

8.1.6 Kanban Control

Toyota considered its system of external and internal processes as connected with invisible conveyor lines (kanbans). The information flow (kanban flow) acts like an invisible conveyor through the entire production system and connects each department together. Figure 8.5 presents a general kanban control system (Lu, 1982).

8.1.6.1 Production Line. Due to different types of material handling systems, there are three types of control (Lu, 1982):

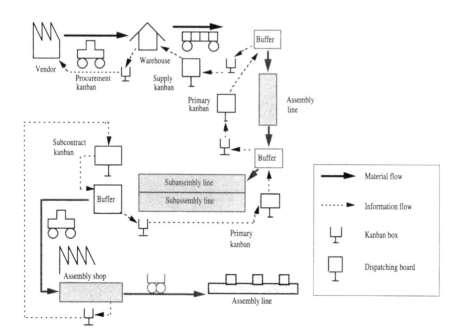

Figure 8.5. General kanban control system.

1. *Single-Kanban System.* The *single-kanban* (single-card) system (SKS) uses *production kanbans* only to block material handling based on the part type. The production is blocked at each stage based on the total queue size (see Figure 8.6). In a single-card system, the size of a station output buffer and part mix might vary. Multiple containers contain the batches to be produced, as long as the total number of full containers in the output buffer does not exceed the buffer output capacity. Note that the single-card system is in no way related to the hybrid push–pull schedule-driven single-card system described by Schonberger (1982a).

   The following conditions are essential for a proper functioning of the single-kanban system:

   - Small distance between any two subsequent stages
   - Fast turnover of kanbans
   - Low WIP
   - Small buffer space and fast turnover of WIP
   - Synchronization between the production rate and speed of material handling

2. *Dual-Kanban System.* The *dual-kanban* (two-card) system (DKS) uses *production and withdrawal kanbans* to implement both the station and material handling blocking by part type. There is a buffer for WIP while transporting the finished parts from a preceding stage to its succeeding stage. The withdrawal kanbans are presented in the buffer area (see Figure 8.7). The most common form

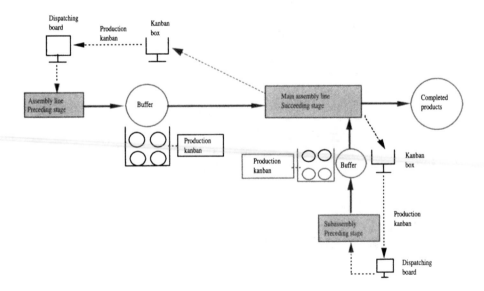

Figure 8.6. Single-kanban system.

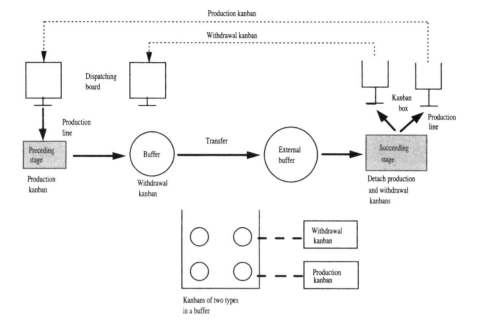

Figure 8.7. Dual-kanban system.

of two-card kanban production control is described in Sugimori et al. (1977), Monden (1983a), and Schonberger (1982a).

This system is appropriate for manufacturers who are not prepared to adopt strict control rules to the buffer inventory. The following conditions are essential for the dual-kanban system:

- Moderate distance between two stages
- Fast turnover of kanbans
- Some WIP in a buffer
- External buffer to the production system
- Synchronization between the production rate and speed of material handling

3. *Semi-Dual-Kanban System.* Figure 8.8 presents the semi-dual-kanban system (SDKS). This kanban system has the following characteristics:

- Large distance between two stages
- Slow turnover of kanbans
- Large WIP needed between subsequent stages
- Slow turnover of WIP

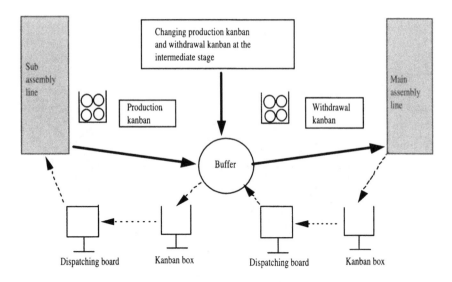

Figure 8.8. Semi-dual-kanban system.

- No need for synchronization between the production rate and speed of material handling

The three types of kanban systems are compared in Table 8.1.

8.1.6.2 *Receiving Area.* Based on different types of receiving, three types of kanban operations are performed:

1. Receiving from a preceding stage in the same facility (see Figure 8.6)
2. Receiving from a storage (see Figure 8.9)
3. Receiving from a vendor (see Figure 8.10)

TABLE 8.1. Comparison of Three Types Kanban Systems

Parameter	SKS	DKS	SDKS
Distance between two stages	Small	Moderate	Large
WIP between two stages	Small	Small	Large
Turnover of kanbans	Fast	Fast	Low
Turnover of WIP	Fast	Moderate	Slow
Synchronization of production and movement of WIP	Necessary	Necessary	Not necessary

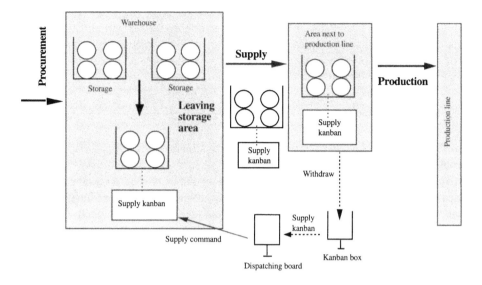

Figure 8.9. Kanban system receiving parts at warehouse.

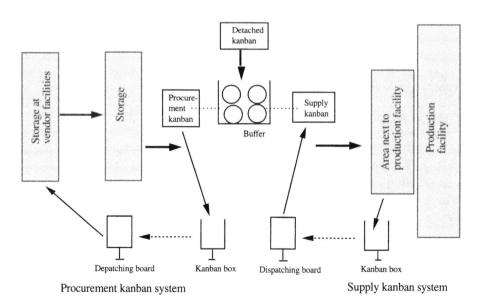

Figure 8.10. Kanban system receiving parts from external vendors.

8.1.6.3 Determining the Number of Kanbans. It is important to have an accurate number of kanbans so that the WIP is minimized and simultaneously the out-of-stock situation is avoided. The number of kanbans between adjacent stations impacts the inventory level between these stations. An estimated number of kanbans issued for a particular part is calculated using the following formula:

$$\text{Number of kanbans} = \frac{\text{unit daily demand} \times \text{order cycle time} \times \text{safety factor}}{\text{lot size}}$$

The unit daily demand refers to the daily production rate of the part. The order cycle time is the time it takes to process the part to procure a purchased item. The safety factor is a percentage increase in the number of kanbans used as a precautionary measure for buffer inventories. The lot size is the number of parts to be fetched by the kanban if it is a withdrawal kanban or to be manufactured if it is a production kanban.

For example, assume that the production requirement for 100-Ω resistors is 6000 units a month. The cycle time for them is 14 days and the lot size used is 1000. Then

$$\text{Unit daily demand} = \frac{6000}{20} = 300 \text{ units/day}$$

Since the process is not stable, a safety factor of 1.25 is used. Using the formula for number of kanbans, we obtain

$$\text{Number of kanbans} = \frac{300 \times 14 \times 1.25}{1000} = 5.25$$

This means that we need four kanbans to run the process and one extra kanban is used as a buffer until the process is steady and predictable.

As shown in the above example, the planner has global information about production levels. This information is used to calculate the number of kanbans required supporting the schedule. Once the kanban system starts to operate, the planner will lose control of the status of the kanbans, and therefore tracking kanban status is one of the main problems that a material planner faces in such a system. As the number of kanbans increases and the traffic becomes intense, the planner would spend considerable time in tracking kanban status.

One of the ways to reduce the additional work load in tracking kanbans is not to keep a kanban uniquely associated with a particular part lot. The only information required is the number of kanbans issued for a part and the size of the lot they represent. It should not matter which kanban numbers are at a particular location. For example, there are two kanbans for 250-Ω resistors in a particular location, and these kanbans represent a lot of 100 resistors each. There is no need to know whether these are kanbans 2 and 4 or kanbans 3 and 6. The only important information is that there are two kanbans in that location with a total of two hundred 250-Ω resistors.

It is important that kanbans meet their estimated lead times, because this impacts supply of parts to subsequent processes. One of the ways to track kanban lead times is to provide every station with an estimate of how long it would take to process a kanban. The operators report any deviation from that estimate. Then, tracking would be based on the default cases. A simple kanban reporting system would require the operators to report the number of kanbans at hand and the number of kanbans that are past due. This reporting should be done at a particular time of the day. Such a report gives the material planner an overview of the status of the materials and provides the opportunity to detect supply problems in advance if the number of kanbans expected in a station are not adequate to meet the production rate.

The calculation of the cycle time used in the formula determining the number of kanbans is illustrated next. For the data in Figure 8.11, the cycle time for kanbans (parts A, B, C) is $0.1 + 0.5 + 0.5 + 0.2 + 0.1 + 0.1 = 1.5$ days.

8.1.6.4 Kanban System Adjustments

1. Insertion Maintenance Action. Insertion maintenance takes place when the number of kanbans used in a month is larger than the number of kanbans used in the previous month. Additional kanbans are introduced to the system immediately after withdrawing the production kanbans and placing them on the dispatching board.

2. Removal Maintenance Action. Removal maintenance, similar to the insertion maintenance, takes place when the number of kanbans used in the current planning period is smaller than the number of kanbans used in the previous planning period. The additional kanbans are always removed immediately after withdrawing the production kanbans and removal of an equivalent number of kanbans from the dispatching board.

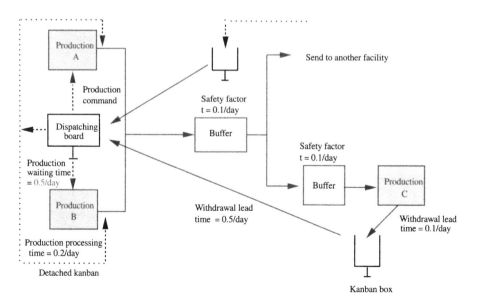

Figure 8.11. Cycle time of kanbans.

8.2 MODELING KANBAN SYSTEMS

Previous research on kanban pull systems included simulation, mathematical, and stochastic models (Uzsoy and Martin-Vega, 1990). The research published has mainly concentrated on modeling kanban systems in a repetitive (job shop) environment, determining the number of kanbans in order to optimize the system performance and comparing kanban systems.

8.2.1 Basic Kanban Models

Simulation Models. The simulation studies of JIT kanban systems can be broadly classified as (1) explorative analysis of pull systems (JIT with kanban) and (2) comparative analysis of push and pull systems. For reviews of simulation models, see, for example, Chu and Shih (1992), Krajewski et al. (1987), and Jothishankar and Wang (1993).

Mathematical Programming Models. Deterministic mathematical programming models are used to optimize some objective functions of the kanban system (e.g., Li and Co, 1991; Bard and Golany, 1991). This approach is suitable for a JIT kanban system since the repetitive environment is deterministic. However, it might not be appropriate in a dynamic environment.

Stochastic Models. In the stochastic approach, the pull demand and the processing time are modeled as random variables. Markov chains are often used to describe the system behavior. The Poisson process arrivals and exponential processing times are the general assumptions (Mitra and Mitrani, 1990; Deleersnyder et al., 1989; Buzacott, 1989).

Design Methodologies. Different methodologies for design of kanban systems have been studied in the literature; for example:

- Design with the server network generator (SNG) in Bouchentouf-Idriss and Zeidner (1991).
- Design with Petri nets in Di Mascolo et al. (1991).

Optimizing the Number of Kanbans. Most studies have concentrated on operational control problems and performance analysis of JIT manufacturing systems, emphasizing the determination of the number of kanbans (Deleersnyder et al., 1989; Berkley, 1987; Bitran and Chang, 1987; Huang et al., 1983; Jordan, 1988; Kim, 1985; Rees et al., 1987; Sarker and Harris, 1988; Sarker and Fitzsimmons, 1989; Villeda et al., 1988; So and Pinault, 1988; Kimura and Terada, 1981; Krajewski et al., 1987; Price et al., 1995).

8.2.2 Control Approaches

Most studies have focused on shop control methodology, allocations of a fixed number of kanbans and buffers, and batch size control.

Chaudhury and Whinston (1990) presented an efficient, decentralized, and adaptive control methodology for flow shops. The methodology is based on stochastic automata methods for modeling learning behavior. It was suggested that such a methodology could be used with a kanban-type control technique to make flow shop systems more flexible and adaptive. The relationship between the control model and computational models such as neural computing was discussed.

Cheng (1993) proved that with a general arrival process and exponential service times, job completion, job departure, and kanban generation processes are increasing concave functions of the initial inventory and kanban counts.

Tayur (1993) studied the structural properties and a heuristic for kanban-controlled serial lines and concluded with:

1. The optimal solutions in the allocation and partitioning problems given a fixed number of kanbans.
2. The reduced computational effort usually required studying these systems.
3. The development of a combinatorial measure as a surrogate for the mean throughput based on structural results; for example, in a five-cell line to be allocated, $(1, C_1, C_2, C_3, 1)$ is better than other allocations.
4. The demonstration of which structure with optimal allocations is insensitive to the variability in the system with balanced lines.

8.2.3 Scheduling Approaches

Most studies have concentrated on leveling the schedules for the mixed model (Miltenburg, 1989; Miltenburg and Sinnamon, 1989; Kubiak and Sethi, 1991). Garey et al. (1988) and Inman and Bulfin (1991) studied the problem of minimizing the total earliness and tardiness of schedules.

8.2.4 Comparing Kanban Systems with Other Systems

Numerous studies have compared kanban systems with the MRP system (Petroff, 1993; Hernandez, 1989; Dürmusoglu, 1991; Rees et al., 1989; Schonberger, 1982a,b, 1983; Grüwald et al., 1987; Sarker and Fitzsimmons, 1989). The stock (Q, r) policy and the tandem queuing model in the generalized semi-Markov processes were also compared with the kanban systems (Axsäter and Rosling, 1993; Berkley, 1987; Glasserman and Yao, 1994).

8.3 MODIFIED KANBAN SYSTEMS

The kanban system approach is difficult to use in certain situations, namely (Monden, 1983a):

- Job orders with short production runs
- Significant setups
- Presence of scrap
- Large, unpredictable fluctuations in demand
- The need for complex information and a hierarchical control system in the shop

Several modified models were developed to overcome those shortcomings of kanban systems.

8.3.1 Constant Work-in-Process Model

Reason. The kanban is intrinsically a system for repetitive manufacturing (Hall, 1981) and it is not appropriate for a shop controlled by job orders.

Model. Spearman et al. (1989) presented a new pull system called CONWIP (CONstant Work In Process). The WIP was kept constant by fixing the total number of kanbans in the system. The purpose of the model was to present a system that possesses the benefits of a pull system and could be used in different production environments.

Model Description. CONWIP is a generalization of the kanban system. Also, it is an integrated system that offers the benefits of JIT systems but is applicable to a broader range of production environments than the traditional JIT approach. CONWIP is focused on the interactions between the planning modules at the different levels in the hierarchy and on the architecture linking them. Like a kanban system, it relies on signals. A card is attached to a standard container of parts at the beginning of the process. When the container approaches the end of the process, the card is removed and sent back to the beginning where it waits in a card queue to eventually be attached to another container of parts. CONWIP production cards are assigned to the production line. Part numbers are assigned to the cards at the beginning of the production line. Figure 8.12 illustrates the operation of the CONWIP system.

Main Difference from the Kanban System

1. Backlog information is used to dictate the part number sequence.
2. Cards are associated with all parts produced on a line rather than individual part numbers.
3. Jobs are pushed between workstations in series once they have been authorized by a card to enter the line.

Results. Many of the benefits of CONWIP can be attributed to be fact that it is a pull-based production system (e.g., shorter flow times and reduced inventory levels). However, the system does offer some distinct advantages over the kanban system.

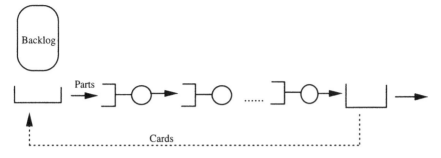

Figure 8.12. The CONWIP system.

One advantage is that it can be used in some production environments where the kanban is not practical due to too many card numbers or because of high setups. By allowing WIP to be collected in front of the bottleneck, CONWIP can function with lower WIPs and less production control personnel than in the kanban system. Spearman et al. (1990, 1992) concluded:

1. CONWIP is more general than a kanban system.
2. CONWIP is more effective than a kanban system.
3. CONWIP not only has better customer service (e.g., less tardy jobs than a pure kanban system) but also solves certain implementation problems (e.g., difficult-to-reduce setups) or optimizes synchronization of parts production.

8.3.2 Generic Kanban System

Reason. With variable demand and processing times, it is difficult to set the master schedule. Also, line balancing and synchronization in the receptive system are impossible to attain. A kanban operation is generally not applicable to a dynamic environment with variable demands and variable processing times (Hall, 1981; Huang et al., 1983; Finch and Cox, 1986; Krajewski et al., 1987).

Models
1. A dynamic environment may be changed (or simplified) toward the repetitive system and adopt the kanban control discipline. However, this requires significant changes in the system (Huang et al., 1983; Finch and Cox, 1986), which is not practical in many cases because many companies cannot afford to do so.
2. Chang and Yih (1994a) proposed a generic kanban system—a modified kanban discipline—for dynamic environments. The generic kanban system does not have all the benefits of the JIT kanban system. However, it is adaptable because

it has advantages over other production systems under the same dynamic conditions.

Model Description. To operate a generic kanban system, the number of kanbans and lot sizes used in the system needs to be determined. The number of kanbans and job lot size directly affect system performance. One approach to determine the number of kanbans at each station and lot sizes of job types to optimize the generic kanban system performance was proposed by Chang and Yih (1994b). This approach included formulating the multiobjective optimization problem with a utility function and searching the maximum utility value with a modified simulated annealing (SA) algorithm.

A generic kanban system includes two phases: acquisition and production.

Acquisition Phase. In the generic kanban system, the demand is unknown due to the dynamic environment. When a demand arrives in the system, kanbans have to be issued for all stages because no parts at any stage are made beforehand. Only when the raw material arrives at the initial station can the actual production of the system begin. Moreover, not every kanban at any stage can be issued immediately since the number of kanbans at each stage is limited. A request may be deferred if at a particular stage kanbans are not available.

Production Phase. When a job finishes processing at one stage, it is moved to the next downstream stage and the attached kanban at this stage is dropped. This kanban is acquired by the next demand request.

Main Difference from the Kanban System. The production phase is different from a JIT kanban system where the free kanban triggers a new production immediately because products are made repetitively in this environment.

Results. To show the adaptability and superiority of such a system, other control methodologies such as push systems, dedicated kanban systems, and CONWIP systems were studied and compared. The simulation results are listed below (Chang and Yih, 1994b):

1. A trade-off exists between cycle time and WIP level in generic kanban systems. The lot size has an impact on the system performance.

2. The generic kanban system behaves similarly to a push system except that a decision maker has more flexibility in relocating resources.

3. The performance of the generic kanban system is preferred to the dedicated kanban system because it provides simpler production control and dominates the performance (e.g., less WIP). It is also preferable to CONWIP because of higher flexibility (e.g., more jobs are allowed to enter the system). The SA algorithm is shown to provide similar solutions in a shorter time in the generic kanban system than in the traditional one.

8.3.3 Modified Kanban System for Semiconductor Manufacturing

Reasons. A conceptually pure kanban system is not suitable for a semiconductor fabrication due to the nature of the manufacturing process. The current systems are designed to prevent uneven line loading resulting from various operational problems. Even after the fundamental problem has been solved, the residual impact on production due to disrupted WIP flow could last for weeks. Thus, poor line loading leads directly to increased cycle times, poor predictability, and more defective products.

Models

1. Otenti (1991) offered a modified kanban WIP control system successfully implemented in a complementary metal–oxide–semiconductor (CMOS) fabrication facility. The approach was to set up a series of kanbans with caps on those lots allowed to enter the system. No additional lots would be allowed to move into a kanban system if the WIP level in that system had reached the maximum allowable limit.

2. Kraft (1992) described a tool that is being currently used at Texas Instruments and with a modified kanban JIT scheduling incorporated to improve the line balancing and WIP flow.

Results

1. Cycle time dropped from 44 days to 30 days, a 32% improvement.
2. Cycle time reduced by more than 36%.

8.3.4 Integrated Push–Pull Manufacturing Strategy

Reasons

1. A pull strategy is not necessarily applicable to all manufacturing environments.
2. Many manufacturing firms using pull systems are interested in attaining the simplicity of push systems.

Model. Olhager and Östlund (1990) combined a push–pull system into a system through three points, the customer order point (i.e., the point where a product is assigned to a specific customer), the bottleneck resources, and the product structure.

Results. In the integrated push–pull system, the major issue is the linkage of the manufacturing strategy with the business strategy. Changing the manufacturing planning and control focus can solve the issue. In the new system, a push principle is applied to the focused machines (bottleneck machines) and succeeding production stages, and incoming parts are pulled. This has resulted in improved dependability of delivery and production flexibility. A case study in a semirepetitive, make-to-order environment illustrates some potential benefits from such an integrated approach.

8.3.5 Periodic Pull System

Using the kanban system, manufacturing factories at Toyota no longer needed to rely on a computer. The reasons for having employed a kanban system instead of a computerized system were as follows:

- Reduction of the cost of processing information
- Rapid and precise acquisition of facts
- Limits on surplus capacity at feeding facilities

Reasons

1. In present management systems, the volume and complexity of information have increased.
2. For some manufacturing environments, computerization is necessary.

Model. Kim (1985) proposed an alternative to the kanban system, a period pull system (PPS), as an operation policy of practicing a pull system. In the PPS, the manual information processing time of a kanban system is replaced with an instant on-line computerized processing.

Model Description. In a computerized material management system, the status of material flow at all stages is reviewed at regular intervals. As the result of the review, only the exact amount of material that has been consumed at a succeeding stage (since the last review time) is allowed to be withdrawn from or produced at a preceding stage. A review interval is called a period. The time for a review is assumed to be nonnegative (i.e., computer processing time). The withdrawal and production start immediately after the review, that is, at the beginning of the period. A PPS is formulated mathematically and a solution approach is provided for target stock levels, as well as the analysis of the fluctuations of in-process material flow, on-hand stock levels, target availability, and so on.

Analogy. One may visualize that a review time is equivalent to a kanban pick-up time, and thus, in a PPS, the imaginary kanbans picked at a review time are delivered to a preceding stage at the same review time, that is, instantly.

Results. The material lead time is much shorter than that of a kanban system, and the system performance improves in terms of less inventory and faster system response.

8.3.6 Case Study

Graham (1992) developed a steady-state Markovian model for calculating the number of kanbans required to control single-stage processes feeding assembly lines. A Markovian model of an alternative JIT system, in which the off-line process is triggered by the passage of vehicle bodies past a point prior to the assembly area, is also described.

Results

1. These models have shown that the use of a trigger system leads to lower inventory levels and a greater pressure for improvement than in the kanban system itself.

2. In a kanban system the level of subassembly inventory required is insensitive to changes in the rate and average duration of body rework, whereas with the triggered system the average level of subassembly inventory is sensitive to both the rate of body rework and the duration of subassembly rework.

3. The only incentive of a kanban system is to reduce the rate and duration of body rework positively correlated with the value of inventory. However, in the triggered system a reduction in the expected level of body rework inventory may reduce the inventory level of all triggered subassemblies.

4. For example, a 50% reduction in the rate of body reworking and a 50% reduction in rework time both lead to a reduction from 69 to 67 in the expected average number of engines of this type in inventory if triggering is being used, whereas with the kanban system, 80 engines would still be needed.

8.4 SUMMARY

- A variety of kanban systems have been proposed in the literature and applied in industry.

- The concept of kanban systems is not a panacea for all industrial problems. It is applicable to a repetitive manufacturing environment. Furthermore, the key to improving manufacturing performance is to consider such factors as lot sizes, setup times, yield losses, workforce flexibility, degree of product customization, and product structure, to *shape* a manufacturing environment with more uniform work flow and flexibility. The kanban system, by itself, is not crucial for improving manufacturing performance.

- The model of kanban operations in its simple form is a stock (Q, r) policy or a tandem queue. However, together with automation (Jidoka), setup reduction, flexibility of workforce, and employment empowerment, the kanban system offers numerous advantages.

- Decreasing the lot size is an effective way to reduce the mean length and waiting time in WIP points at all kanban levels that combine kanbans and production stations.

- The optimal allocation structure of a fixed number of kanbans is insensitive to the variability in the system with balanced production lines.
- The inventory function in the kanban system is to stabilize the demand rather than balance the setup cost.
- For a kanban system to operate effectively, it is crucial that the delivery times and quality of the downstream suppliers are reliable.
- In most practical approaches, the product/process design was not modified before implementing the JIT kanban system concept.

The issues that need further research are categorized as follows:

- Design of products and processes for a JIT kanban system.
- Development of a general model that has the advantages of kanban systems can be integrated with manufacturing systems of different types and applicability of the integrated concept to a nonrepetitive manufacturing environment.
- The problem of production leveling through scheduling is crucial in kanban systems. Selecting the proper scheduling rules becomes even more important in case of high product variety and uncertainty of processing times.
- Introduction of feeder lines into the pull system configuration adds flexibility in adjusting to the lumpy demand and flow synchronization.
- Development of optimal bounding schemes for the *sum (minmax) objective function* in the leveling schedule problem.
- The trade-off cost between more frequent material handling and benefits of reduced WIP when the optimal number of kanbans is to be determined. Most previous studies only considered the minimization of the throughput/WIP and ignored the minimization of total cost when the optimal number of kanbans was determined.

REFERENCES

Albino, V., G. Carella, and O. G. Okogbaa (1992), Maintenance policies in just-in-time manufacturing lines, *International Journal of Production Research*, Vol. 30, No. 2, pp. 369–382.

Axsüter, S., and K. Rosling (1993), Notes: Installation vs. echelon stock policies for multilevel inventory control, *Management Science*, Vol. 39, No. 10, pp. 1274–1280.

Bard, J. F., and B. Golany (1991), Determining the number of kanbans in a multiproduct, multistage production system, *International Journal of Production Research*, Vol. 29, No. 5, pp. 881–895.

Berkley, B. J. (1987), Stochastic kanban-controlled lines and tandem queues, Working Paper, Department of Decision Sciences, School of Business Administration, University of Southern California, Los Angels, CA.

Berkley, B. J. (1993), Effect of buffer capacity and sequencing rules on single-card kanban system performance, *International Journal of Production Research*, Vol. 31, No. 12, pp. 2875–2894.

Bitran, G. R., and L. Chang (1985), An optimization approach to the kanban system, Working Paper, No. 1635-85, Sloan School of Management, Massachusetts Institute of Technology, Cambridge, MA.

Bitran, G. R., and L. Chang (1987), A mathematical programming approach to a deterministic kanban system, *Management Science*, Vol. 33, No. 4, pp. 427–441.

Bouchentouf-Idriss, A., and L. Zeidner (1991), Design a kanban manufacturing system using the server network generator (SNG) CASE tool, *APL Quote QUAD*, Vol. 21, No. 4, pp. 62–70.

Buzacott, J. A. (1989), Queuing models of kanban and MRP controlled production systems, Working Paper, Department of Management Sciences, University of Waterloo, Waterloo, Ontario, Canada.

Chang, T. M., and Y. Yih (1994a), Generic kanban systems for dynamic environments, *International Journal of Production Research*, Vol. 32, No. 4, pp. 889–902.

Chang, T. M., and Y. Yih (1994b), Determining the number of kanbans and lot sizes in a generic kanban systems: A simulated annealing approach, *International Journal of Production Research*, Vol. 32, No. 8, pp. 1991–2004.

Chase, R. B., and N. J. Aquilano (1985), *Production and Operations Management: A Life Cycle Approach*, Richard D. Irwin, Homewwod, IL.

Chaudhury, A., and A. B. Whinston (1990), Toward an adaptive Kanban system, *International Journal of Production Research*, Vol. 28, No. 3, pp. 437–458.

Cheng, D. W. (1993), Concavity of system throughput and information transfer in a PAC system, *Operations Research Letters*, Vol. 14, No. 3, pp. 143–146.

Chu, C. H., and W. L. Shih (1992), Simulation studies in JIT production, *International Journal of Production Research*, Vol. 30, No. 11, pp. 2573–2586.

Davis, W. J., and S. J. Stubitz (1987), Configuring a kanban system using a discrete optimization of multiple stochastic responses, *International Journal of Production Research*, Vol. 25. No. 5, pp. 721–740.

Deleersnyder, J. L., T. J. Hodgson, H. Muller (-Malek), and P. J. O'Grady (1989), Controlled pull systems: An analytic approach, *Management Science*, Vol. 35, No. 9, pp. 1079–1091.

Di Mascolo, M., Y. Frein, Y. Dallery, and R. David (1991), A unified modeling of kanban systems using Petri nets, *International Journal of Flexible Manufacturing Systems*, Vol. 3, Nos. 3/4, pp. 275–307.

Dürmusoglu, B. M. (1991), Comparison of push and pull systems in a cellular manufacturing environment, in A. Satir (Ed.), *Just-in-Time Manufacturing Systems: Operational Planning and Control Issues*, Proceedings of Manufacturing Research and Technology Conference, Elsevier, Amsterdam, The Netherlands, pp. 115–132.

Ebrahimpour, M., and B. M. Fathi (1985), Dynamic simulation of a kanban production inventory system, *International Journal of Operations and Production Management*, Vol. 5, No. 1, pp. 5–14.

Elmaghraby, S. E. (1978), The economic lot scheduling problem (ELSP): Review and extensions, *Management Science*, Vol. 24, pp. 587–598.

Finch, B. J., and J. F. Cox (1986), An examination of just-in-time management for the small manufacturing: With an illustration, *International Journal of Production Research*, Vol. 24, No. 2, pp. 329–342.

Garey, M., R. Tarjan, and G. Wilfong (1988), One-processor scheduling with symmetric early and tardiness penalties, *Mathematics Operations Research*, Vol. 13, No. 2, pp. 330–344.

Glasserman, P., and D. D. Yao (1994), *Monotone Structure in Discrete-Event Systems*, John Wiley, New York.

Graham, I. (1992), Comparing trigger and kanban control of flow-line manufacture, *International Journal of Production Research*, Vol. 30, No. 10, pp. 2351–2362.

Gravel, M., and W. L. Price (1988), Using the kanban in a job shop environment, *International Journal of Production Research*, Vol. 26, No. 6, pp. 1105–1118.

Grüwald, H., P. E. T. Striekwood, and P. J. Weeda (1987), A frame-work for quantitative comparison of production control concepts, *International Journal of Production Research*, Vol. 27, No. 2, pp. 281–292.

Gupta, Y. P., and M. C. Gupta (1989a), A system dynamic model of a JIT-kanban system, *Engineering Costs and Production Economics*, Vol. 18, No. 2, pp. 117–130.

Gupta, Y. P., and M. C. Gupta (1989b), A system dynamics model for a multi-stage-line dual-card JIT kanban system, *International Journal of Production Research*, Vol. 27, No. 2, pp. 309–352.

Gupta, Y. P., W. G. Mangold, and S. C. Lonial (1991), An empirical examination of the characteristics of JIT manufacturers versus non-JIT manufacturers, *Manufacturing Review*, Vol. 4, No. 2, pp. 78–86.

Hall, R. W. (1981), *Driving the Productivity Machine: Production and Control in Japan*, American Production and Inventory Control Society, Alexandria, VA.

Hall, R. W. (1983), *Zero Inventory*, Dow-Jones-Irwin, Homewood, IL.

Harvey, D. A., and S. W. Jones (1989), Design a kanban system in an aerospace environment, *Proceeding of the International Conference on Implementing Flexible Manufacturing—Challenges for Organization and Education in a Changing Europe*, Amsterdam, Netherlands, pp. 151–161.

Hernandez, A. (1989), *Just-in-Time Manufacturing: A Practice Approach*, Prentice-Hall, Englewood Cliffs, NJ, pp. 14–20.

Huang, C. C., and A. Kusiak (1996), Overview of kanban systems, *International Journal of Computer Integrated Manufacturing*, Vol. 9, No. 3, pp. 169–189.

Huang, P. Y., L. P. Rees, and B. W. Taylor III (1983), A simulation analysis of the Japanese JIT techniques (with kanban) for multiline, multistage production systems, *Decision Sciences*, Vol. 14, No. 3, pp. 326–344.

Ichihashi, E. (1990), An example of CIM in Nippondenso Kota plant, *System Control and Information*, Vol. 34, No. 3, pp. 185–187.

Inman, R., and R. Bulfin (1991), Sequencing JIT mixed-model assembly lines, *Management Science*, Vol. 37, No. 7, pp. 192–207.

Japanese Management Association (1986), *Kanban: Just-in-Time at Toyota*, Productivity Press, Stamfort, CT.

Jordan, S. (1988), Analysis and approximation of a JIT production line, *Decision Sciences*, Vol. 19, No. 3, pp. 672–681.

Jothishankar, M. C., and H. P. Wang (1992), Determination of optimal number of kanban using stochastic Petric Nets, *Journal of Manufacturing Systems*, Vol. 11, No. 6, pp. 449–461.

Jothishankar, M. C., and H. P. Wang (1993), Metamodelling a just-in-time kanban system, *International Journal of Operations and Production Management*, Vol. 13, No. 8, pp. 18–36.

Karmarkar, U. S., and S. Kekre (1989), Batching policy in kanban systems, *Journal of Manufacturing Systems*, Vol. 8, No. 4, pp. 317–328.

Kim, T. (1985), Just-in-time manufacturing system: A periodic pull system, *International Journal of Production Research*, Vol. 23, No. 3, pp. 553–562.

Kimura, O., and H. Terada (1981), Design and analysis of pull system, a method of multistage production control, *International Journal of Production Research*, Vol. 19, No. 3, pp. 241–253.

Kraft, C. (1992), Dynamic kanban semiconductor inventory management system, paper presented at the 1992 IEEE/SEMI Advanced Semiconductor Manufacturing Conference & Workshop, Cambridge, MA, pp. 30–35.

Krajewski, L. J., B. E. King, L. P. Ritzman, and D. S. Wang (1987), Kanban, MRP, and shaping the manufacturing environment, *Management Science*, Vol. 33, No. 1, pp. 39–57.

Kubiak, W., and S. Sethi (1989), A note on level schedules for mixed-model assembly lines in just-in-time production system, *Management Science*, Vol. 37, No. 1, pp. 901–904.

Li, A., and H. C. Co (1991), A dynamic programming model for the kanban assignment problem in a multistage multi-period production system, *International Journal of Production Research*, Vol. 29, No. 1, pp. 1–16.

Lu, Y. Y. (1982), *Introduction to Industrial Engineering*, Hua-Tai, Taipei, Taiwan.

Miltenburg, J. (1989), Level schedules for mixed-model assembly lines in just-in-time production system, *Management Science*, Vol. 32, No. 2, pp. 121–122.

Miltenburg, J., and G. Sinnamon (1989), Scheduling mixed-model, multilevel assembly lines in just-in-time production system, *International Journal of Production Research*, Vol. 27, No. 9, pp. 1487–1510.

Mitra, D., and I. Mitrani (1990), Analysis of a kanban discipline for cell coordination in production lines, *Management Science*, Vol. 36, No. 12, pp. 1548–1566.

Mitra, D., and I. Mitrani (1991), Analysis of a kanban discipline for cell coordination in production lines. II. Stochastic demands, *Operations Research*, Vol. 39, No. 5, pp. 807–823.

Mitwasi, M. G., and R. G. Askin (1994), Production planning for a multi-item, single-stage kanban system, *International Journal of Production Research*, Vol. 32, No. 5, pp. 1173–1195.

Monden, Y. (1981a), What makes the Toyota production system really tick, *Industrial Engineering*, Vol. 13, No. 1, pp. 36–46.

Monden, Y. (1981b), Adaptable kanban system helps Toyota maintain just-in-time production, *Industrial Engineering*, Vol. 13, No. 5, pp. 29–46.

Monden, Y. (1983a), *Toyota Production System: Practical Approach to Production Management*, Industrial Engineering and Management Press, Norcross, GA.

Monden, Y. (1983b), *Applying Just in Time*, Industrial Engineering and Management Press, Norcross, GA.

Ohno, K., K. Nakashima, and M. Kojima (1995), Optimal numbers of two kinds of kanbans in a JIT production system, *International Journal of Production Research*, Vol. 33, No. 5, pp. 1387–1402.

Ohno, T. (1988), *Toyota Production System: Beyond Large-Scale Production*, Productivity Press, Cambridge, MA.

Olhager, J., and B. Östlund (1990), An integrated push-pull manufacturing strategy, *European Journal of Operational Research*, Vol. 45, No. 2/3, pp. 135–142.

Otenti, S. (1991), A modified kanban system in a semiconductor manufacturing environment, paper presented at the 1991 IEEE/SEMI Advanced Semiconductor Manufacturing Conference & Workshop, Boston, MA, pp. 43–45.

Pervozvanskiy, A. A., and I. Y. Sheynis (1994), A kanban system as a manufacturing control system with feedback, *International Journal of Computer and Systems Science*, Vol. 32, No. 4, pp. 153–157.

Petroff, J. N. (1993), *Handbook of MRP II and JIT: Strategies for Total Manufacturing Control*, Prentice-Hall, Englewood Cliffs, NJ, pp. 5–13.

Philipoom, P. R., L. P. Rees, B. W. Taylor III, and P. Y. Huang (1987), An investigation of the factors influencing the number of kanbans required in the implementation of the JIT technique with kanbans, *International Journal of Production Research*, Vol. 25, No. 3, pp. 457–472.

Price, W., M. Gravel, A. L. Nsakanda, and F. Cantin (1995), Modeling the performance of a kanban assembly shop, *International Journal of Production Research*, Vol. 33, No. 4, pp. 1171–1177.

Rees, L. P., P. Y. Huang, and B. W. Taylor III (1989), A comparative analysis of an MRP lot-for-lot system and a Kanban system for a multistage production operation, *International Journal of Production Research*, Vol. 27, No. 8, pp. 1427–1443.

Rees, L. P., P. R. Philipoom, B. W. Taylor III, and P. Y. Huang (1987), Dynamically adjusting the kanbans in a just-in-time production system using estimated values of lead time, *IIE Transactions*, Vol. 19, No. 2, pp. 199–137.

Sarker, B. R. (1989), Simulation a just-in-time production system, *Computers and Industrial Engineering*, Vol. 16, No. 1, pp. 127–137.

Sarker, B. R., and J. A. Fitzsimmons (1989), The performance of push and pull systems: A simulation and comparative study, *International Journal of Production Research*, Vol. 27, No. 10, pp. 1715–1731.

Sarker, B. R., and R. D. Harris (1988), The effect of imbalance in a just-in-time production system: A simulation study, *International Journal of Production Research*, Vol. 26, No. 1, pp. 1–18.

Schonberger, R. J. (1982a), *Japanese Manufacturing Techniques*, Free Press, New York.

Schonberger, R. J. (1982b), Inventory control in Japanese industry, in S. M. Lee and T. Schwendiman (Eds.), *Management by Japanese Systems*, Praeger, New York.

Schonberger, R. J. (1983), Application of single-card and dual-card kanban, *Interfaces*, Vol. 13, No. 4, pp. 56–67.

Shingo, S. (1987), *The Saying of Shigeo Shingo: Key Strategies for Plant Improvement*, Productivity Press, Cambridge, MA.

Shipper, D., and R. Shapira (1989), JIT vs. WIP—a trade-off analysis, *International Journal of Production Research*, Vol. 27, No. 6, pp. 903–914.

Siha, S. (1994), The pull production system: Modeling and characteristics, *International Journal of Production Research*, Vol. 32, No. 4, pp. 933–950.

Singh, N., and D. R. Falkenburg (1994), Kanban systems, in R. C. Dorf and A. Kusiak, (Eds.), *Handbook of Design, Manufacturing, and Automation*, John Wiley, New York, pp. 567–585.

So, K. C., and S. C. Pinault (1988), Allocating buffer storages in a pull system, *International Journal of Production Research*, Vol. 26, No. 12, pp. 1959–1980.

Sohal, A. S., and D. Naylor (1992), Implementation of JIT in a small manufacturing firm, *Production and Inventory Management Journal*, Vol. 33, No. 1, pp. 20–26.

Sohal, A. S., G. Lewis, and D. Samson (1993), Integrating CNC technology and the JIT kanban system: A case study, *International Journal of Technology Management*, Vol. 8, No. 3/5, pp. 422–431.

Spearman, M. L. (1992), Customer service in pull production systems, *Operations Research*, Vol. 40, No. 5, pp. 948–958.

Spearman, M. L., D. L. Woodruff, and W. J. Hopp (1989), A hierarchical architecture for production systems, *Journal of Manufacturing and Operations Management*, Vol. 2, No. 3, pp. 147–171.

Spearman, M. L., D. L. Woodruff, and W. J. Hopp (1990), CONWIP: A pull alternative to kanban, *International Journal of Production Research*, Vol. 28, No. 5, pp. 879–894.

Sugimori, Y., K. Kusunoki, F. Cho, and K. Uchikawa (1977), Toyota production system and kanban system: Materialization of just-in-time and respect-for-human system, *International Journal of Production Research*, Vol. 15, No. 6, pp. 553–564.

Tayur, S. R. (1993), Structural properties and a heuristic for kanban-controlled serial lines, *Management Science*, Vol. 39, No. 11, pp. 1347–1368.

Uzsoy, R., and L. A. Martin-Vega (1990), Modeling kanban-based demand-pull systems: A survey and critique, *Manufacturing Review*, Vol. 3, No. 3, pp. 155–160.

Villeda, R., R. Dudek, and M. L. Smith (1988), Increasing the production rate of a just-in-time production system with variable operations times, *International Journal of Production Research*, Vol. 26, No. 11, pp. 1749–1768.

QUESTIONS

8.1. What is a kanban system?

8.2. What kanban types do you know?

8.3. What basic information is contained in a kanban?

8.4. What is the difference between single- and dual-kanban systems?

8.5. What is a semi-dual-kanban system?

8.6. What data are needed to estimate the number of kanbans?

8.7. What are some approaches used to model kanban systems?

8.8. Why are kanban systems called pull systems?

8.9. What type of production system is suitable for kanban implementation?

8.10. What is the relationship between the kanban and MRP concepts?

PROBLEMS

8.1. Consider the kanban system in Figure 8.13.

(a) What type of kanban system is it?

(b) What do W and P mean?

8.2. An assembly system is to process 100 circuits per hour. It takes 20 minutes to receive the necessary components from the preceding workstation. Completed circuits are placed on a rack that holds 10 boards. The rack must be full before it is sent to the next station. If the factory uses a safety factor of 10%, how many kanbans are needed for the circuit board assembly process?

8.3. Corporation AA produces three models of chemical analyzers—labeled A, B, and C—using a three-stage batch flow process. The company uses a JIT production system that produces each model in lots of 50 units at a time. A one-hour setup is required at each stage while changing the system from one model to another. The company can process 10 units per hour at each stage when the stage is operating. Table 8.2 shows the operation of the factory during the past eight-hour shift. In the table fill in the demands listed at the next shift, indicating what is being done at each production stage, where and when kanbans are issued, how materials flow between stages, and the inventory level at each stage and for the final products. Assume that production completed at one stage is available at the next stage in the next hour. Raw material requested one day is received from the supplier at the beginning of the next day (regardless of when they were requested during the previous day).

8.4. An assembly system is to process 100 circuits per hour. It takes 20 minutes to receive the necessary components from the preceding workstation. Completed circuits are placed on a rack that holds 10 boards. The rack must be full before it is sent to the next station. If the factory uses a safety factor of 10%, how many kanbans are needed for the circuit board assembly process?

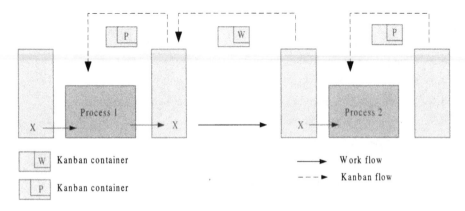

Figure 8.13. Kanban system.

TABLE 8.2. Production Data

Day-Hour	Raw Matl	Stage 1 Activ.	Stage 1 Kanban	Stage 1 Inv.	Stage 2 Activ.	Stage 2 Kaban	Stage 2 Inv.	Stage 3 Activ.	Stage 3 Kanban	Beginning Product Inventory A	B	C	Demand A	B	C
1-1	BAC	Run B	CA	ABC	Su C	BA	BC	Run A	CB	60	80	90	5	10	0
1-2	BAC	Run B	CAC	AB	Run C	BA	BC	Run A	CB	55	70	90	10	5	10
1-3	BAC	Run B	CAC	AB	Run C	BA	BC	Run A	CB	45	65	80	5	5	5
1-4	BAC	SU C	AC	BAB	Run C	BA	BC	Run A	CBA	40	60	75	5	10	10
1-5	BA	Run C	AC	BAB	Run C	BA	BC	SU C	BA	85	50	65	5	0	0
1-6	BA	Run C	AC	BAB	Run C	BAC	B	Run C	BAB	80	50	65	5	5	5
1-7	BA	Run C	AC	BAB	SU B	AC	CB	Run C	BAB	75	45	60	0	0	5
1-8	BA	Run C	ACB	BA	Run B	AC	CB	Run C	BAB	75	45	55	5	5	0
2-1	—	—	—	—	—	—	—	—	—	—	—	—	0	5	5
2-2	—	—	—	—	—	—	—	—	—	—	—	—	5	0	5
2-3	—	—	—	—	—	—	—	—	—	—	—	—	5	5	0
2-4	—	—	—	—	—	—	—	—	—	—	—	—	5	0	5
2-5	—	—	—	—	—	—	—	—	—	—	—	—	5	5	5
2-6	—	—	—	—	—	—	—	—	—	—	—	—	0	0	5
2-7	—	—	—	—	—	—	—	—	—	—	—	—	5	5	0
2-8	—	—	—	—	—	—	—	—	—	—	—	—	5	0	5

Activ. = Activity
Inv. = Inventory

Note: Letters A–C represent 59 units of materials for products A–C or kanbans requesting 50 units of production for that product. ——→ material movement; ┈┈┈ information movements (kanban issued). SU = production setup. Run = products being processed.

271

CHAPTER 9

SELECTION OF MANUFACTURING EQUIPMENT

9.1 DESIGN OF MANUFACTURING SYSTEMS

The manufacturing industry has witnessed dramatic changes in recent years. The number of new products designed and new manufacturing systems installed are a measure of the developments that have taken place. An important factor to the successful implementations of modern manufacturing systems is proper selection and effective use of manufacturing equipment (i.e., machines, tools, fixtures, and material handling systems). Design of a manufacturing system is a complex task involving constraints, for example, space and proximity of the equipment to dedicated outlets.

A manufacturing system can be designed following the four phases given in Figure 9.1 (Kusiak, 1990).

9.1.1 Manufacturing Equipment Selection

Selecting the right type and number of manufacturing equipment may result in:

- Reduced procurement cost
- Reduced operating and maintenance cost
- Increased machine utilization

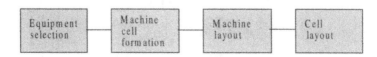

Figure 9.1. Basic phases in design of a manufacturing system.

- Improved layout of equipment
- Increased efficiency of a manufacturing system

The issues related to selection of manufacturing equipment have been analyzed and modeled using a number of approaches. Miller and Davis (1977) surveyed, classified, and compared a number of deterministic and stochastic approaches for the equipment selection problem. Some existing approaches consider assumptions that make them unsuitable for industrial applications. For example, the dynamic programming approach (Hayes et al., 1981) assumed that the machines in a manufacturing system are homogeneous.

9.1.2 Machine Cell Formation

Having selected the equipment in a newly designed manufacturing system, an attempt is made to form machine (product) cells. If product cells cannot be formed, a functional layout of machines is considered.

The concept of group technology (GT) is also often applied to the existing manufacturing facilities. Of course, the major concern here is whether physical rearrangement of machines or even organizational changes can be justified with the benefits of cellular manufacturing. The benefits of the functional and cellular approach as well as details of the layout of machines are discussed in subsequent chapters.

9.1.3 Machine Layout

Machine layout aims at determining the best arrangement of machines in each product cell. The minimization of material handling cost is an often used objective in determining the layout of machines in a cell. Constraints related to the availability of space, material handling system type, and so on are considered.

The type of operations and parts are not the only factors that impact the layout of machines. The type of material handling system to be used also needs to be considered; for example, the articulated robot (R) in Figure 9.2a implies a circular arrangement of machines. If an AGV had been selected to tend the same machines, it would have been necessary to use the layout in Figure 9.2b.

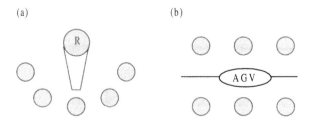

Figure 9.2. Layout patterns of machines.

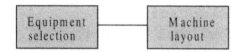

Figure 9.3. Two steps in the design of manufacturing systems.

9.1.4   Machine Cell Layout

The goal of machine cell layout is to arrange the product or functional cells formed on the factory floor. Determining the layout of machine cells involves locating the cells in order to minimize the total material handling cost subject to some constraints (e.g., shape of the facility). If all cells were square in shape and of the same size, then the cell layout could be modeled as the quadratic assignment problem (QAP). The cell layout problem can be viewed as a machine layout problem, where each machine represents a cell.

Although cellular manufacturing offers numerous benefits, it is not always implemented due to the following:

1. Parts and machines may not form mutually exclusive clusters.

2. The data required from the formation of cells might not be available.

In such cases, the four steps in the design of manufacturing systems reduce to two steps, shown in Figure 9.3, possibly more than once.

In practice, it is often necessary to repeat all or some of the steps illustrated in Figures 9.1 and 9.3.

9.2   SELECTION OF MACHINES AND MATERIAL HANDLING EQUIPMENT

9.2.1   Machine Selection

To date, numerous approaches for the selection of machines have been developed. The simplest way of selecting machines is based on the formula (8.1), which can be found in engineering handbooks:

$$M_i = \frac{\sum_{j=1}^{p} t_j}{hu} \tag{9.1}$$

where

M_i = number of machines of type i

t_j = processing time of part j in hours

p = number of parts

h = number of standard hours per day

u = scrap factor

Rounding off the number of machines M_i obtained from formula (9.1), the number of machines of type i would be determined. The formula (9.1) is simple to use; however, the decision made is only for one machine at a time.

To consider the interactions among parts and machines, the previously discussed set-covering model can be used. The application of this model for the selection of machines is illustrated in Example 9.1.

Example 9.1.   From the set of six machine types, select machines that are able to process five different operations. The relationship between operations and machines is defined in the matrix

$$
[a_{ij}] =
\begin{array}{c}
\\
o_1 \\
o_2 \\
o_3 \\
o_4 \\
o_5
\end{array}
\begin{array}{c}
\begin{array}{cccccc}
M_1 & M_2 & M_3 & M_4 & M_5 & M_6
\end{array} \\
\left[
\begin{array}{cccccc}
1 & 1 & & & 1 & 1 \\
1 & & & 1 & & \\
& 1 & 1 & & 1 & \\
& & & 1 & & \\
& & & & 1 &
\end{array}
\right]
\end{array}
\tag{9.2}
$$

$$
[t_j]\ [0.5 \quad 0.6 \quad 0.4 \quad 0.95 \quad 1.1 \quad 0.42]
$$

For the data in matrix (9.2), the model (9.3)–(9.5) is formulated. The objective function (9.3) of the set-covering model minimizes the total processing time of all operations:

$$
\min \sum_{j=1}^{6} t_j x_j
\tag{9.3}
$$

$$
\sum_{j=1}^{6} a_{ij} x_{ij} \geq 1 \qquad i = 1, \ldots, 5
\tag{9.4}
$$

$$
x_j = 0,\ 1 \qquad j = 1, \ldots, 6
\tag{9.5}
$$

Constraint (9.4) implies that each operation has to be performed at least once. The integrality of variable x_j is imposed by constraint (9.5).

Solving this model results in the selection of machines M_4 and M_5, i.e., $x_4 = x_5 = 1$, with the total processing time of 2.05.

The model (9.3)–(9.5) allows selecting a set of machines that satisfies the functional machining requirements of all operations.

9.2.2   Selection of Machines and Material Handling Systems

Matrix (9.2) can be generalized to consider machines and material handling equipment at the same time, the case which applies to many computer-integrated manufacturing systems. Consider the following generalized operation–manufacturing equipment incidence matrix (Kusiak, 1990):

$$R_1 = M_1 + H_1 \quad R_2 = M_2 + H_2 \quad R_3 = M_1 + H_3 \quad \cdots \quad R_n = M_p + M_p + H_r$$

$$[a_{ij}] = \begin{array}{c} o_1 \\ o_2 \\ \vdots \\ \vdots \\ o_m \end{array}
\left[\begin{array}{ccccc}
1 & & 2 & & \\
& 1 & 1 & & 1 \\
\cdot & \cdot & \cdot & & \cdot \\
\cdot & \cdot & \cdot & & \cdot \\
\cdot & \cdot & \cdot & \cdots & \\
1 & \cdots & 2 & \cdots &
\end{array} \right]$$

$$c_1 \qquad\qquad c_2 \qquad\qquad c_3 \qquad \cdots \qquad c_n \qquad (9.6)$$

Each row in the matrix corresponds to an operation and each column denotes a module. A module is a combination of machines and material handling equipment. The modules can be generated in groups by a knowledge-based system. Each group of modules (set of columns) is based on elements from a subset of machines and material handling carriers. This approach reduces the number of columns generated. The first three columns in the matrix (group of modules $F_1 = \{R_1, R_2, R_3\}$) can be interpreted as follows:

Column 1: lathe M_1 and robot H_1

Column 2: lathe M_2 and robot H_2

Column 3: lathe M_1, lathe M_2, and robot H_3

Each nonzero entry in the matrix has a value $a_{ij} \geq 1$. The value $a_{ij} = 1$ means that operation o_i can be performed by only one machine from module j. The value $a_{ij} > 1$ means that operation o_i can be performed by $a_{ij} > 1$ machines from module j, which is one of the basic requirements of modern manufacturing systems. Multiple machines allow rerouting parts in the case where one of them is down. In a typical manufacturing system, only a subset of all operations is assigned multiple machines.

To formulate the manufacturing equipment selection model, denote:

$m =$ number of operations

$n =$ number of equipment modules

$c_j =$ cost of equipment module j

$$a_{ij} = \begin{cases} l \geq 1 & \text{if operation } o_i \text{ can be performed } l \text{ times on all machines of module } j \\ 0 & \text{otherwise} \end{cases}$$

n_i = required (minimum) number of machines that can perform operation o_i

F_k = group k of modules

p = number of groups of modules

$$x_j = \begin{cases} 1 & \text{if equipment module } j \text{ is selected} \\ 0 & \text{otherwise} \end{cases}$$

Model (9.7)–(9.10) minimizes the total procurement cost of manufacturing equipment:

$$\min \sum_{=1}^{n} c_j x_j \tag{9.7}$$

$$\sum_{j=1}^{n} a_{ij} x_j \geq n_i \qquad \text{for all } i = 1, \ldots, m \tag{9.8}$$

$$\sum_{j \in F_k} x_j \leq 1 \qquad \text{for all } k = 1, \ldots, p \tag{9.9}$$

$$x_j = 0, 1 \qquad \text{for all } j = 1, \ldots, n \tag{9.10}$$

Constraint (9.8) ensures that each operation is performed at least the required number of times. Constraint (9.9) means that from each group of equipment modules, at most one module is selected. Constraint (9.10) ensures integrality.

Additional constraints can be incorporated into model (9.7)–(9.10), for example, machine capacity constraints (see Kusiak, 1990).

Model (9.7)–(9.10) is illustrated in Example 9.2.

Example 9.2. Determine the machines and material handling carriers required to build a manufacturing system for the following data:

1. Vector O of operations:

$$O = [o_i] = [o_1, o_2, o_3, o_4, o_5, o_6]$$

2. Vector R of equipment modules:

$$R = [R_j] = [R_1 = M_1 + H_1, R_2 = M_2 + H_2, R_3 = M_1 + M_2 + H_3, R_4 = M_4 + H_4,$$

$$R_5 = M_5 + H_5, R_6 = M_6 + H_6, R_7 = M_6 + M_7 + H_7]$$

3. Groups of modules: $F_1 = \{M_1, M_2, M_3\}$, $F_2 = \{M_4\}$, $F_3 = \{M_5\}$, $F_4 = \{M_6, M_7\}$

3. Vector C of costs of equipment modules:

$$C = [c_j] = [150{,}000 \quad 130{,}000 \quad 220{,}000 \quad 190{,}000 \quad 90{,}000 \quad 110{,}000 \quad 210{,}000]$$

4. Vector of the required number of machines that can perform operation o_i:

$$N = [n_i] = [2, 2, 1, 2, 2, 1]^{\mathrm{T}}$$

5. Generalized operation–equipment module incidence matrix:

	1	2	3	4	5	6	7
	$M_1 + H_1$	$M_2 + H_2$	$M_1 + M_2 + H_3$	$M_4 + H_4$	$M_5 + H_5$	$M_6 + H_6$	$M_6 + M_7 + H_7$
o_1	1	1	2				
o_2	1	1	2	1			
o_3	1		1	1			
o_4		1	1	1		1	1
o_5					1	1	2
o_6						1	2
	150,000	130,000	220,000	190,000	90,000	110,000	210,000

$[a_{ij}] = $ (matrix above)

$$(9.11)$$

For the data provided in this example, the following optimal solution has been generated (the model listing is provided in Appendix 9.1): $x_3 = 1$, $x_5 = 1$, $x_6 = 1$, corresponding to columns 3, 5, and 6 in matrix (9.11).

Based on matrix (9.11), the following manufacturing equipment is selected: machines M_1, M_2, M_5, M_6 and materials handling carriers H_3, H_5, H_6. The total cost of this equipment is $420,000.

9.2.3 Special Case of the Equipment Selection Model

Solving model (9.7)–(9.10) results in the selection of both machines and material handling carriers. In some cases, a purchasing decision regarding only either of these two has to be made. For example, consider the expansion of a machining system with AGVs. In this case, it is more than likely that the same form of material carriers (i.e., AGVs) will be used in the redesigned system.

The selection model for machines (or material handling carriers) can be formulated as a special case model (9.12)–(9.14):

$$\min \sum_{j=1}^{n} c_j x_j \tag{9.12}$$

$$\sum_{j=1}^{n} a_{ij}x_j \geq n_i \qquad \text{for all } i = 1, \ldots, m \qquad (9.13)$$

$$x_j = 0, 1 \qquad \text{for all } j = 1, \ldots, n \qquad (9.14)$$

This model [(9.12)–(9.14)] is known as the generalized (on coefficients $a_{ij} \geq 0$) multiple set-covering model.

9.3 SELECTION OF MANUFACTURING RESOURCES BASED ON PROCESS PLANS

In this section, manufacturing equipment is selected based on the information included in process plans. The previous two models utilized basically the relationship between operations and machines [the set-covering model (9.3)–(9.5)] and operations and machines and material handling equipment in model (9.7)–(9.10)]. The model discussed in this section includes more information in the process plans than the previous models. All resources included in the process plans may be considered. The model and solution approach discussed can be used for an existing manufacturing system as well as for a system being designed. In the case of an existing manufacturing system, the model allows us to determine the core resources required. When the model is used at the design stage of a manufacturing system, it selects the minimal set of resource types needed to deliver the production planned for the system. The manufacturing resources of concern here, besides the machines, include auxiliary equipment, such as (see Figure 9.4):

- Tools to machine or assemble parts
- Fixtures for holding parts to be machined or assembled
- Grippers for handling parts
- Feeders for presenting parts

Parts are machined or assembled using tools whose designs are different from the ones used in traditional manufacturing systems. For many parts to be manufactured in

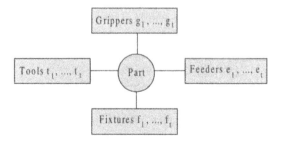

Figure 9.4. Auxiliary manufacturing equipment associated with a part.

a computer-integrated manufacturing system, a number of different process plans might be generated. Each process plan imposes requirements not only for machines but also for the auxiliary devices. One is interested in selecting from the set of available process plans a subset of process plans with the minimum cost of associated manufacturing resources (i.e., machines, fixtures, grippers, feeders, and tools).

There are at least three reasons for solving the equipment selection problem besides the selection of the equipment itself:

(a) Reduction of manufacturing cost

(b) Limited capacity of tool magazines

(c) Reduction of number of types of machines and auxiliary devices

While the first two reasons seem obvious, the last requires some elaboration. It may appear that the reduction of the number of different types of machines and auxiliary devices decreases the scheduling flexibility; however, it is likely to have an opposite effect. A manufacturing system with a smaller number of machines and auxiliary equipment is easier to manage than a large system. Since in such a system the bottleneck tools and devices are relatively easy to determine, if required, using multiple tools and auxiliary devices can easily increase the scheduling flexibility. This option is always less costly than the development of unique tools and devices.

To simplify the equipment selection, in this section, rather than all machines and auxiliary devices, only tools and fixtures are considered.

9.3.1 Model Background

In this section, a model is formulated for the selection of manufacturing equipment (tools and fixtures) based on the information included in process plans. The model for m parts is equivalent to the problem of finding the maximum clique in an m-partite graph with the minimum corresponding cost (Kusiak and Finke, 1988). The model minimizes the number of different types of equipment and the corresponding manufacturing cost.

The following notation is adopted in defining a process plan:

$V_l = (e_{l_1}, \ldots, e_{l_n})$, set of features to be removed in one setup, i.e., without resetting a part on the currently used fixture or changing the fixture

$V = $ set of all features to be removed for a given part (note that $V = \cup_{l=1}^m V_l$, where m is the total number of setups)

$T_l = (t_{l_1}, \ldots, t_{l_n})$ is a set of tools for removing feature V_l, $l = 1, \ldots, m$

$f_l = $ fixture for holding a part while removing feature V_l, $l = 1, \ldots, m$

Based on this notation, a process plan PP_i is defined as a vector of triplets:

$$PP_i = \{(V_l; T_l; f_l), \ldots, (V_m; T_m; f_m)\}$$

Without the loss of generality, machines have not been included in this definition. A machine or any other resource can be added as an element to any of the triplets $(V_k; T_k; f_k)$. The vector of resources associated with a process plan can be made as general as necessary. Of course, the above considerations and the model presented in the next section apply to other processes (e.g., electronics assembly) and services where the notion of a process plan is used.

The above process plan definition is illustrated in Example 9.3.

Example 9.3.   Consider the rotational part in Figure 9.5 with features $v_1, \ldots, v_{15}$ removed. The following two process plans can be generated for this part:

$$PP_1 = \{(v_6, v_{11}, v_{15}; t_1; f_1), (v_5, v_4, v_3; t_2; f_1), (v_{10}; t_1; f_1),$$

$$(v_1, v_7, v_{12}, v_{14}; t_1; f_2), \{(v_2, v_8, v_{13}; t_1; f_2), (v_9; t_1; f_2)\}$$

$$PP_2 = \{(v_6, v_5, v_4, v_3; t_3; f_1), (v_{10}, v_{11}; t_3; f_1), (v_{15}; t_3; f_1),$$

$$(v_1, v_7, v_{12}, v_{14}; ; f_2), (v_2, v_8, v_{13}; t_1; f_2), (v_9; t_1; f_2)\}$$

The costs of removing features $v_1, \ldots, v_{15}$ in Figure 9.5 are $c_1, \ldots, c_{15}$, respectively. For each process plan PP_i, $i \in N$, define the following incidence column vector:

$$a_i = [a_{li}, \ldots, a_{ri}, a_{si}, \ldots, a_{zi}]^T$$

where

$$a_{ti} = \begin{cases} 1 & \text{if tool } t \text{ is used in } PP_i, \ t \in \{1, \ldots, r\} \\ 0 & \text{otherwise} \end{cases}$$

$$a_{fi} = \begin{cases} 1 & \text{if fixture } f \text{ is used in } PP_i, \ f \in \{s, \ldots, z\} \\ 0 & \text{otherwise} \end{cases}$$

For any two process plans PP_i and PP_j define the weighted Hamming distance:

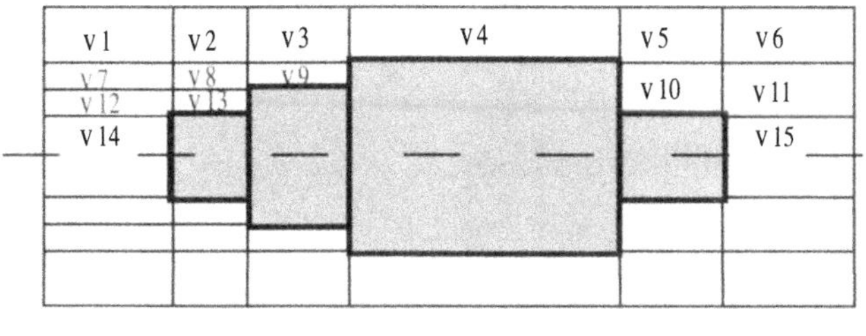

Figure 9.5. Rotational part with features $v_1, \ldots, v_{15}$ removed.

$$d_{ij} = \sum_{q=1}^{z} w_q \delta(a_{qi}, a_{qj}) \quad \text{for all } i, j \qquad (9.15)$$

where

$$\delta(x_{qi}, x_{qj}) = \begin{cases} 1 & \text{if } a_{qi} \neq a_{qj} \\ 0 & \text{otherwise} \end{cases}$$

w_q = weight coefficient of auxiliary device q

The weighted Hamming distance d_{ij} measures dissimilarity (distance) between process plans PP_i and PP_j. Note that assuming $w_q = 1$ for all q, the weighted Hamming distance (9.15) becomes the Hamming distance.

As the importance of each machine or auxiliary device varies, introducing the weight coefficient w_q for each q has modified the Hamming distance. For example, the weight assigned to a fixture typically has a higher value than the weight assigned to a tool. A way to determine the value of the weight w_q for all q is to set it proportional to the cost of auxiliary device q.

The weighted Hamming distance defined in (9.15) and the Hamming distance are illustrated in Example 9.4.

Example 9.4. Consider the vector of weight coefficients $[w_q]$ and two process plans PP_1 and PP_5:

$$[w_q] = \begin{bmatrix} 1 \\ 2 \\ 4 \\ 1 \\ 3 \\ 5 \end{bmatrix} \qquad PP_1 = \begin{bmatrix} 1 \\ 1 \\ 0 \\ 1 \\ 0 \\ 1 \end{bmatrix} \qquad PP_5 = \begin{bmatrix} 0 \\ 1 \\ 1 \\ 0 \\ 1 \\ 1 \end{bmatrix}$$

First calculate the Hamming distance between the two process plans PP_1 and PP_5 as the total number of "mismatches" of the corresponding elements of the two vectors:

$$d_{15} = 1 + 0 + 1 + 1 + 1 + 0 = 4$$

The weighted Hamming distance between the two process plans is

$$d_{15} = 1 \times 1 + 2 \times 0 + 4 \times 1 + 1 \times 1 + 3 \times 1 + 5 \times 0 = 9$$

9.3.2   Integer Programming Model

To formulate the equipment selection model, in addition to the notation introduced in the previous section, the following are defined:

$K = \{1, 2, \ldots, m\}$ is the set of parts to be manufactured

N_k = set of process plans, where

$$\cup_{k \in K} N_k = N$$

A = set of arcs connecting process plans from set N_k to set N_l, $(k, l) \in K \times K$ and $k \neq l$

c_i = cost of process plan PP_i, for all $i \in N$

$$x_i = \begin{cases} 1 & \text{if process plan } PP_i, \text{ is selected} \\ 0 & \text{otherwise} \end{cases}$$

$$y_{ij} = \begin{cases} 1 & \text{if process plans } PP_i \text{ and } PP_j \text{ are selected} \\ 0 & \text{otherwise} \end{cases}$$

The equipment selection model is to select, for each set of process plans (a part), only one representative process plan and the corresponding equipment in such a way that the total distances among the selected process plans and the corresponding costs are minimized:

$$\min \frac{1}{2} \sum_{(i,j) \in A} d_{ij} y_{ij} + \sum_{i \in N} c_i x_i \tag{9.16}$$

subject to

$$\sum_{i \in N_k} x_i = 1 \qquad k \in K \tag{9.17}$$

$$x_i + x_j - 1 \leq y_{ij} \qquad (i, j) \in A \tag{9.18}$$

$$x_i = 0, 1 \qquad i \in N \tag{9.19}$$

$$y_{ij} = 0, 1 \qquad (i, j) \in A \tag{9.20}$$

Constraint (9.17) ensures that for each part exactly one process plan and the corresponding equipment are selected. The consistency among variables is imposed by constraint (9.18). Constraints (9.19) and (9.20) ensure integrality.

The definition of cost c_i of process plan i requires some elaboration. This cost could be defined as the cost of manufacturing a part according to process plan i, the cost of equipment used by process plan i, and so on. In addition to the weights w_q associated with each resource in the weighted Hamming distance (9.15), the two components in the objective function (9.16), that is, total distance and total cost, could be assigned different weights to balance their impact on the solution.

From the definition, $y_{ij} = y_{ji}$ and the term $d_{ij}y_{ij} = d_{ji}y_{ji}$ appears twice in (9.16), so the factor $\frac{1}{2}$ has been introduced. Furthermore, due to the nature of constraint (9.19), the integrality requirement for constraint (9.20) is redundant. Relaxing the integrality of the latter constraint transforms model (9.16)–(9.20) into the following mixed-integer programming program:

$$\min \frac{1}{2} \sum_{(i,j) \in A} d_{ij}y_{ij} + \sum_{i \in N} c_i x_i \qquad (9.21)$$

subject to (9.17), (9.18), (9.19), and

$$y_{ij} \geq 0 \qquad (i, j) \in A \qquad (9.22)$$

Due to the high computational complexity of the models (9.16)–(9.20) and (9.21)–(9.22), an efficient heuristic, named the construction algorithm, is presented (Kusiak and Finke, 1988).

9.3.3 Construction Algorithm

Consider a sequence of process plan sets $N_1, N_2, \ldots, N_m$. One wants to construct the set $S = \{S_i\}$ following this sequence: first $S_1 \in N_1$ is selected, then $S_2 \in N_2, \ldots$, and finally $S_m \in N_m$.

Let $S^{(k)} = \{S_1, S_2, \ldots, S_k\}$ denote the elements selected at stage k of the following algorithm:

Step 0. Generate a random permutation of process plans.

Step 1. $S^{(0)} = \varnothing$ and $k = 1$.

Step 2. Calculate the cost

$$c_i + \sum_{s \in S^{(k-1)}} d_{is} \qquad \text{for all } i \in N_k$$

Let S_k be the process plan with the minimum corresponding cost.

Step 3. $S^{(k)} = S^{(k-1)} + \{S_k\}$. If $k = m$, stop ($S^{(m)} = S$ is the final set of process plans); else set $k = k + 1$ and go to Step 2.

In Step 0 of the algorithm a random permutation of $N_1, N_2, \ldots, N_m$ is generated.

In Step 2, the best possible element $S_k \in N_k$ is selected, leaving all previous elements $S^{(k-1)}$ fixed. Initially, $S_1 \in N_1$ is simply a process plan with the corresponding cost $c_i, i \in N_1$. In general, the minimum cost represents exactly the expansion cost for a clique in the k-partite graph spanned over $N_1, N_2, \ldots, N_k$.

The construction algorithm is illustrated in Example 9.5.

Example 9.5. Consider the following data sets:

1. Set of parts $K = \{1, 2, 3, 4\}$.
2. Sets of process plans $N_1 = \{1, 2\}$, $N_2 = \{3, 4, 5\}$, $N_3 = \{6, 7\}$, $N_4 = \{8, 9, 10\}$.
3. The incidence matrix for five tools, three fixtures, four parts, and 10 process plans:

$$
\begin{array}{l}
\\
\text{Tools}\\
\\
\\
\\
\text{Fixtures}\\
\\
\end{array}
\begin{array}{l}
t_1\\
t_2\\
t_3\\
t_4\\
t_5\\
f_1\\
f_2\\
f_3\\
\end{array}
\left[
\begin{array}{cccccccccc}
1 & & 1 & 1 & & 1 & 1 & 1 & & 1\\
 & 1 & & & 1 & & & 1 & & 1\\
1 & 1 & 1 & 1 & 1 & 1 & & & 1 & \\
1 & & & & 1 & & 1 & 1 & 1 & 1\\
 & 1 & & & & 1 & & & 1 & \\
1 & & 1 & 1 & 1 & 1 & 1 & 1 & 1 & 1\\
 & 1 & & & 1 & & & 1 & & \\
1 & 1 & & & 1 & & 1 & & 1 & 1\\
\end{array}
\right]
\tag{9.23}
$$

$$
\begin{array}{cccccccccc}
\text{Part 1} & & \text{Part 2} & & & \text{Part 3} & & \text{Part 4}\\
1\ \ 2 & & 3\ \ 4\ \ 5 & & & 6\ \ 7 & & 8\ \ 9\ \ 10
\end{array}
$$

4. Vector of process plan costs:

$$[c_i] = [5.8, 9.4, 11.6, 5.7, 3.4, 4.3, 5.1, 6.4, 5.2, 5.3] \tag{9.24}$$

5. Vector of weights:

$$[w_q] = [1, 1, 1, 1, 1, 1, 1, 1]^{\mathrm{T}} \tag{9.25}$$

For the incidence matrix (9.23) and weight vector (9.25), calculate matrix $D = [d_{ij}]$ of Hamming distances:

$$
D = [d_{ij}] =
\begin{array}{l}
1\\2\\3\\4\\5\\6\\7\\8\\9\\10
\end{array}
\left[
\begin{array}{cccccccccc}
\infty & \infty & 2 & 1 & 3 & 3 & 1 & 4 & 1 & 4\\
 & \infty & 5 & 6 & 4 & 4 & 8 & 5 & 6 & 5\\
 & & \infty & \infty & \infty & 1 & 3 & 4 & 3 & 4\\
 & & & \infty & \infty & 2 & 2 & 5 & 1 & 5\\
 & & & & \infty & 6 & 4 & 3 & 2 & 5\\
 & & & & & \infty & \infty & 5 & 4 & 3\\
 & & & & & & \infty & 3 & 2 & 3\\
 & & & & & & & \infty & \infty & \infty\\
 & & & & & & & & \infty & \infty\\
 & & & & & & & & & \infty\\
\end{array}
\right]
\tag{9.26}
$$

$$
\begin{array}{cccccccccc}
1 & 2 & 3 & 4 & 5 & 6 & 7 & 8 & 9 & 10
\end{array}
$$

The following observations can be made regarding Example 9.5:

1. The distance matrix (9.26) is symmetric. The factor $\frac{1}{2}$ should not appear in the objective function (21) in the model (9.21)–(9.22) instantiated with the data from this example.

2. For all entries $(i,j) \in A$ of matrix (9.26), $d_{ij} \geq 0$ and for entries $(i,j) \notin A$, $d_{ij} = \infty$. The latter implies that distances between the same process plans as well as distances between process plans belonging to the same part are excluded from the model (9.21)–(9.22). The set A can define the valid connections among process plans, e.g., (1, 3) is included in A while (1, 2) is not as process plans 1 and 2 belong to the same part.

3. An easy way to formulate the model (9.21)–(9.22) is to list in the objective function (9.21) only variables y_{ij} that correspond to the entries $(i,j) \in A$ of matrix (9.26). The same applies to the variables in constraint (9.18). For example, for the entry $d_{12} = \infty$ of (9.26) the variable y_{12} would appear neither in the objective function (9.21) nor jointly with the variables x_1 and x_2 in the constraint (9.18). However, the variables x_1 and x_2 would appear in the objective function (9.21).

4. The term process plan can be replaced with the term resource vector (e.g., resource vector 1 for product 2).

For the above data, the equipment selection problem is solved with the construction algorithm. The steps of this algorithm are presented next.

Step 0. A random permutation of process plan sets $\{N_1, N_2, N_3, N_4\}$ is generated.

Step 1. Set $S^{(0)} = \emptyset$, $k = 1$.

Step 2. Since

$$\sum_{s \in \emptyset} d_{is} = 0$$

the costs c_i for all $i \in N_k$ are compared, that is, $c_1 = 5.8$, $c_2 = 9.4$; hence, $S_1 = 1$.

Step 3. $S^{(1)} = \{1\}$, $k = 2$.

Step 2. Calculate

$$c_i + \sum_{s \in S^{(1)}} d_{is} \quad \text{for all } i \in N_2$$

that is, $c_3 + d_{31} = 11.6 + 2 = 13.6$, $c_4 + d_{41} = 5.7 + 1 = 6.7$, $c_5 + d_{51} = 3.4 + 3 = 6.4$. The minimum cost defines $S_2 = 5$.

Step 3. $S^{(2)} = \{1, 5\}$, $k = 3$.

Step 2. Calculate

$$c_i + \sum_{s \in S^{(2)}} d_{is} \quad \text{for all } i \in N_3$$

that is, $c_6 + d_{61} + d_{65} = 13.3$, $c_7 + d_{71} + d_{75} = 10.1$; hence $S_3 = 7$.

Step 3. $S^{(3)} = \{1, 5, 7\}$, $k = 4$.

Step 2. Calculate

$$c_i + \sum_{s \in S^{(3)}} d_{is} \quad \text{for all } i \in N_4$$

that is, $c_8 + d_{81} + d_{85} + d_{87} = 16.4$, $c_9 + d_{91} + d_{95} + d_{97} = 10.2$, $c_{10} + d_{10.1} + d_{10.5} + d_{10.7} = 17.3$; hence $S_4 = 9$.

Step 3. The selected set of process plans $S = S^{(4)} = \{1, 5, 7, 9\}$ with the corresponding total cost $c_1 + c_5 + c_7 + c_9 + d_{15} + d_{17} + d_{19} + d_{57} + d_{59} + d_{79} = 32.5$.

A solution of better quality $x_1 = x_4 = x_7 = x_9 = 1$; that is, process plans 1, 4, 7, 9 with the cost of 29.8, has been obtained by solving the equipment selection model with an optimal algorithm. From matrix (9.23) the following resources corresponding to this solution are determined: tools t_1, t_3, and t_4 and fixture f_3, which represents only a portion of all resources considered. The reduced number of different types of equipment significantly reduces manufacturing cost. The desired level of system flexibility can be accomplished by procuring multiple copies of the same piece of equipment. A listing of the model (9.21)–(9.22) for the data in Example 9.5 is given in Appendix 9.2.

9.4 SUMMARY

In this chapter, various models for the selection of manufacturing equipment were discussed. The simplest model involves selecting one machine at a time. The more complex models may select all machines to meet the part processing requirements as well as material handling and any other equipment required. It should be stressed that the approach presented in this chapter is of a static nature and emphasizes meeting the functional requirements for manufacturing equipment. The equipment capacity is not considered. The static models with the equipment capacity considered are discussed in Kusiak (1990); however, the solutions generated by such models are in general not accurate. To determine the exact capacity of the equipment required, a dynamic analysis needs to be conducted. Such analysis can be performed with discrete-event simulation software, for example, ARENA and SIMPLE++.

APPENDIX 9.1:   INPUT FILE OF EXAMPLE 9.2

```
MIN 150X1+130X2+220X3+190X4+90X5+110X6+210X7
SUBJECT TO
1X1+1X2+2X3>=2
```

```
1X1+1X2+2X3+1X4>=2
1X1+1X3+1X4>=1
1X2+1X3+1X4+1X6+1X7>=2
1X5+1X6+2X7>=2
1X6+2X7>=1
1X1+1X2+1X3<=1
1X4<=1
1X5<=1
1X6+1X7<=1
END
INTEGER 7
```

APPENDIX 9.2: INPUT FILE OF EXAMPLE 9.5

```
MIN
2Y13+1Y14+3Y15+3Y16+1Y17+4Y18+1Y19+4Y110+
5Y23+6Y24+4Y25+4Y26+8Y27+5Y28+6Y29+5Y210
+1Y36+3Y37+4Y38+3Y39+4Y310
+2Y46+2Y47+5Y48+1Y49+5Y410
+6Y56+4Y57+3Y58+2Y59+5Y510
+5Y68+4Y69+3Y610
+3Y78+2Y79+3Y710
5.81+9.4X2+11.6X3+5.7X4+3.4X5+4.3X6+5.1X7+6.4X8+5.2X9+
5.3X10
SUBJECT TO
X1+X2=1
X3+X4+X5=1
X6+X7=1
X8+X9+X10=1
X1+X3-Y13<=1
X1+X4-Y14<=1
X1+X5-Y15<=1
X1+X6-Y16<=1
X1+X7-Y17<=1
X1+X8-Y18<=1
X1+X9-Y19<=1
X1+X10-Y110<=1
X2+X3-Y23<=1
X2+X4-Y24<=1
X2+X5-Y25<=1
X2+X6-Y26<=1
X2+X7-Y27<=1
X2+X8-Y28<=1
X2+X9-Y29<=1
```

```
X2+X10-Y210<=1
X3+X6-Y36<=1
X3+X7-Y37<=1
X3+X8-Y38<=1
X3+X9-Y39<=1
X3+X10-Y310<=1
X4+X6-Y46<=1
X4+X7-Y47<=1
X4+X8-Y48<=1
X4+X9-Y49<=1
X4+X10-Y410<=1
X5+X6-Y56<=1
X5+X7-Y57<=1
X5+X8-Y58<=1
X5+X9-Y59<=1
X5+X10-Y510<=1
X6+X8-Y68<=1
X6+X9-Y69>=1
X6+X10-Y610<=1
X7+X8-Y78<=1
X7+X9-Y79<=1
X7+X10-Y710<=1
END
INTEGER X1
INTEGER X2
INTEGER X3
INTEGER X4
INTEGER X5
INTEGER X6
INTEGER X7
INTEGER X8
INTEGER X9
INTEGER X10
```

REFERENCES

Hayes, G. M., R. P. Davis, and R. A. Wysk (1981), A dynamic programming approach to machine requirements planning, *AIIE Transactions*, Vol. 13, pp. 175–181.

Kusiak, A. (1990), *Intelligent Manufacturing Systems*, Prentice-Hall, Englewood Cliffs, NJ.

Kusiak, A., and G. Finke (1988), Selection of process plans in automated manufacturing systems, *IEEE Journal of Robotics and Automation*, Vol. 4, No. 4, pp. 397–402.

Lee, R. C. T. (1981), Clustering analysis and its applications, in J. T. Tou (Ed.), *Advances in Information Systems Science*, Plenum, New York.

Miller, D. M., and R. P. Davis (1977), The machine requirements problem, *International Journal of Production Research*, Vol. 15, No. 2, pp. 219–231.

QUESTIONS

9.1. What are the basic phases in designing a manufacturing system?

9.2. What is the role of material handling systems in the layout pattern of a manufacturing system?

9.3. What are the advantages of selecting the manufacturing equipment necessary to implement a system one piece at a time?

9.4. What are the advantages of selecting machine tools and material handling equipment at the same time?

9.5. Can a static model determine the exact number of manufacturing equipment needed?

9.6. What is the role of dynamic models (e.g., discrete simulation) in equipment selection?

9.7. What are the advantages and disadvantages of minimizing the number of distinct pieces of machines, tools, material handling carriers, and so on.

PROBLEMS

9.1. For the data below, find the minimum value of n_2 and n_6 that produce a feasible solution to the equipment selection model. Select the lowest cost manufacturing equipment required to produce all six batches of parts?

(a)   Vector P of part batches: $P = [i] = [1, 2, 3, 4, 5, 6]$.

(b)   Vector R of equipment modules (machines and material handling equipment):

$$R = [R_j] = [R_1 = M_1 + H_1, R_2 = M_2 + H_1, R_3 = M_3 + M_4 + H_3,$$
$$R_4 = M_3 + H_4, R_5 = M_5 + H_5, R_6 = M_6 + H_5, R_7 = M_5 + M_6 + H_6]$$

(c)   Group of modules: $F_1 = \{R_1, R_2\}$, $F_2 = \{R_3, R_4\}$, $F_3 = \{R_5, R_6, R_7\}$.

(d)   Vector of module costs:

$$C = [c_j] = [85,000 \quad 100,000 \quad 95,000 \quad 65,000 \quad 31,000 \quad 60,000 \quad 95,000]$$

(e)   Vector of required number of machines to process each batch i of parts:

$$N = [n_i] = [1, n_2, 1, 1, 1, n_6]$$

(f)   Part batch–manufacturing module incidence matrix:

$$
[a_{ij}] = \begin{array}{c} \\ 1 \\ 2 \\ 3 \\ 4 \\ 5 \\ 6 \end{array}
\begin{array}{cccccccc}
1 & 2 & 3 & 4 & 5 & 6 & 7 \\
\left[\begin{array}{ccccccc}
1 & & 1 & & & & \\
1 & 1 & 2 & 2 & & & \\
1 & & & 1 & 1 & & \\
 & 1 & 1 & 1 & 1 & 1 & 2 \\
1 & 1 & 2 & 1 & 1 & & 1 \\
1 & & & 1 & & 1 &
\end{array}\right]
\end{array}
$$

9.2. Modify the equipment selection model (9.7)–(9.10) so that:

  (a) The total equipment budget does not exceed B.

  (b) The total number of equipment modules is minimized.

9.3. Formulate (write the objective function and constraints in a compact form) an equipment selection model to minimize the number of modules as to not exceed budget B and assuming that for each group of modules one module is selected. Use the following notation:

  m = number of operations

  n = number of equipment modules

  c_j = cost of equipment module j

  $a_{ij} = \begin{cases} 1 & \text{if operation } c_i \text{ can be performed } l \text{ times on machines of module } j \\ 0 & \text{otherwise} \end{cases}$

  n_i = required (minimum) number of machines that can perform operation o_i

  F_k = group of modules k

  p = number of groups of modules

  B = total budget available

  $x_j = \begin{cases} 1 & \text{if module } j \text{ has been selected} \\ 0 & \text{otherwise} \end{cases}$

9.4. A systems engineer has collected the following data for the selection of assembly equipment:

  (a) Assembly operations: $O = [o_i] = [o_1, o_2, o_3, o_4, o_5, o_6]$.

  (b) Assembly modules:

$$
R = [R_j] = [R_1 = M_1, R_2 = M_2, R_3 = M_3 + M_4,
$$
$$
R_4 = M_3, R_5 = M_5, R_6 = M_6, R_7 = M_5 + M_6]
$$

  (c) Module costs:

$$
C = [c_j] = [110{,}000 \quad 135{,}000 \quad 120{,}000 \quad 165{,}000 \quad 48{,}000 \quad 90{,}000 \quad 125{,}000]
$$

  (d) Required number of assembly stations needed to perform each operation o_i:

$$N = [n_i] = [1, 2, 1, 2, 2, 1]$$

(e) Assembly operation–equipment incidence matrix:

$$
[a_{ij}] = \begin{matrix} & \begin{matrix} 1 & 2 & 3 & 4 & 5 & 6 & 7 \end{matrix} \\ \begin{matrix} o_1 \\ o_2 \\ o_3 \\ o_4 \\ o_5 \\ o_6 \end{matrix} & \begin{bmatrix} 1 & 1 & 2 & & & & \\ 1 & 1 & 2 & 3 & & & \\ 1 & & 1 & 2 & & & \\ & 1 & 1 & 1 & & 1 & 1 \\ & 1 & & 1 & 1 & 1 & 2 \\ 1 & & & 1 & & 1 & \end{bmatrix} \end{matrix}
$$

Using the data collected, the engineer has formulated the model (9.27)–(9.35), which is needed in a staff meeting:

$$\min x_1 + x_2 + x_3 + x_4 + x_5 + x_5 + x_6 + x_7 \tag{9.27}$$

$$x_1 + x_2 + 2x_3 \geq 1 \tag{9.28}$$

$$x_1 + x_2 + 2x_3 + 3x_4 \geq 2 \tag{9.29}$$

$$x_1 + x_3 + 2x_4 \geq 1 \tag{9.30}$$

$$x_2 + x_3 + x_4 + x_6 + x_7 \geq 2 \tag{9.31}$$

$$x_2 + x_4 + x_5 + x_6 + 2x_7 \geq 2 \tag{9.32}$$

$$x_1 + x_4 + x_6 + x_7 \geq 1 \tag{9.33}$$

$$110x_1 + 135x_2 + 120x_3 + 165x_4 + 48x_5 + 90x_6 + 125x_7 \leq 475 \tag{9.34}$$

$$x_i = 0, 1 \qquad i = 1, \ldots, 7 \tag{9.35}$$

Explain the objective function and constraints of the model (9.27)–(9.35).

9.5. For the data below solve the process plan selection problem.

(a) Incidence matrix:

$$
\begin{array}{c}
 \\
\text{Tools} \\
 \\
 \\
 \\
\text{Grippers} \\

\end{array}
\begin{array}{c}
t_1 \\
t_2 \\
t_3 \\
t_4 \\
g_1 \\
g_2 \\
g_3 \\
g_4
\end{array}
\begin{bmatrix}
1 & 1 & 0 & 1 & 1 & 1 & 0 & 1 \\
0 & 0 & 1 & 0 & 0 & 1 & 0 & 1 \\
1 & 1 & 1 & 1 & 0 & 0 & 1 & 0 \\
0 & 0 & 1 & 0 & 1 & 1 & 1 & 1 \\
0 & 0 & 0 & 1 & 0 & 0 & 0 & 1 \\
1 & 1 & 1 & 1 & 1 & 1 & 1 & 1 \\
0 & 0 & 1 & 0 & 0 & 1 & 0 & 0 \\
1 & 0 & 0 & 1 & 0 & 1 & 0 & 1
\end{bmatrix}
$$

Part 1   Part 2   Part 3

(b)   Vector of process plan costs:

$$[c_{ij}] = [\ 2.8,\ 1.6,\ 1.3,\ 4.4,\ 5.4,\ 6.6,\ 8.1,\ 7.2\ ]$$

(c)   Vector of costs of tools and grippers:

$$[w_i] = [\ 7.5,\ 7.6,\ 5.0,\ 3.0,\ 220,\ 300,\ 250,\ 540]$$

Find the following:

(i)   Optimal set of process plans

(ii)   Optimal set of tools and grippers

CHAPTER 10

GROUP TECHNOLOGY

10.1 INTRODUCTION

The basic idea of group technology (GT) is to decompose a manufacturing system into subsystems. Implementation of GT in manufacturing has the following advantages:

- Reduced production lead time
- Reduced work-in-process
- Reduced labor
- Reduced tooling
- Reduced rework and scrap materials
- Reduced setup time
- Reduced order time delivery
- Improved human relations
- Reduced paper work

This chapter focuses on models and algorithms developed for GT. There are two basic methods used for solving the GT problem:

- Classification
- Cluster analysis

The classification method is used to group parts into part families based on their design features. There are two variations of the classification method:

- Visual method
- Coding method

10.1.1 Visual Method

In the visual method parts are grouped according to their similarities in the geometric shape, as shown in Figure 10.1, where 10 parts have been grouped into 4 part families. Grouping parts using the visual method is dependent on personal preference. However, this method is applicable when the number of parts is rather limited.

10.1.2 Coding Method

In the coding method, parts are first classified based on features (characteristics), for example:

- Geometric shape and complexity
- Dimensions
- Type of material
- Shape of raw material
- Required accuracy of the finished part

Using a coding system, each part is assigned a numerical or alphanumerical code. Each digit of this code represents a feature of a part. The currently available coding systems differ with respect of the depth of coverage of the five features. For example, a coding system may provide more information on the shape and dimension of a part, whereas another may emphasize the accuracy of a part. There are three basic types of coding systems:

- Monocode
- Polycode
- Hybrid

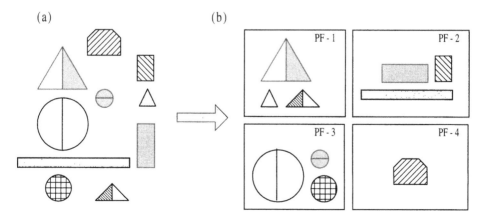

Figure 10.1. Parts grouped into families using a visual method: (*a*) ungrouped parts; (*b*) grouped parts.

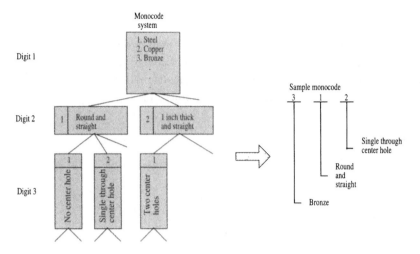

Figure 10.2. Monocode system and part monocode.

An example of a monocode system is illustrated in Figure 10.2. Features of each part are matched with the list of features corresponding to each node of the tree, and a part code is generated. Since a monocode system has a tree structure, a digit selected at a particular node depends on the digit selected at the preceding node. To fully interpret a part, all digits of its monocode are required. For a given part, the length of its monocode is rather short compared to other coding systems (Ingram, 1982).

10.2 CLUSTER ANALYSIS METHOD

Another approach used in group technology is cluster analysis, which is concerned with grouping objects into homogeneous clusters (groups) based on the object features. The application of cluster analysis in manufacturing leads to grouping parts into part families (PFs) and machines into machine cells (MCs). The result of this grouping is implemented as:

1. Physical cell
2. Logical (virtual) cell

The physical implementation of a GT cell requires rearrangement of machines so that the shop floor is altered, as shown in Figure 10.3. Using the logical machine layout, machines are grouped into logical machine cells and the positions of machines are not altered (see Figure 10.4). Logical grouping can be applied when the manufacturing content (e.g., part geometry, part mix) is changing frequently so that the physical machine layout is not justified.

To model the GT problem, three clustering formulations are used:

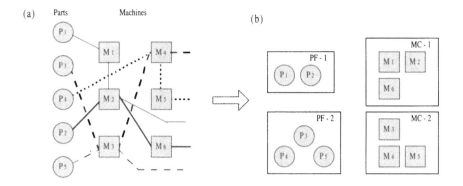

Figure 10.3. Physical machine layout (*a*) ungrouped parts and machines; (*b*) grouped parts and machines.

- Matrix formulation
- Mathematical programming formulation
- Graph formulation

In this chapter, the first two formulations will be considered.

10.2.1 Matrix Formulation

In the matrix formulation, a machine–part incidence matrix $[a_{ij}]$ is constructed. The machine–part incidence matrix $[a_{ij}]$ consists of 0, 1 entries, where an entry 1 (0) indicates that machine i is used (not used) to process part j. Typically when an initial machine–part incidence matrix $[a_{ij}]$ is constructed, clusters of machines and parts are not visible. A clustering algorithm transforms an initial incidence matrix into a more

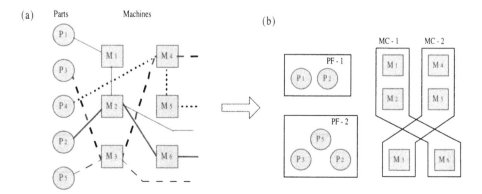

Figure 10.4. Logical machine layout: (*a*) ungrouped parts and machines; (*b*) grouped parts and machines.

structured (possibly block diagonal) form. To illustrate the clustering concept applied to GT, consider the following machine–part incidence matrix:

$$
\begin{array}{cc}
 & \text{Part number} \\
 & \begin{array}{ccccc} 1 & 2 & 3 & 4 & 5 \end{array}
\end{array}
$$

$$
\begin{array}{cc}
\begin{array}{c} \\ \text{Machine} \\ \text{number} \\ \\ \end{array}
\begin{array}{c} 1 \\ 2 \\ 3 \\ 4 \end{array}
\end{array}
\left[
\begin{array}{ccccc}
1 & & 1 & 1 \\
1 & & 1 & & \\
 & 1 & & 1 & \\
1 & & 1 & &
\end{array}
\right]
\tag{10.1}
$$

Rearranging rows and columns in matrix (10.1) results in the matrix

$$
\begin{array}{cc}
 & \begin{array}{ccc} \text{PF-1} & & \text{PF-2} \end{array} \\
 & \begin{array}{ccccc} 1 & 3 & 2 & 4 & 5 \end{array}
\end{array}
$$

$$
\begin{array}{c}
\text{MC-1} \begin{array}{c} 2 \\ 4 \end{array} \\
\text{MC-2} \begin{array}{c} 1 \\ 3 \end{array}
\end{array}
\left[
\begin{array}{ccccc}
1 & 1 & & & \\
1 & 1 & & & \\
 & & 1 & 1 & 1 \\
 & & 1 & 1 &
\end{array}
\right]
\tag{10.2}
$$

Two machine cells (clusters) MC-1 = {2, 4} and MC-2 = {1, 3} and two corresponding part families PF-1 = {1, 3} and PF-2 = {2, 4, 5} are visible in matrix (10.2).

Clustering a binary incidence matrix may result in the following two classes of clusters:

1. Mutually separable clusters
2. Partially separable clusters

The mutually separable clusters are shown in matrix (10.2), while the partialy separable clusters are presented in the matrix

$$
\begin{array}{cc}
 & \begin{array}{ccccc} 1 & 2 & 3 & 4 & 5 \end{array}
\end{array}
$$

$$
\begin{array}{c}
\text{MC-1} \begin{array}{c} 1 \\ 2 \end{array} \\
\text{MC-2} \begin{array}{c} 3 \\ 4 \end{array}
\end{array}
\left[
\begin{array}{ccccc}
1 & 1 & & & 1 \\
1 & 1 & & & \\
 & & 1 & 1 & 1 \\
 & & 1 & 1 &
\end{array}
\right]
\tag{10.3}
$$

Matrix (10.3) cannot be separated into two disjoint clusters because of part 5, which is to be machined in two cells, MC-1 and MC-2. Removing part 5 from matrix (10.3) results in the decomposition of matrix (10.3) into two separable machine cells, MC-1 = {1, 2} and MC-2 = {3, 4}, and two part families, PF-1 = {1, 2} and PF-2 = {3, 4}. The two clusters are called partially separable clusters and the overlapping part is called a bottleneck part. To deal with the bottleneck part 5, one of the following three actions can be taken:

1. It can be machined in one machine cell and transferred to the other machine cell by a material handing carrier.
2. It can be machined in a functional facility.
3. It can be subcontracted.

Analogous to the bottleneck part, a bottleneck machine is defined. A bottleneck machine prevents the decomposition of a machine–part incidence matrix into disjoint submatrices. For example, machine 3 in matrix (10.4) does not permit decomposition of that matrix into two machine cells and two part families:

$$
\begin{array}{c}
\\
\text{Machine} \\
\text{number}
\end{array}
\begin{array}{c}
\\
1 \\
2 \\
3 \\
4 \\
5
\end{array}
\begin{array}{cccccc}
\multicolumn{6}{c}{\text{Part number}} \\
1 & 2 & 3 & 4 & 5 & 6 \\
\left[\begin{array}{cccccc}
1 & 1 & & & & \\
1 & 1 & & & & \\
1 & 1 & 1 & & 1 & 1 \\
& & 1 & 1 & 1 & 1 \\
& & 1 & & 1 & 1
\end{array}\right]
\end{array}
\tag{10.4}
$$

A way to decompose matrix (10.4) into two disjoint submatrices is to use an additional copy of machine 3. The latter leads to the transformation of matrix (10.4) into matrix (10.5):

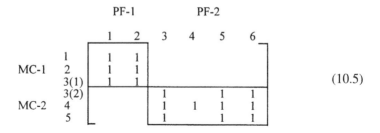

$$\tag{10.5}$$

Two machine cells, MC-1 = {1, 2, 3(1)} and MC-2 = {3(2), 4, 5}, and two corresponding part families, PF-1 = {1, 2} and PF-2 = {3, 4, 5, 6}, are shown in matrix (10.5).

To solve the matrix formulation of the group technology problem, the following solution methods have been developed (Kusiak, 1990):

- Similarity coefficient methods
- Sorting-based algorithms
- Bond energy algorithm
- Cost-based method
- Cluster identification algorithm
- Extended cluster identification algorithm

Some of these solution methods are discussed next.

10.2.1.1 Similarity Coefficient Methods. In this section, the single linkage cluster analysis (SLCA) presented by McAuley (1972) is discussed. The SLCA is based on the similarity coefficient s_{ij} measure between two machines i and j computed as follows:

$$s_{ij} = \frac{\sum_{k=1}^{n} \delta_1(a_{ik}, a_{jk})}{\sum_{k=1}^{n} \delta_2(a_{ik}, a_{jk})}$$

where

$$\delta_1(a_{ik}, a_{jk}) = \begin{cases} 1 & \text{if } a_{ik} = a_{jk} = 1 \\ 0 & \text{otherwise} \end{cases}$$

$$\delta_2(a_{ik}, a_{jk}) = \begin{cases} 0 & \text{if } a_{ik} = a_{jk} = 1 \\ 1 & \text{otherwise} \end{cases}$$

$$n = \text{number of parts}$$

To solve the group technology problem using SLCA, similarity coefficients for all possible pairs of machines are computed. Machine cells are generated using a threshold value for the similarity coefficient.

To illustrate SLCA, consider matrix (10.3). The similarity coefficients s_{ij} for this matrix are computed next and depicted in Figure 10.5:

$$s_{12} = s_{34} = \frac{2}{3} = 0.66$$

$$s_{13} = \frac{1}{4} = 0.25$$

$$s_{14} = s_{24} = s_{23} = \frac{0}{5} = 0$$

Assuming the threshold value of the similarity coefficient $s_{ij} = 60\%$, the following machine cells are obtained from Figure 10.5:

$$\text{MC-1} = \{1, 2\} \qquad \text{MC-2} = \{3, 4\}$$

A value of s_{ij} greater than 66% suggests that the grouping of machines does not place, that is, each machine acts as a cell.

A disadvantage of SLCA is that it fails to recognize the chaining problem resulting from the duplication of bottleneck machines.

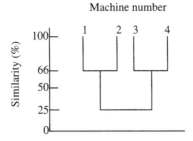

Figure 10.5. Tree of machines and similarity coefficients.

10.2.1.2 Sorting-Based Algorithms. Many authors have studied clustering methods based on sorting rows and columns of the machine–part incidence matrix. King (1980) developed the rank-order clustering (ROC) algorithm, which is presented next.

Rank Order Clustering Algorithm

Step 1. For each row of the machine–part incidence matrix, assign binary weight and calculate its decimal equivalent (a weight).

Step 2. Sort rows of the binary matrix in decreasing order of the corresponding decimal weights.

Step 3. Repeat the preceding two steps for each column.

Step 4. Repeat the preceding steps until the position of each element in each row and column does not change.

A weight for each row i and column j is calculated as follows:

$$\text{Row } i: \sum_{k=1}^{n} a_{ik} 2^{n-k}$$

$$\text{Column } j: \sum_{k=1}^{m} a_{kj} 2^{n-k}$$

In the final matrix generated by the ROC algorithm, clusters are identified visually. The ROC algorithm for the previously considered matrix (10.1) is illustrated in Example 10.1.

Example 10.1

Step 1. Assign binary weights to each row and calculate their decimal equivalents:

$$
\begin{array}{c}
\text{Part number} \\
\begin{array}{ccccc}
1 & 2 & 3 & 4 & 5
\end{array}
\end{array}
$$

Binary weight $\longrightarrow$

2^4	2^3	2^2	2^1	2^0	Decimal equivalent	
	1		1	1	11	1
1		1			20	2 Machine
	1		1		10	3 number
1		1			20	4

Step 2. Sorting the decimal weights in decreasing order results in the matrix

Part number

1	2	3	4	5		
1		1			2	
1		1			4	Machine
	1		1	1	1	number
	1		1		3	

Step 3. Repeating the preceding steps for each column produces the matrix

1	3	2	4	5		
1	1				2	
1	1				4	Machine
		1	1	1	1	number
		1	1		3	

In this matrix two separable clusters are visible. The matrix obtained is identical to matrix (10.2).

The ROC algorithm was further extended in King and Nakornchai (1982) and Chandrasekharan and Rajagopalan (1986).

10.2.1.3 *Cluster Identification Algorithm.* Kusiak and Chow (1987) applied the concept presented in Iri (1968) to develop the *cluster identification (CI) algorithm.* The CI algorithm checks the existence of mutually separable clusters in a binary machine–part incidence matrix, provided that they exist. The CI algorithm presented in Chapter 2 (page 52) is illustrated in Example 10.2.

Example 10.2. Determine machine cells and part families for the machine–part incidence matrix

$$
\begin{array}{c}
\hspace{3.3cm}\text{Part number}\\[2pt]
\begin{array}{cccccccc}
1 & 2 & 3 & 4 & 5 & 6 & 7 & 8
\end{array}\\
\begin{array}{c}
1\\2\\3\\4\\5\\6\\7
\end{array}
\left[
\begin{array}{cccccccc}
 & 1 & 1 &   & 1 &   &   &   \\
1 &   &   &   &   & 1 &   &   \\
 &   &   & 1 &   &   & 1 &   \\
1 &   &   &   &   & 1 &   &   \\
 &   & 1 &   & 1 &   &   & 1 \\
 &   &   & 1 &   &   &   &   \\
 & 1 & 1 &   & 1 &   &   & 1
\end{array}
\right]
\begin{array}{l}
\\ \\ \\ \text{Machine}\\ \text{number}\\ \\ \\
\end{array}
\end{array}
\qquad (10.6)
$$

Step 0. Set iteration number $k = 1$.

Step 1. Row 1 of matrix (10.6) is selected and horizontal line h_1 is drawn [see matrix (10.7)].

Step 2. Three vertical lines v_2, v_3, and v_5 are drawn in matrix (10.7) for each single-crossed entry 1:

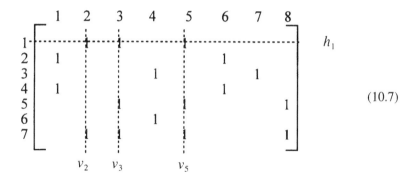

$$(10.7)$$

As a result of drawing the three vertical lines in Step 2, five new single-crossed entries 1 are created in matrix (10.7), that is, entries (5, 3), (5, 5), (7, 2), (7, 3), and (7, 5).

Step 3. Two horizontal lines h_5 and h_7 are drawn through the single-crossed entries of matrix (10.7):

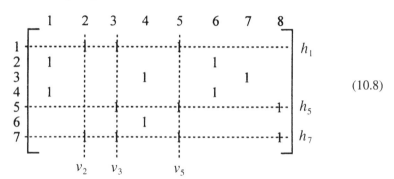

$$(10.8)$$

Step 4. Since the entries (5, 8) and (7, 8) of matrix (10.8) are single crossed, the vertical line v_8 is drawn:

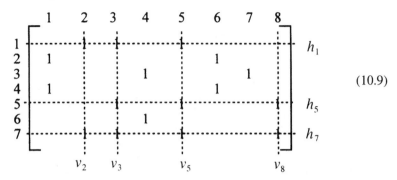

$$(10.9)$$

As there are no more single-crossed entries, all double-crossed entries 1 of matrix (10.9) form:

- Machine cell MC-1 = {1, 5, 7}
- Part family PF-1 = {2, 3, 5, 8}

Step 5. Matrix (10.9) is transformed into matrix (10.10) by removing all double-crossed entries:

$$
\begin{array}{c}
\begin{array}{cccc} 1 & 4 & 6 & 7 \end{array} \\
\begin{array}{c} 2 \\ 3 \\ 4 \\ 6 \end{array}
\left[
\begin{array}{cccc}
1 & & 1 & \\
 & 1 & & 1 \\
1 & & 1 & \\
 & 1 & &
\end{array}
\right]
\end{array}
\qquad (10.10)
$$

In the second iteration ($k = 2$), Steps 1–4 are performed for matrix (10.10). This iteration results in the incidence matrix

$$
\begin{array}{c}
\begin{array}{cccc} 1 & 4 & 6 & 7 \end{array} \\
\begin{array}{c} 2 \\ 3 \\ 4 \\ 6 \end{array}
\left[
\begin{array}{cccc}
 & & & \\
 & 1 & & 1 \\
 & & & \\
 & 1 & &
\end{array}
\right]
\begin{array}{c} h_2 \\ \\ h_4 \\ \end{array}
\end{array}
\qquad (10.11)
$$

The double-crossed elements of matrix (10.11) form:

- Machine cell MC-2 = {2, 4}
- Part family PF-2 = {1, 6}

In the third iteration ($k = 3$), matrix (10.12) is generated by removing double-crossed entries from matrix (10.11):

$$
\begin{array}{c}
\quad\;\; 4 \quad\;\; 7 \\
\begin{array}{c} 3 \\ 6 \end{array}
\left[\begin{array}{cc} \vdots & \vdots \\ \vdots & \vdots \end{array} \right]
\begin{array}{c} h_3 \\ h_6 \end{array} \\
\quad\;\; v_4 \quad\; v_7
\end{array}
\qquad (10.12)
$$

From this matrix MC-3 = {3, 6} and PF-3 = {4, 7} are obtained. The final clustering result is shown in matrix (10.13) with three machine cells and part families:

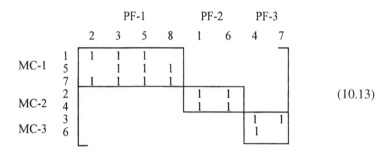

$$(10.13)$$

The conceptual design of a cellular manufacturing system corresponding to matrix (10.13) is illustrated in Figure 10.6. The computational experience showed that the CI algorithm is very efficient. For example, it takes only 0.06 seconds to identify clusters in an $m \times n = 60 \times 80$ matrix.

The CI algorithm determines optimal machine cells and part families when the machine–part incidence matrix partitions into mutually separable submatrices. The extended CI algorithm presented next heuristically solves the group technology problem with bottleneck parts or bottleneck machines.

10.2.1.4   *Extended CI Algorithm*

Step 0. Set iteration number $k = 1$.

Step 1. Select those machines (rows of matrix $[a_{ij}]^{(k)}$) that based on the user's expertise are potential candidates for inclusion in machine cell MC-k. Draw a

Manufacturing system	
Machine cell MC - 1	
Machines: 1, 5, 7	Parts: 2, 3, 5, 8
Machine cell MC - 2	Machine cell MC - 3
Machines: 2, 4	Machines: 3, 6
Parts: 1, 6	Parts: 4, 7

Figure 10.6. Conceptual design of a cellular manufacturing system from Example 10.2.

horizontal line h_i through each row of matrix $[a_{ij}]^{(k)}$ corresponding to these machines. In the absence of the user's expertise any machine can be selected.

Step 2. For each column in $[a_{ij}]^{(k)}$ corresponding to entry 1 that is single crossed by any of the horizontal lines h_i, draw a vertical line v_j.

Step 3. For each row in $[a_{ij}]^{(k)}$ corresponding to the entry 1 that is single crossed by the vertical line v_j drawn in Step 2, draw a horizontal line h_i. Based on the machines corresponding to all the horizontal lines drawn in Steps 1 and 3, a temporary machine cell MC′-k is formed. If either a constraint limiting the number of machines in a cell or the user's expertise indicates that some machines cannot be included in the temporary machine cell MC′-k, erase the corresponding horizontal lines in the matrix $[a_{ij}]^{(k)}$. Removal of these horizontal lines results in machine cell MC-k. Delete from matrix $[a_{ij}]^{(k)}$ parts (columns) that are to be manufactured on at least one of the machines already included in MC-k. Place these parts on the list of parts to be manufactured in a functional machining facility. Draw a vertical line v_j through each single-crossed entry 1 in $[a_{ij}]^{(k)}$ that involves only those machines included in MC-k.

Step 4. For all the double-crossed entries 1 in $[a_{ij}]^{(k)}$, form a machine cell MC-k and a part family PF-k.

Step 5. Transform the incidence matrix $[a_{ij}]^{(k)}$ into $[a_{ij}]^{(k+1)}$ by removing all the rows and columns included in MC-k and PF-k, respectively.

Step 6. If matrix $[a_{ij}]^{(k+1)} = \mathbf{0}$ (where $\mathbf{0}$ denotes a matrix with all elements equal to zero), stop; otherwise set $k = k + 1$ and go to step 1.

As one can see, the computational complexity of the extended CI algorithm is slightly higher than that of the CI algorithm, mainly due to Step 3. When the clustered matrix has block diagonal structure, the computational time complexity of the extended CI algorithm reduces to $O(mn)$; that is, it is equal to the complexity of the CI algorithm. The extended CI algorithm is illustrated in Example 10.3.

Example 10.3. Given the machine–part incidence matrix (10.14), determine mutually separable machine cells and part families:

Part number

	1	2	3	4	5	6	7	8	9	10	
1	1	1	1				1				
2	1				1					1	
3									1		Machine
4	1		1			1					number
5					1			1			
6	1			1				1	1	1	
7			1	1			1	1			
	1										

$$(10.14)$$

The user's expertise indicates that machines 1 and 4 should be included in machine cell MC-1 and machines 3 and 5 should be included in machine cell MC-2. The maximum number of machines in a cell is limited to 4.

Step 0. Set iteration number $k = 1$.

Step 1. Since the user's expertise indicates that machines 1 and 4 should be included in machine cell MC-1, two horizontal lines h_1 and h_4 are drawn:

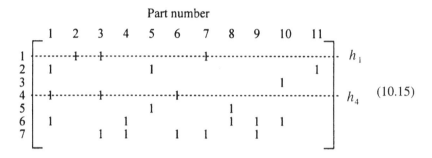

(10.15)

Step 2. For columns 1, 2, 3, 6, and 7 crossed by the horizontal lines h_1 and h_4 in matrix (10.15), five vertical lines v_1, v_2, v_3, v_6, and v_7 are drawn:

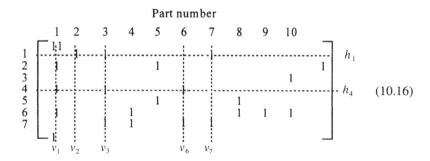

(10.16)

Step 3. Three horizontal lines h_2, h_6, and h_7 are drawn through rows 2, 5, and 7 corresponding of the single-crossed elements 1 of matrix (10.16):

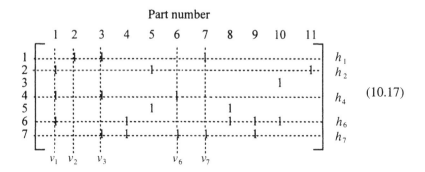

(10.17)

Based on all the double-crossed elements of matrix (10.17), temporary machine cell MC'-1 with machines {1, 2, 4, 6, 7} is formed. The user evaluates this temporary machine cell. Obviously, the constraint limiting the number of machines in a cell to 4 is violated. The user decided to remove machines 2 and 6 from the temporary machine cell MC'-1 as each of them includes less double-crossed (committed) elements than machine 7. As machines 2 and 6 are excluded, the horizontal lines h_2 and h_6 are erased in matrix (10.17). This leads to the deletion of columns (parts) 1, 4, and 9 from matrix (10.17). Note that each of the three parts 1, 4, and 9 is to be manufactured on at least one of the machines 2 and 6 [lines h_2 and h_6 in (10.17)] and at least one of the machines 1, 4, and 7 that form machine cell MC-1 [see matrix (10.18) and Step 4].

The three parts 1, 4, and 9 are placed on the list of parts to be manufactured in the functional machining facility. Since there are no more single-crossed entries 1, no vertical line is drawn. The above transformations lead to the matrix

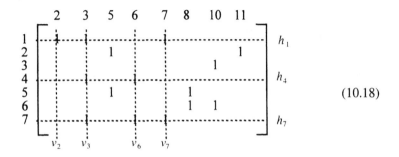

$$(10.18)$$

Step 4. The double-crossed entries 1 of matrix (10.18) indicate:

- Machine cell MC-1 = {1, 4, 7}
- Part family PF-1 = {2, 3, 6, 7}

Step 5. Matrix (10.18) is transformed into

$$
\begin{array}{c c}
 & \begin{array}{c c c c} 5 & 8 & 10 & 11 \end{array} \\
\begin{array}{c} 2 \\ 3 \\ 5 \\ 6 \end{array} &
\left[\begin{array}{c c c c}
1 & & & 1 \\
 & & 1 & \\
1 & 1 & & \\
 & 1 & 1 &
\end{array} \right]
\end{array}
\qquad (10.19)
$$

Step 6. Set $k = k + 1 = 2$ and go to Step 1.
The second iteration ($k = 2$) results in:

- Machine cell MC-2 = {2, 3, 5, 6}
- Part family PF-2 = {5, 8, 10, 11} as shown in the matrix

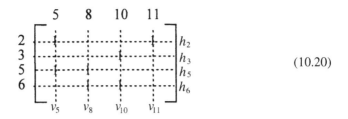

$$(10.20)$$

The final result generated by the extended CI algorithm is presented in matrix (10.21)

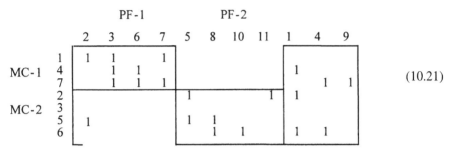

$$(10.21)$$

Machine cells MC-1 and MC-2 and corresponding part families PF-1 and PF-2 are visible in matrix (10.21). The corresponding conceptual layout of a manufacturing system is illustrated in Figure 10.7. The functional machining facility in Figure 10.7 should involve a set of machines that can perform operations on parts 1, 4, and 9. However, these machines do not have to be identical to machines 4, 7, 2, and 6 [see matrix (10.21)], initially designated for machining these three parts. Apart from the approach for manufacturing parts 1, 4, and 9, which is illustrated in Figure 10.7, three other alternative approaches are possible:

1. Process plans for parts 1, 4, and 9 might be modified, so that each of these parts could be machined in one of the two existing machine cells MC-1 or MC-2.

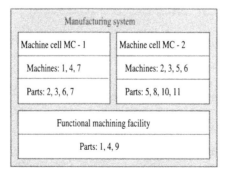

Figure 10.7. Conceptual layout of the manufacturing system from Example 10.3.

2. Designs of parts 1, 4, and 9 might be modified, so that the resulting process plans could fit the existing machine cell MC-1 or MC-2.

3. Parts 1, 4, and 9 might be manufactured in the two existing machine cells MC-1 and MC-2 without any changes of their process plans. This approach is applicable as long as the flow of parts among different machine cells (in the preceding case, MC-1 and MC-2) is relatively low.

Finally, if a user is not satisfied with the machine cells and part families generated, the entire computational process might be repeated. Initiating the extended CI algorithm with another machine (or machines) in Step 1 may result in a different configuration of machine cells and part families.

10.2.2 Mathematical Programming Formulation

Most mathematical programming models developed for group technology applications consider a distance measure d_{ij} between parts i and j. The distance measure d_{ij} is a real-valued symmetric function obeying the following axioms:

- Reflexivity $d_{ii} = 0$
- Symmetry $d_{ij} = d_{ji}$
- Triangular inequality $d_{iq} \leq d_{ip} + d_{pq}$

For a machine–part incidence matrix $[a_{ij}]$ the most commonly used distance measures are presented:

1. Minkowski distance measure (Arthanari and Dodge, 1981):

$$d_{ij} = \left[\sum_{k=1}^{m} |a_{ki} - a_{kj}|^r \right]^{1/r}$$

where

r = a positive integer

m = number of machines

Two special cases of the above measure are widely used:

(a) Absolute metric (for $r = 1$)
(b) Euclidean metric (for $r = 2$)

2. Weighted Minkowski distance measure (Arthanari and Dodge, 1981):

$$d_{ij} = \left[\sum_{k=1}^{m} w_k |a_{ki} - a_{kj}|^r \right]^{1/r}$$

There are two special cases:

(a)   Weighted absolute metric (for $r = 1$)
(b)   Weighted Euclidean metric (for $r = 2$)

3. Hamming distance (Lee, 1981):

$$d_{ij} = \sum_{l=1}^{m} \delta(a_{ki}, a_{kj})$$

where

$$\delta(a_{ki}, a_{kj}) = \begin{cases} 1 & \text{if } a_{ki} \neq a_{kj} \\ 0 & \text{otherwise} \end{cases}$$

Distance measures are also referred to as dissimilarity measures. In this section, the following mathematical programming models are discussed:

- p-Median model
- Generalized p-median model

10.2.2.1   *The p-Median Model.* The p-median model is used to group n parts into p part families. In order to consider the p-median model, the following notation needs to be defined:

$m =$ number of machines
$n =$ number of parts
$p =$ number of part families
$x_{ij} = \begin{cases} 1 & \text{if part } i \text{ belongs to part family } j \\ 0 & \text{otherwise} \end{cases}$
$d_{ij} =$ distance measure between parts i and j

The objective function of the p-median model is to minimize the total distance between any two parts i and j:

$$\min \sum_{i=1}^{n} \sum_{j=1}^{n} d_{ij} x_{ij} \tag{10.22}$$

subject to

$$\sum_{j=1}^{n} x_{ij} = 1 \quad \text{for all } i = 1, \ldots, n \tag{10.23}$$

$$\sum_{j=1}^{n} x_{jj} = p \tag{10.24}$$

$$x_{ij} \leq x_{jj} \quad \text{for all } i = 1, \ldots, n, j = 1, \ldots, n \tag{10.25}$$

$$x_{ij} = 0,1 \quad \text{for all } i = 1, \ldots, n, j = 1, \ldots, n \tag{10.26}$$

Note that in this model subscript j has a double meaning, part number and part family number. Constraint (10.23) ensures that each part belongs to exactly one part family. Constraint (10.24) specifies the required number of part families. Constraint (10.25) ensures that part i belongs to part family j only when this part family is formed. The last constraint, (10.26), ensures integrality. Note that in the p-median model, the required number of part families, p, is specified in advance. The p-median model (10.22)–(10.26) is illustrated in Example 10.4.

Example 10.4.   Given the machine–part incidence matrix (10.1), form $p = 2$ part families and their corresponding machine cells. For the matrix (10.1), the following matrix $[d_{ij}]$ of Hamming distances is obtained:

$$
[d_{ij}] =
\begin{array}{c}
 \\
 \\
 \\
 \\
 \\
\end{array}
\begin{array}{cc}
 & \text{Part number} \\
 & \begin{array}{ccccc} 1 & 2 & 3 & 4 & 5 \end{array} \\
\begin{array}{c} 1 \\ 2 \\ 3 \\ 4 \\ 5 \end{array} &
\left[
\begin{array}{ccccc}
0 & 4 & 0 & 4 & 4 \\
4 & 0 & 4 & 0 & 1 \\
0 & 4 & 0 & 4 & 3 \\
4 & 0 & 4 & 0 & 1 \\
4 & 1 & 3 & 1 & 0
\end{array}
\right]
\end{array}
\begin{array}{c}
 \\
 \\
\text{Part} \\
\text{number}
\end{array}
$$

Note that all diagonal elements in the matrix $[d_{ij}]$ are 0's. Solving the p-median model (10.22)–(10.26) for the above matrix and $p = 2$ results in the following solution (see Appendix 10.1 for the model listing):

$$x_{11} = 1, \qquad x_{31} = 1$$

and

$$x_{24} = 1, \qquad x_{44} = 1 \qquad x_{54} = 1$$

Based on the definition of x_{ij}, two part families are formed:

$$PF\text{-}1 = \{1, 3\} \qquad PF\text{-}2 = \{2, 4, 5\}$$

Recall that the part family number is indicated by the second subscript of x, while the part number is indicated by the first subscript. For convenience, PF-4 indicated by the solution has been changed to PF-2.

For the two part families, the corresponding two machine cells are determined from matrix (10.1):

$$MC\text{-}1 = \{2, 4\} \qquad MC\text{-}2 = \{1, 3\}$$

One can easily observe that the solution to the p-median model is identical to that obtained using the matrix formulation [see matrix (10.2)].

10.2.2.2 *Generalized p-Median Model.* The model (10.22)–(10.26) has been developed under the assumption that for each part i only one process plan is developed. To relax this constraint, Kusiak (1987) modified the model (10.22)–(10.26) so that it permits consideration of more than one process plan for each part. In addition, a production cost was associated with each process plan.

To present the generalized p-median model, define:

$n =$ total number of process plans
$F_k =$ set of process plans for part number k, $k = 1, \ldots, l$, where $\cup_{k=1}^{l}|F_k| = n$
$p =$ minimum required number of part (process) families
$d_{ij} =$ distance measure between process plans i and j
$c_j =$ cost of process plan j

The objective of the generalized p-median model is to minimize the total distance measure and cost of the process plans:

$$\min \sum_{i=1}^{n} \sum_{j=1}^{n} d_{ij} x_{ij} + \sum_{j=1}^{n} c_j x_{jj} \tag{10.27}$$

subject to

$$\sum_{i \in F_k} \sum_{j=1}^{n} x_{ij} = 1 \quad \text{for all } k = 1, \ldots, l \tag{10.28}$$

$$\sum_{j=1}^{n} x_{jj} \leq p \tag{10.29}$$

$$x_{ij} \le x_{jj} \quad \text{for all } i = 1, \ldots, n, \, j = 1, \ldots, n \tag{10.30}$$

$$x_{ij} = 0, 1 \quad \text{for all } i = 1, \ldots, n, \, j = 1, \ldots, n \tag{10.31}$$

Constraint (10.28) ensures that for each part (a set of process plans) exactly one process plan is selected. Constraint (10.29) imposes a lower bound on the number of part families. Constraints (10.30) and (10.31) correspond to constraints (10.25) and (10.26) in the model (10.22)–(10.26), respectively. Model (10.27)–(10.31) is illustrated in Example 10.5.

Example 10.5. Given the machine–part process incidence matrix (10.32) with the corresponding vector of process plan costs (10.33), form $p = 2$ process families and two machine cells using the model (10.27)–(10.31):

		Part number									
	1			2		3		4		5	
				Process plan number							
Machine number	1	2	3	4	5	6	7	8	9	10	11
1			1		1	1		1	1		1
2		1	1	1			1				
3	1			1	1				1	1	
4	1	1				1	1	1		1	

$$[a_{ij}] \tag{10.32}$$

For simplicity, the vector of process plan costs is

$$[c_j] = [1, 1, 1, 1, 1, 1, 1, 1, 1, 1, 1] \tag{10.33}$$

The Hamming distances d_{ij} for matrix (10.32) are shown in the matrix

Process plan number	1	2	3	4	5	6	7	8	9	10	11
1	0	∞	∞	2	2	2	2	2	2	0	3
2	∞	0	∞	2	4	2	0	2	4	2	3
3	∞	∞	0	2	2	2	2	2	2	4	1
4	2	2	2	0	∞	4	2	4	2	2	3
5	2	4	2	∞	0	2	4	2	0	2	1
6	2	2	2	4	2	0	∞	0	2	2	1
7	2	0	2	2	4	∞	0	2	4	2	3
8	2	2	2	4	2	0	2	0	∞	2	1
9	2	4	2	2	0	2	4	∞	0	2	1
10	0	2	4	2	2	2	2	2	2	0	∞
11	3	3	1	3	1	1	3	1	1	∞	0

$$[d_{ij}] \tag{10.34}$$

All entries in matrix (10.34) with value other than ∞ indicate that for each F_k, only one process plan is to be selected. Solving the model (10.27)–(10.31) for the above data with the LINDO software produces the following solution (see Appendix 10.2 for the model listing):

$$x_{27} = 1, \qquad x_{77} = 1$$

and

$$x_{59} = 1, \qquad x_{99} = 1, \qquad x_{11,9} = 1$$

This solution can be interpreted as follows:

$$PF\text{-}1 = \{2, 7\}(\text{relabeled PF-7})$$

$$PF\text{-}2 = \{5, 9, 11\}(\text{relabeled PF-9})$$

For these two process families from matrix (10.32), two machine cells are obtained:

$$MC\text{-}1 = \{2, 4\} \qquad MC\text{-}2 = \{1, 3\}$$

The final result is shown in matrix (10.35), where two machine cells and two process families are visible:

		Process family PF-1		Process family PF-2		
		2	7	5	9	11
Machine cell MC-1	2	1	1			
	4	1	1			
Machine cell MC-2	1			1	1	1
	3			1	1	

(10.35)

10.2.3 Innovative Applications of Group Technology

The decomposition concept used in GT can also be used in numerous innovative applications, for example to form tool–part families. Consider a set of five parts that are to be machined in a given time period. Each part requires a number of tools and fixtures. Assume that for the set of parts and tools the following incidence matrix has been formed:

		Part number				
		1	2	3	4	5
	1	1		1		
	2		1			1
	3		1		1	1
Tool	4	1		1		
number	5				1	1
	6	1				
	7			1		
	8		1		1	1

(10.36)

Applying the cluster identification algorithm to matrix (10.36) results in the matrix

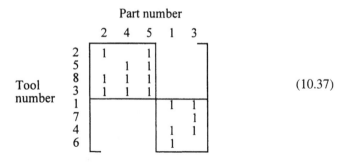

$$(10.37)$$

Matrix (10.37) consists of two clusters of tools, TF-1 = {2, 5, 8, 3} and TF-2 = {1, 7, 4, 6}, and two corresponding part families, PF-1 = {2, 4, 5} and PF-2 = {1, 3}.

Clustering parts and tools simplifies tool management and part scheduling.

10.2.3.1 *Data Mining.* The recently emerged concepts of data mining and data farming have created enormous potential for applications of the models and algorithms discussed in this chapter. Rather than grouping physical entities such as parts, machines, or tools, data items are grouped to determine their similarity, trends, and so on. Replacing a machine with an object and a part with a feature (attribute, characteristic) leads to an object–feature incidence matrix. A group of objects and the corresponding group of features represent a large number of items included in these groups, thus clearly expressing the data dependency in a structured form, in particular a block diagonal.

For other methods of data mining see Chapter 17.

10.3 BRANCHING ALGORITHMS

In this section, heuristic algorithms using different branching schemes are developed for solving the GT problem with bottleneck parts and bottleneck machines. The main features of the algorithms developed here are as follows:

- High-quality solutions
- Detection of bottleneck parts and bottleneck machines
- Ease of incorporation of constraints of any type, for example, cell capacity constraints
- Simplicity of algorithms and ease of coding.

Each of the three heuristics uses the *cluster identification algorithm* from page 52 of Chapter 2.

The implicit enumeration algorithm presented next uses the CI algorithm at each node.

Algorithm 10.1

Step 0. (Initialization): Begin with the incidence matrix $[a_{ij}]$ at level 0. Solve the GT problem represented with $[a_{ij}]$ with the CI algorithm.

Step 1. (Branching): Using the breadth-first search strategy, select an active node (not fathomed) and solve the corresponding GT problem with the CI algorithm.

Step 2. (Fathoming): Exclude a new node from further consideration if:

    Test 1: Cluster size is not satisfactory.

    Test 2: Cluster structure is not satisfactory.

Step 3. (Backtracking): Return to an active node.

Step 4. (Stopping rule): Stop when there are no active nodes remained; the current incumbent solution is optimal; otherwise go to Step 1.

Branching. Branching is performed in one of the following two ways:

1. By the CI algorithm when the incidence matrix partitions into mutually separable submatrices.
2. By removal of one column at a time from the corresponding incidence matrix when the matrix does not partition into mutually separable submatrices.

Fathoming. Fathoming is based on the following tests:

    Test 1: Cluster size is not satisfactory.

    Test 2: Cluster structure is not satisfactory.

One of the parameters of grouping might be an upper limit on the number of machines or parts in a cluster. This requirement is checked in test 1. Fathoming can also be performed based on the structure of the incidence matrix (e.g., the matrix sparcity). The branching algorithm is illustrated in Example 10.6.

Example 10.6. Solve the GT problem represented by the following machine–part incidence matrix:

$$
\begin{array}{c}
\text{Part number}
\end{array}
$$

	1	2	3	4	5	6	7
1	1		1		1		
2		1				1	
3				1			
4			1			1	
5			1				
6		1					
7	1			1			1
8			1	1			

Machine number   (10.38)

Assume that the size of each machine cell is to be not smaller than 2 and not greater than 3.

Applying the CI algorithm, matrix (10.38) partitions (Step 1: branching) into the following two submatrices:

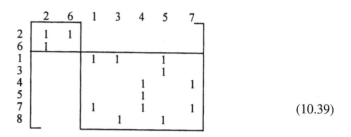

$$(10.39)$$

with the corresponding machine cells

$$MC\text{-}1 = \{2, 6\}, \qquad MC'\text{-}2 = \{1, 3, 4, 5, 7, 8\}$$

and part families

$$PF\text{-}1 = \{2, 6\}, \qquad PF'\text{-}2 = \{1, 3, 4, 5, 7\}.$$

Since none of the columns has been removed, each branch is assigned a value 0 (see Figure 10.8). Also, since the size of the matrix representing MC-1 is satisfactory (Step 2: fathoming), the corresponding matrix becomes an incumbent solution.

Further branching (Step 1) is performed on the active matrix. The resulting submatrices are presented in Figure 10.8. The five matrices at level 2 in Figure 10.8 are generated by removing one column at a time from the matrix at level 1. Only the matrix with the column (part) 1 removed has partitioned into two submatrices (level 3 in Figure 10.8). Each of the two submatrices satisfies test 1. The final solution of the GT problem is shown in the matrix

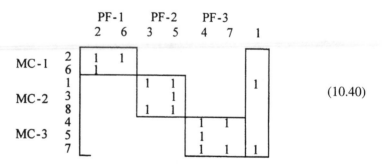

$$(10.40)$$

Three machine cells MC-1 = {2, 6}, MC-2 = {1, 3, 8}, MC-3 = {4, 5, 7}, three part families PF-1 = {2, 6}, PF-2 = {3, 5}, PF-3 = {4, 7}, and the bottleneck part 1 are visible in matrix (10.40).

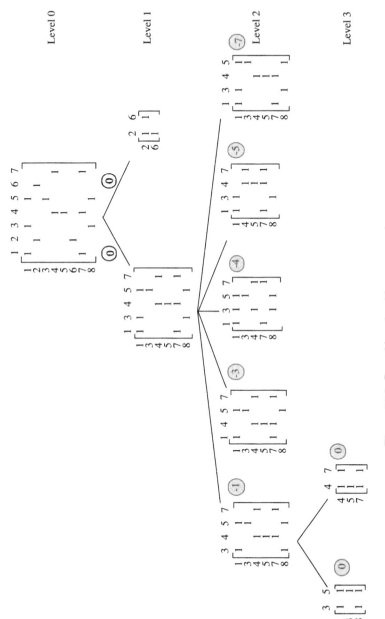

Figure 10.8. Branching algorithm enumeration tree.

Algorithm 10.2.   Algorithm 10.1 can be used to solve the GT problem with bottleneck machines provided that its Step 1 (branching) is modified. Rather than removing columns, rows representing multiple machines are added. The branching scheme is discussed next.

Branching Scheme.   Consider the group technology problem represented in the matrix

$$
\begin{array}{c}
\text{Part number}\\
\begin{array}{cccccc}
 & 1 & 2 & 3 & 4 & 5
\end{array}\\
\begin{array}{c}
\text{Machine} \\
\text{number}
\end{array}
\begin{array}{c}
1\\2\\3\\4
\end{array}
\left[
\begin{array}{ccccc}
 & 1 & 1 & & 1\\
1 & & & 1 & \\
 & 1 & 1 & & 1\\
1 & & 1 & 1 &
\end{array}
\right]
\end{array}
\qquad (10.41)
$$

As each of the four machines in matrix (10.41) could potentially be a bottleneck machine, the maximum number of multiple copies that one may consider for each machine is equal to the number of parts processed on that machine.

Assuming that machine 1 is the bottleneck machine, matrix (10.41) is transformed into matrix (10.42), where each copy of machine 1 corresponds to one part:

$$
\begin{array}{c}
\text{Part number}\\
\begin{array}{cccccc}
 & 1 & 2 & 3 & 4 & 5
\end{array}\\
\begin{array}{c}
\text{Machine} \\
\text{number}
\end{array}
\begin{array}{c}
1(1)\\1(2)\\1(3)\\2\\3\\4
\end{array}
\left[
\begin{array}{ccccc}
 & 1 & & & \\
 & & 1 & & \\
 & & & & 1\\
1 & & & 1 & \\
 & 1 & 1 & & 1\\
1 & & 1 & 1 &
\end{array}
\right]
\end{array}
\qquad (10.42)
$$

Applying the CI algorithm to matrix (10.42) does not lead to the decomposition of the matrix. This proves that machine 1 is not a bottleneck machine. The same is true for machines 2 and 3.

Considering machine 4 as the bottleneck machine results in the matrix

$$
\begin{array}{c}
\begin{array}{cccccc}
 & 1 & 2 & 3 & 4 & 5
\end{array}\\
\begin{array}{c}
1\\2\\3\\4(1)\\4(2)\\4(3)
\end{array}
\left[
\begin{array}{ccccc}
 & 1 & 1 & & 1\\
1 & & & 1 & \\
 & 1 & 1 & & 1\\
1 & & & & \\
 & & 1 & & \\
 & & & 1 &
\end{array}
\right]
\end{array}
\qquad (10.43)
$$

In the first iteration of the CI algorithm the following matrix is obtained:

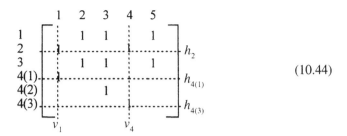

$$(10.44)$$

Two copies 4(1) and 4(3) of machine 4 are involved in the first machine cell. However, for the first cell it is enough to consider only one (the first copy) as illustrated in the matrix

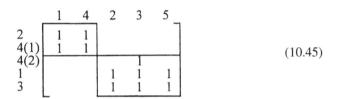

$$(10.45)$$

The second copy of machine 4 is used to form the second cell. In the case considered only two of three copies (one for each of the two cells) of machine 4 are needed to decompose incidence matrix (10.43).

Algorithm 10.1 removes one part (column) at a time and applies the CI algorithm to the corresponding matrix. It uses an explicit enumeration scheme that could potentially lead to the computation of feasible solutions. Algorithm 10.2 inserts machines (rows), one at a time, and applies the CI algorithm. The two branching schemes are simple but not efficient. The algorithm presented next uses a branching scheme that does not guarantee optimality, as do the complete enumeration schemes in Algorithms 10.1 and 10.2; however, it is computationally more efficient (Kusiak and Park, 1990).

Algorithm 10.3

Step 0. Begin with the incidence matrix $[a_{ij}]$. Solve the GT problem represented with $[a_{ij}]$ using the CI algorithm.

Step 1. If the incidence matrix does not partition into mutually separable submatrices, apply the branching scheme below; otherwise, stop.

Step 2. Using the selection scheme below, select a child node, solve the corresponding GT problem with the CI algorithm, and go to Step 1.

Branching Scheme. The branching is performed as follows:

- Child nodes are generated by removing one column at a time from the corresponding incidence matrix.

The node "–*j*" is read: the child node is generated by removing the *j*th column from the incidence matrix [a_{ij}].

Selection Scheme. For each row of the incidence matrix (child node) generated by the branching scheme above, calculate a row index (the number of 1's in the corresponding row). Figure 10.9 shows the row index for the incidence matrix (child node). The maximum value of the row index (MRI) is computed at each child node. For example, the MRI for the matrix in Figure 10.9 is 3.
    The selection scheme is implemented as follows:

1. Select a node with the minimum value of MRI among the child nodes at the same level.
2. If there exist a tie, select an arbitrary node.

The heuristic Algorithm 10.3 is illustrated in Example 10.7.

Example 10.7. Solve the GT problem represented by the machine–part incidence matrix (10.46). Assume that the size of each machine cell is not restricted:

$$
\begin{array}{c}
\text{Part number}\\
\begin{array}{ccccc}
1 & 2 & 3 & 4 & 5
\end{array}\\
\begin{array}{c}
1\\
\text{Machine } 2\\
\text{number } 3\\
4
\end{array}
\left[
\begin{array}{ccccc}
1 & 1 & & & 1\\
1 & & & 1 & \\
& 1 & 1 & & 1\\
1 & & & 1 & 1
\end{array}
\right]
\end{array}
\qquad (10.46)
$$

Step 0. Matrix (10.46) does not partition into mutually separable submatrices with the CI algorithm.

Step 1. Performing branching by removing one column at a time generates five child nodes –1, –2, –3, –4, and –5 shown in Figure 10.10.

$$
\begin{array}{c}
\begin{array}{c}
\text{Part}\\
\text{number}
\end{array}\\
\begin{array}{cccc}
1 & 4 & 5 & 7
\end{array}\quad\text{Row index}\\
\begin{array}{c}
1\\
3\\
\text{Machine } 4\\
\text{number } 5\\
7\\
8
\end{array}
\left[
\begin{array}{cccc}
1 & & 1 & \\
& & 1 & \\
& 1 & & 1\\
& 1 & & \\
1 & 1 & & 1\\
& & 1 &
\end{array}
\right]
\begin{array}{c}
2\\
1\\
2\\
1\\
3\\
1
\end{array}
\end{array}
$$

Figure 10.9. Row index.

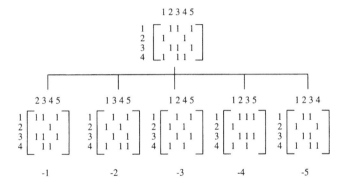

Figure 10.10. Child nodes obtained from matrix (10.46).

Step 2. For each child node, the row index is calculated as shown in Table 10.1. Then the MRI for each matrix is computed (see Table 10.2). Child node −3 [the column (part) 3 removed] with the lowest MRI of 2 is selected.

Applying the CI algorithm to the matrix corresponding to child node −3 results in matrix (10.47), where two machine cells and two part families are visible:

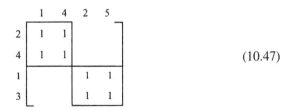

(10.47)

The final solution of the GT problem is shown in matrix (10.48), that is, machine cells, two part families, and one bottleneck part 3:

TABLE 10.1. Row Index for Child Nodes

Row	Child Nodes				
	−1	−2	−3	−4	−5
1	3	2	2	3	2
2	1	2	2	1	2
3	3	2	2	3	2
4	2	3	2	2	3

TABLE 10.2. MRIs for Child Nodes

Child Node	−1	−2	−3	−4	−5
MRI	3	3	2	3	3

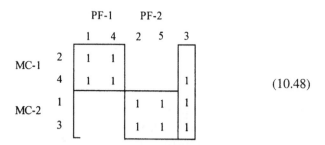

(10.48)

Applying Algorithm 10.3 may result in undesirable partitioning of the incidence matrix; for example, some of the clusters created may be too large. Thus it is desirable to limit the size of each machine cell generated by the algorithm.

Algorithm 10.4 solves the GT problem where the size of each machine cell is limited.

Algorithm 10.4

Step 0. Begin with the incidence matrix $[a_{ij}]$. Solve the GT problem represented with $[a_{ij}]$ using the CI algorithm.

Step 1. For each submatrix identified by the CI algorithm: If the size of machine cell is satisfactory, store the corresponding matrix as a machine cell; otherwise, store it as a subproblem.

Step 2. If there is no subproblem to solve, stop; otherwise, arbitrarily select a subproblem to be solved.

Step 3. Perform branching using the strategy of Algorithm 10.3.

Step 4. Using the selection strategy of Algorithm 10.3, select a child node and solve the corresponding GT problem with the CI algorithm and go to Step 1.

Algorithm 10.4 is illustrated in Example 10.8.

Example 10.8. Solve the GT problem represented by the machine–part incidence matrix (10.49). Assume that the size of each machine cell is not smaller than 2 and not greater than 3:

Part number

Machine number	1	2	3	4	5	6	7
1	1		1		1		
2		1				1	
3					1		
4				1			1
5				1			
6		1					
7	1			1			1
8			1	1			

(10.49)

Step 0. Begin with the incidence matrix (10.49). Solve the GT problem with the CI algorithm. Applying the CI algorithm, matrix (10.49) partitions into the following two submatrices:

	2	6	1	3	4	5	7
2	1	1					
6	1						
1			1	1		1	
3						1	
4					1		1
5					1		
7			1		1		1
8				1	1		

(10.50)

with the corresponding machine cells

$$MC\text{-}1 = \{2, 6\} \qquad MC\text{-}2 = \{1, 3, 4, 5, 7, 8\}$$

and part families

$$PF\text{-}1 = \{2, 6\} \qquad PF\text{-}2 = \{1, 3, 4, 5, 7\}.$$

Step 1. Since the size of the matrix representing MC-1 is satisfactory, the corresponding matrix is stored as a machine cell. As the size of the matrix representing MC-2 is not satisfactory, the corresponding matrix (10.51) is stored as a subproblem:

	1	3	4	5	7
1	1	1		1	
3				1	
4			1		1
5			1		
7	1		1		1
8		1	1		

(10.51)

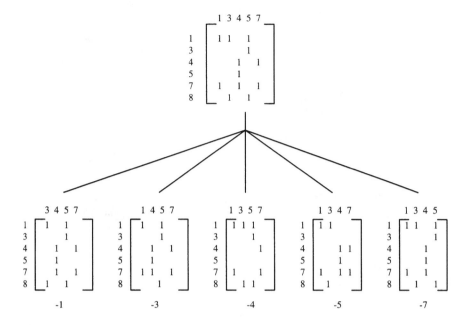

Figure 10.11. Child nodes obtained from matrix (10.51).

Step 2. Matrix (10.51) representing MC-2 is selected for solving.

Step 3. The five child nodes −1, −3, −4, −5, and −7 are generated by removing one column at a time from matrix (10.51), as shown in Figure 10.11.

Step 4. For each matrix, the row index is calculated as shown in Table 10.3. Then the MRI for each matrix is computed as shown in Table 10.4. Child node −1 (column (part) 1 removed) with the lowest MRI of 2 is selected. The CI algorithm partitions the matrix corresponding to the child node −1 into the following two submatrices:

TABLE 10.3. Row Index for Child Nodes

	Child Nodes				
Row	−1	−3	−4	−5	−7
1	2	2	3	2	3
3	1	1	1	0	1
4	2	2	1	2	1
5	1	1	1	1	1
7	2	3	2	3	2
8	2	1	2	1	1

TABLE 10.4. MRI of Child Nodes

Child Nodes	−1	−3	−4	−5	−7

$$
\begin{array}{c|cc|cc}
 & 3 & 5 & 4 & 7 \\
\hline
1 & 1 & 1 & & \\
3 & & 1 & & \\
8 & 1 & 1 & & \\
4 & & & 1 & 1 \\
5 & & & 1 & \\
7 & & & 1 & 1 \\
\end{array}
\qquad (10.52)
$$

Step 1. Since the size of each the two submatrices is satisfactory, store the resulting two submatrices as machine cells.

Step 2. As there is no subproblem left, stop.

The final solution of the GT problem is shown in the matrix

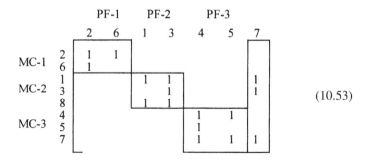

$$(10.53)$$

Three machine cells MC-1 = {2, 6}, MC-2 = {1, 3, 8}, MC-3 = {4, 5, 7}, three part families PF-1 = {2, 6}, PF-2 = {1, 3}, PF-3 = {4, 5}, and the bottleneck part 1 are visible in matrix (10.53).

In a machine–part incidence matrix, some machines (rows) might be incident to a large number of parts (have many 1's). Such machines may be potentially bottleneck machines. One would remove them from the incidence matrix prior to grouping.

Some parts (columns) considered as possible bottleneck parts can be removed from the incidence matrix. If the number of machines required by a part is greater than the maximum size of machine cells allowed, that part can be considered as a possible bottleneck part. Algorithm 10.5 incorporates screening of bottleneck parts and machines.

Algorithm 10.5

Step 0′. (Screening) Remove rows (columns) with a larger number of 1's than the threshold value for rows (columns).

Steps 0–4 are identical to Algorithm 10.4.

Threshold Value

1. The threshold value for screening of rows for possible bottleneck machines depends on the maximum number of parts allowed in a machine cell.
2. The threshold value for screening of columns for possible bottleneck parts is equal to the maximum-size machine cell.

Algorithm 10.5 is illustrated in Example 10.9.

Example 10.9. Solve the GT problem represented with the machine–part incidence matrix (10.54) (see Burbidge, 1975). Assume the maximum number of parts allowed in a machine cell is 15. Thus the threshold value for rows is 15. Assume that the size of each machine cell is not to be smaller than 2 or greater than 3. Thus the threshold value for columns is 3:

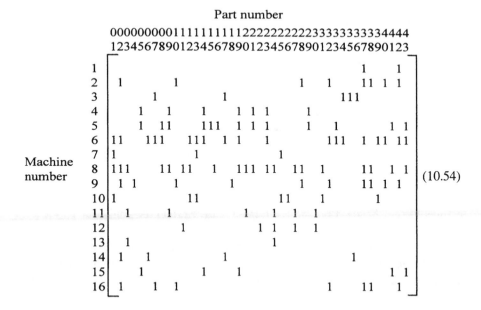

Step 0′. (Screening) As the number of entries 1 in rows 6 and 8 is 19 and 20, respectively, and the threshold value is 15, rows 6 and 8 are removed and matrix (10.55) is formed:

Part number

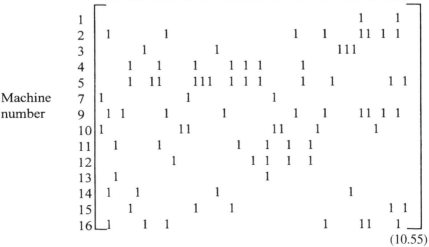

$$(10.55)$$

Since the number of entries 1 in columns 2, 37, and 42 is 4 and the threshold value is 3, remove columns 2, 37, and 42 and form matrix (10.56):

Part number

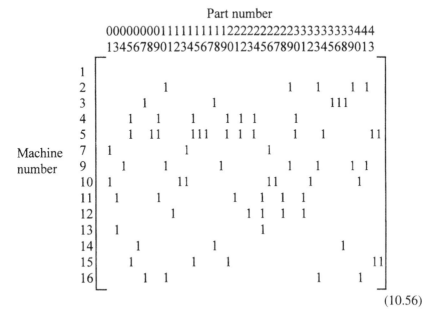

$$(10.56)$$

Step 0. Begin with the incidence matrix (10.56). Applying the CI algorithm to matrix (10.56), it partitions into the following three submatrices:

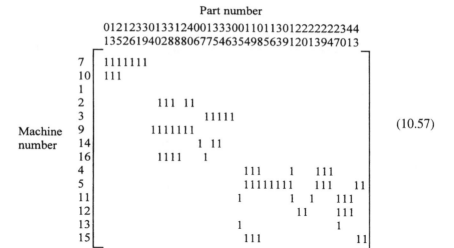

$$(10.57)$$

with the corresponding machine cells

MC-1 = {7, 10}   MC-2 = {1, 2, 3, 9, 14, 16}   MC-3 = {4, 5, 11, 12, 13, 15}

and part families

$$PF\text{-}1 = \{1, 13, 25, 12, 26, 31, 39\}$$

$$PF\text{-}2 = \{4, 10, 32, 38, 18, 28, 40, 6, 7, 17, 35, 34, 36\}$$

$$PF\text{-}3 = \{3, 5, 14, 19, 8, 15, 16, 33, 9, 11, 22, 20, 21, 23, 29, 24, 27, 30, 41, 43\}$$

Step 1. As the size of the matrix representing MC-1 is satisfactory, the corresponding matrix is stored as a machine cell. Since the size of the matrix representing MC-2 is not satisfactory, the corresponding (10.58) is stored as a subproblem:

```
            0 1 3 3 1 2 4 0 0 1 3 3 3
            4 0 2 8 8 8 0 6 7 7 5 4 6
        1 ┌                             ┐
        2 │   1 1 1   1 1                │
        3 │             1 1 1 1 1 1      │        (10.58)
        9 │ 1 1 1 1 1 1 1                │
       14 │             1   1 1          │
       16 └   1 1 1 1                    ┘
```

Also, the size of the matrix representing MC-3 is not satisfactory and the corresponding matrix (10.59) is stored as a subproblem:

$$
\begin{array}{c}
\begin{array}{l}
0\ 0\ 1\ 1\ 0\ 1\ 1\ 3\ 0\ 1\ 2\ 2\ 2\ 2\ 2\ 2\ 2\ 3\ 4\ 4 \\
3\ 5\ 4\ 9\ 8\ 5\ 6\ 3\ 9\ 1\ 2\ 0\ 1\ 3\ 9\ 4\ 7\ 0\ 1\ 3
\end{array} \\
\begin{array}{c}
4 \\ 5 \\ 11 \\ 12 \\ 13 \\ 15
\end{array}
\left[
\begin{array}{l}
1\ 1\ 1 \qquad\quad 1 \qquad 1\ 1\ 1 \\
1\ 1\ 1\ 1\ 1\ 1\ 1 \qquad 1\ 1\ 1 \qquad\quad 1\ 1 \\
1 \qquad\qquad\quad 1 \quad 1 \qquad 1\ 1\ 1 \\
\qquad\qquad\qquad\quad 1\ 1 \qquad 1\ 1\ 1 \\
1 \qquad\qquad\qquad\qquad\qquad 1 \\
\quad 1\ 1\ 1 \qquad\qquad\qquad\qquad 1\ 1
\end{array}
\right]
\end{array}
\qquad (10.59)
$$

Step 2. Matrix (10.58) representing MC-1 is selected for the decomposition.

Step 3. The 13 child nodes −4, −10, −32, −38, −18, −28, −40, −7, −6, −17, −35, −34, and −36 are generated by removing one column at a time from matrix (10.58).

Then for each child node, the row index is calculated as shown in Table 10.5. Child nodes −10, −32, −38 are selected for decomposition and merged into a single node (−10, −32, −38). The matrix corresponding to the child node (-10, -32, -38) does not partition into mutually separable submatrices.

After column 7 has been removed, matrix (10.58) partitions into two submatrices with one bottleneck part shown in matrix (10.60). The removed columns 10, 32, and 38 belong to the previously obtained submatrix:

$$
\begin{array}{c}
\begin{array}{l}
0\ 1\ 2\ 4\ 1\ 3\ 3\ 0\ 1\ 3\ 3\ 3\ 0 \\
4\ 8\ 8\ 0\ 0\ 2\ 8\ 6\ 7\ 5\ 4\ 6\ 7
\end{array} \\
\begin{array}{c}
2 \\ 9 \\ 16 \\ 3 \\ 14 \\ 1
\end{array}
\left[
\begin{array}{l}
\qquad 1\ 1\ 1\ 1\ 1 \\
1\ 1\ 1\ 1\ 1\ 1\ 1 \\
1 \qquad 1\ 1\ 1 \qquad\qquad\qquad 1 \\
\qquad\qquad\qquad\qquad 1\ 1\ 1\ 1\ 1 \\
\qquad\qquad\qquad\qquad 1\ 1\ 1
\end{array}
\right]
\end{array}
\qquad (10.60)
$$

Step 2. Matrix (10.59) representing MC-2 is solved next.

Step 3. The nineteen child nodes −3, −5, −14, 19, −8, −15, −16, −33, −9, −11, −22, −20, −21, −23, −29, −24, −27, −41, and −43 are generated by removing

TABLE 10.5. Row Index for Child Nodes

Row	Child Nodes												
	−4	−10	−32	−38	−18	−28	−40	−7	−6	−17	−35	−34	−36
1	0	0	0	0	0	0	0	0	0	0	0	0	0
2	5	4	4	4	5	4	4	5	5	5	5	5	5
3	6	6	6	6	6	6	6	5	5	5	5	5	5
9	6	6	6	6	6	6	6	7	7	7	7	7	7
14	3	3	3	3	3	3	3	3	2	2	2	3	3
16	5	4	4	4	4	5	5	4	5	5	5	5	5

one column at a time from matrix (10.59). The child node −9 is selected for decomposition. Applying the CI algorithm, matrix (10.59) partitions into the following two submatrices with bottleneck part 9:

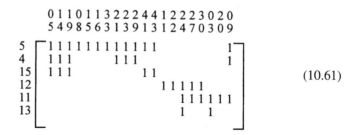

$$(10.61)$$

The final solution of the GT problem is presented in the matrix

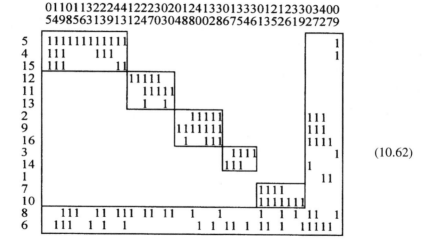

$$(10.62)$$

The solution in matrix (10.62) has the following structure:

- Machine cells: MC-1 = {4, 5, 15}, MC-2 = {11, 12, 13}, MC-3 = {2, 9, 16}, MC-4 = {3, 14}, MC-5 = {7, 10}
- Part families: PF-1 = {5, 8, 14, 15, 16, 19, 21, 23, 29, 41, 43}, PF-2 = {3, 11, 22, 20, 24, 27, 30}, PF-3 = {4, 18, 28, 40, 10, 32, 38}, PF-4 = {6, 17, 35, 34, 36}, PF-5 = {1, 13, 25, 12, 26, 31, 39}
- Bottleneck parts: 2, 37, 42, 7, 9
- Bottleneck machines: 6, 8

The question arises of which of the modeling and solution approaches discussed in this chapter should be used. Some of the representation approaches and algorithms are more suitable for demonstrating the principles of group technology, while others are appropriate for industrial applications. The two integer programming models offer flexibility in capturing numerous objective functions and constraints. The only difficulty with using these models for industrial applications is that they cannot be solved optimally for a large number of parts and machines, especially with standard software. The matrix representation appears to be the most suitable representation of industrial group technology problems. The cluster identification algorithm should always be used as the first solution approach as some large-scale group technology matrices may decompose into mutually separable submatrices. If one would like to maintain a certain degree of interaction with the computational process, the extended cluster identification algorithm can be used for further decomposition of the resulting submatrices. The branching algorithms discussed can perform the same task as the cluster identification algorithm, however without any user interaction. The basic difference between the branching algorithms is in the computational complexity. Algorithms 10.1 and 10.2 can generate optimal machine cells and part families but at the expense of computational time. Algorithm 10.3 and 10.4 provide heuristic solutions in a short time. The quality of the solutions generated by Algorithms 10.3 and 10.4 are acceptable for numerous industrial applications.

10.4   ASSIGNMENT OF PARTS TO THE EXISTING MACHINE CELLS

In some cases, machine cells might have been formed and one is interested in allocating new parts to the existing cells. This is often possible, as machines of the same type or at least with the same capability might exist in different cells. To present the model discussed in Stanfel (1989) for the assignment of parts to machine cells, the following need to be defined:

N = maximum number of newly assigned parts per cell

n = number of parts

m = number of machines

k = number of cells

K_l = set of cells that include machine l

M_i = set of machines that have to be visited by part i (included in process plan of part i)

$$x_{ij} = \begin{cases} 1 & \text{if } part i \text{ is assigned to machine cell } j \\ 0 & \text{otherwise} \end{cases}$$

The model (10.63)–(10.66) minimizes the total number of assignments (visits) of parts to cells:

$$\min \sum_{i=1}^{n} \sum_{j=1}^{k} x_{ij} \tag{10.63}$$

subject to

$$\sum_{j \in K_l} x_{ij} \geq 1 \quad \text{all } l \in M_i; \, i = 1, \ldots, n \tag{10.64}$$

$$\sum_{i=1}^{n} x_{ij} \leq N \qquad j = 1, \ldots, k \tag{10.65}$$

$$x_{ij} = 0,1 \quad i = 1, \ldots, n; \, j = 1, \ldots, k \tag{10.66}$$

Constraint (10.64) ensures that a part is assigned to at least one cell. The total number of assignments of parts per each cell is limited to N by constraint (10.65). Constraint (10.66) ensures integrality.

The model (10.63)–(10.66) is illustrated in Example 10.10 (Stanfel, 1989).

Example 10.10. Five parts are to be assigned to the existing four machine cells with the corresponding machine–part incidence matrix (10.67). This matrix indicates the machines required for processing each of the five parts:

$$
\text{Machine}
\begin{array}{c}
 \\
1 \\
2 \\
3 \\
4 \\
5 \\
6
\end{array}
\begin{array}{c}
\text{Part} \\
\begin{array}{ccccc}
1 & 2 & 3 & 4 & 5
\end{array} \\
\left[
\begin{array}{ccccc}
  &   & 1 & 1 &   \\
1 & 1 &   &   &   \\
1 &   & 1 &   & 1 \\
  &   &   & 1 & 1 \\
  & 1 & 1 &   &   \\
1 &   &   &   & 1
\end{array}
\right]
\end{array}
\tag{10.67}
$$

The existing four cells include machine types 1, . . . , 6 as shown in the matrix

	Machine number		
Machine cell 1	1	2	
2	1	3	
3	2	4	6
4	1	2	5

(10.68)

Assume that the maximum number of assignments (visits) of parts per cell is $N = 3$.

The model (10.63)–(10.66) is easy to formulate once all sets K_l and M_i have been defined. The formation of sets K_l and M_i for part 1 and the corresponding constraints are illustrated next:

The set M_1 of machines to be visited by part 1 is determined from matrix (10.67):

$$M_1 = \{2, 3, 6\}$$

The sets of cells that include each of the three machines 2, 3, and 6 in the set M_1 are determined from (10.68) as follows:

$$K_2 = \{1, 3, 4\} \quad K_3 = \{2\} \quad K_6 = \{3\}$$

Based on the sets M_1, K_2, K_3, and K_6, constraints (10.64) for part 1 are defined:

$$x_{11} + x_{13} + x_{14} \geq 1$$

$$x_{12} \geq 1 \quad \text{for part 1}$$

$$x_{13} \geq 1$$

For the data above, the model (10.63)–(10.66) is formulated as follows:

$$\min x_{11} + x_{12} + \cdots + x_{14} + x_{21} + \cdots + x_{24}$$

$$+ x_{31} + \cdots + x_{34} + x_{41} + \cdots + x_{44} + x_{51} + \cdots + x_{54}$$

$$x_{11} + x_{13} + x_{14} \geq 1$$

$$x_{12} \geq 1 \text{ for part1}$$

$$x_{13} \geq 1$$

$$x_{21} + x_{22} + x_{24} \geq 1$$

$$x_{21} + x_{23} + x_{24} \geq 1 \text{ for part2}$$

$$x_{24} \geq 1$$

$$x_{32} \geq 1$$

$$x_{34} \geq 1 \text{ for part3}$$

$$x_{41} + x_{42} + x_{44} \geq 1$$

$$x_{43} \geq 1 \text{ for part 4}$$

$$x_{52} \geq 1$$

$$x_{53} \geq 1 \text{ for part 5}$$

The constraints (10.65) that not more than $N = 3$ parts may visit a cell are as follows:

$$x_{11} + x_{21} + x_{31} + x_{41} + x_{51} \leq 3$$

$$x_{12} + x_{22} + x_{32} + x_{42} + x_{52} \leq 3$$

$$x_{13} + x_{23} + x_{33} + x_{43} + x_{53} \leq 3$$

$$x_{14} + x_{24} + x_{34} + x_{44} + x_{54} \leq 3$$

$$x_{ij} = 0,1 \qquad i = 1, \ldots, 5 \qquad j = 1, \ldots, 4$$

Solving the above program with an integer programming code produces the following solution: $x_{12} = x_{13} = x_{24} = x_{32} = x_{34} = x_{43} = x_{44} = x_{52} = x_{53} = 1$.

Recalling that the first subscript in the above solution denotes part number and the second denotes cell number, the five parts have been assigned as follows:

Machine cell 2: parts 1, 3, 5
Machine cell 3: parts 1, 4, 5
Machine cell 4: parts 2, 3, 4

The solution indicates that each of the parts 1, 3, and 5 visit two machine cells. None of the five new parts visits machine cell 1.

10.5 SUMMARY

In this chapter, the following two basic methods of solving the group technology problem were discussed:

- Classification
- Cluster analysis

The classification method is used to group parts into part families based on their design features. Two variations of the classification method, the visual and coding methods, were outlined. The visual method is applicable when the number of parts is rather limited. Therefore, the coding methods, which are more commonly used, were discussed.

The cluster analysis approach was used to group machines into machine cells and parts into part families. To model the group technology problem, two clustering formulations were presented: the matrix formulation and mathematical programming formulation. Since the GT problem is NP complete, heuristic algorithms are most likely to be used for solving large-scale industrial problems. The results of grouping leads to the physical machine layout or logical machine layout. The latter is used when the production content changes rather frequently. Group technology applied to tools and parts simplifies tool management and part scheduling.

The cluster identification algorithm and its extensions are appropriate for solving large-scale group technology problems.

Five grouping algorithms for solving the GT problem with bottleneck parts and bottleneck machines were developed. They take advantage of the efficiency and simplicity of the cluster identification algorithm.

The key features of the grouping algorithms presented in this chapter are summarized as follows:

- Algorithms can handle GT problems with bottleneck parts and bottleneck machines.

- The machine-cell-size constraint can be imposed.

- Screening used in Algorithm 10.5 identifies possible bottleneck parts and bottleneck machines.

APPENDIX 10.1: MODEL LISTING FOR EXAMPLE 10.4

```
MIN
4X12+4X14+3X15+4X21+4X23+1X25+4X32+4X34+3X35+4X41+4X43
+1X45+3X51+1X52+3X53+1X54

SUBJECT TO
X11+X12+X13+X14+X15=1
X21+X22+X23+X24+X25=1
X31+X32+X33+X34+X35=1
X41+X42+X43+X44+X45=1
X51+X52+X53+X54+X55=1

X11+X22+X33+X44+X55=2

X21-X11<=0
```

```
X31-X11<=0
X41-X11<=0
X51-X11<=0
X12-X22<=0
X32-X22<=0
X42-X22<=0
X52-X22<=0
X13-X33<=0
X23-X33<=0
X43-X33<=0
X53-X33<=0
X14-X44<=0
X24-X44<=0
X34-X44<=0
X54-X44<=0
X15-X55<=0
X25-X55<=0
X35-X55<=0
X45-X55<=0

END
INTEGER 25
```

APPENDIX 10.2:   MODEL LISTING FOR EXAMPLE 10.5

```
MIN
999X12+999X13+2X14+2X15+2X16+2X17+2X18+2X19+3X1a1+999X21+
999X23+2X24+4X25+2X26+2X28+2X210+3X211+999X31+999X32+2X34
+2X35+2X36+2X37+2X38+2X39+4X310+1X311+2X41+2X42+2X43+999X
45+4X46+2X47+4X48+2X49+2X410+3X411+2X51+4X52+2X53+999X54+
2X56+4X57+2X58+2X510+1X511+2X61+2X62+2X63+4X64+2X65+999X6
7+2X69+2X610+1X611+2X71+2X73+2X74+4X75+999X76+2X78+4X79+2
X710+3X711+2X81+2X82+2X83+4X84+2X85+2X87+999X89+2X810+1X8
11+2X91+2X93+2X94+2X96+4X97+999X98+2X910+1X911+2X102+4X10
3+2X104+2X105+2X106+2X107+2X108+2X109+999X1011+3X11a1+
3X112+1X113+3X114+1X115+1X116+3X117+1X118+1X119+999X1110+
X11+X22+X33+X44+X55+X66+X77+X88+X99+X1010+X1111

SUBJECT TO
X11+X12+X13+X14+X15+X16+X17+X18+X19+X110+X1a11+
X21+X22+X23+X24+X25+X26+X27+X28+X29+X210+X211+
X31+X32+X33+X34+X35+X36+X37+X38+X39+X310+X311=1
X41+X42+X43+X44+X45+X46+X47+X48+X49+X410+X411+
X51+X52+X53+X54+X55+X56+X57+X58+X59+X510+X511=1
```

```
X61+X62+X63+X64+X65+X66+X67+X68+X69+X610+X611+
X71+X72+X73+X74+X75+X76+X77+X78+X79+X710+X711=1
X81+X82+X83+X84+X85+X86+X87+X88+X89+X810+X811+
X91+X92+X93+X94+X95+X96+X97+X98+X99+X910+X911=1
X101+X102+X103+X104+X105+X106+X107+X108+X109+X1010+X1011+
X11a1+X112+X113+X114+X115+X116+X117+X118+X119+X1110+X1111
=1

X11+X22+X33+X44+X55+X66+X77+X88+X99+X1010+X1111<=2

X21-X11<=0
X31-X11<=0
X41-X11<=0
X51-X11<=0
X61-X11<=0
X71-X11<=0
X81-X11<=0
X91-X11<=0
X101-X11<=0
X11a1-X11<=0
X12-X22<=0
X32-X22<=0
X42-X22<=0
X52-X22<=0
X62-X22<=0
X72-X22<=0
X82-X22<=0
X92-X22<=0
X102-X22<=0
X112-X22<=0
X13-X33<=0
X23-X33<=0
X43-X33<=0
X53-X33<=0
X63-X33<=0
X73-X33<=0
X83-X33<=0
X93-X33<=0
X103-X33<=0
X113-X33<=0
X14-X44<=0
X24-X44<=0
X34-X44<=0
X54-X44<=0
```

```
X64-X44<=0
X74-X44<=0
X84-X44<=0
X94-X44<=0
X104-X44<=0
X114-X44<=0
X15-X55<=0
X25-X55<=0
X35-X55<=0
X45-X55<=0
X65-X55<=0
X75-X55<=0
X85-X55<=0
X95-X55<=0
X105-X55<=0
X115-X55<=0
X16-X66<=0
X26-X66<=0
X36-X66<=0
X46-X66<=0
X56-X66<=
X76-X66<=0
X86-X66<=0
X96-X66<=0
X106-X66<=0
X116-X66<=0
X17-X77<=0
X27-X77<=0
X37-X77<=0
X47-X77<=
X57-X77<=0
X67-X77<=0
X87-X77<=0
X97-X77<=0
X107-X77<=0
X117-X77<=0
X18-X88<=0
X28-X88<=0
X38-X88<=0
X48-X88<=0
X58-X88<=0
X68-X88<=0
X78-X88<=0
X98-X88<=0
```

```
X108-X88<=0
X118-X88<=0
X19-X99<=0
X29-X99<=0
X39-X99<=0
X49-X99<=0
X59-X99<=0
X69-X99<=0
X79-X99<=0
X89-X99<=0
X109-X99<=0
X119-X99<=0
X110-X1010<=0
X210-X1010<=0
X310-X1010<=0
X410-X1010<=0
X510-X1010<=0
X610-X1010<=0
X710-X1010<=0
X810-X1010<=0
X910-X1010<=0
X1110-X1010<=0
X1a10-X1010<=0
X211-X1111<=0
X311-X1111<=0
X411-X1111<=0
X511-X1111<=0
X611-X1111<=0
X711-X1111<=
X811-X1111<=0
X911-X1111<=0
X1011-X1111<=0

END
INTEGER 121
```

REFERENCES

Arthanari, T. S., and Y. Dodge (1981), *Mathematical Programming in Statistics*, John Wiley, New York.

Burbidge, J. L. (1971), Production flow analysis, *The Production Engineer*, April/May, pp. 139–152.

Burbidge, J. L. (1975), *The Introduction to Group Technology*, Heinemann, London.

Chan, H. M., and D. A. Milner (1982), Direct clustering algorithm for group formation in cellular manufacturing, *Journal of Manufacturing Systems*, Vol. 1, No. 1, pp. 65–74.

Chandrasekharan, M. P., and R. Rajagopalan (1986), MODROC: An extension of rank order clustering for group technology, *International Journal of Production Research*, Vol. 24, No. 5, pp. 1221–1233.

Ingram, F. B. (1982), Group technology, *Production and Inventory Management Journal*, 4th quarter pp. 71–84.

Iri, M. (1968), On the synthesis of the loop cut set matrices and the related problem, in K. Kondo (Ed.), *RAAG Memoirs*, Vol. 4, pp. 376–410.

King, J. R. (1980), Machine-component group formation in production flow analysis: An approach using a rank order clustering algorithm, *International Journal of Production Research*, Vol. 18, No. 2, pp. 213–232.

King, J. R., and V. Nakornchai (1982), Machine-component group formation in group technology: Review and extension, *International Journal of Production Research*, Vol. 20, No. 1, pp. 117–133.

Kusiak, A. (1987), The generalized group technology concept, *International Journal of Production Research*, Vol. 25, No. 4, pp. 561–569.

Kusiak, A. (1990), *Intelligent Manufacturing Systems*, Prentice-Hall, Englewood Cliffs, NJ.

Kusiak, A., and W. S. Chow (1987), Efficient solving of the group technology problem, *Journal of Manufacturing Systems*, Vol. 6, No. 2, pp. 117–124.

Kusiak, A., and K. Park (1990), Group technology, Working Paper No. 90-14, Department of Industrial Engineering, The University of Iowa, Iowa City, IA.

Lee, R. C. T. (1981), Clustering analysis and its applications, in J. T. Tou (Ed.), *Advances in Information Systems Science*, Plenum, New York.

McAuley, J. (1972), Machine grouping for efficient production, *The Production Engineer*, February, pp. 53–57.

Stanfel, L. E. (1989), A successive approximations method for a cellular manufacturing problem, *Annals of Operations Research*, Vol. 17, pp. 13–30.

QUESTIONS

10.1. What is group technology?

10.2. What is the relationship between the concepts of aggregation and decomposition and group technology?

10.3. What are the classes of methods used in group technology?

10.4. What are the advantages and disadvantages of the classification and coding method in forming machine cells and part families?

10.5. What types of clustering approaches are used in group technology?

10.6. What are the advantages and disadvantages of King's algorithm?

10.7. What is a bottleneck part and a bottleneck machine?

10.8. What type of group technology problem can be solved with the cluster identification algorithm?

10.9. What is accomplished with branching in the decomposition of machine–part matrices?

10.10. What is the relationship between the generalized p-median model and the process plan selection model (9.16)–(9.20) from Chapter 9?

PROBLEMS

10.1. Consider a manufacturing system that includes four machines and five parts and is represented by the following machine–part incidence matrix:

$$
\text{Machines}
\begin{array}{c}
\text{Parts}\\
\begin{bmatrix}
1 & 0 & 1 & 0 & 0\\
0 & 1 & 0 & 1 & 1\\
1 & 0 & 1 & 0 & 0\\
0 & 1 & 1 & 0 & 1
\end{bmatrix}
\end{array}
\qquad (10.69)
$$

(a) Determine machine cells and part families with the rank-order algorithm (King's algorithm).

(b) Identify the bottleneck parts and bottleneck machines, if any.

10.2. Consider a manufacturing system with five machines and seven parts represented by the following incidence matrix:

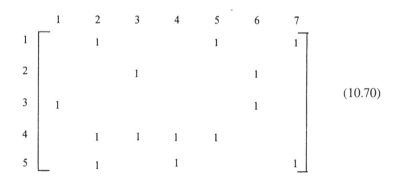

$$(10.70)$$

1. Identify two machine cells and two part families by solving the p-median problem in two different ways:

    (a) Considering distances between parts

    (b) Considering distances between machines

    Draw the solution matrix for 1(a) and 1(b).

2. Generate two machine cells and two part families with the branching algorithm.

    (a) Identify bottleneck part(s)

  (b) Draw the solution matrix.

 3. Comment on the differences between the solutions in 1(a), 1(b), and 2(b), if any.

10.3. Consider the following group technology problem with five machines and eight parts represented in the matrix (10.71)

Part

		1	2	3	4	5	6	7	8
	1	1					1		1
	2	1		1				1	1
Machine	3		1		1				
	4			1		1		1	
	5	1	1						

$$(10.71)$$

(a) Formulate a $p = 2$ median model and determine machine cells and part families.

(b) Does the p-median model generate an infeasible solution when the machine–part incidence matrix cannot be separated into mutually exclusive p submatrices? If yes, explain why.

(c) Does it make a difference whether the Hamming distances in model (a) are calculated for the parts or the machines?

(d) Identify bottleneck parts with an efficient branching algorithm. The number of machine cells must not be smaller than 2.

(e) Identify bottleneck machines with an efficient branching algorithm. The number of machine cells must not be smaller than 2.

(f) Explain the relationship between the number of machine cells and the number of bottleneck machines. If necessary, use an example.

Part

		1	2	3	4	5	6	7	8
	1	1					1		1
	2	1					1	1	1
Machine	3		1		1				
	4	1	1	1	1	1			
	5	1						1	

$$(10.72)$$

10.4. Consider the manufacturing system represented with five machines and eight parts in the matrix (10.72)

   (a) Determine two machine cells with the minimum number of bottleneck (overlapping) parts.

   (b) Determine two machine cells with the minimum number of bottleneck (overlapping) machines.

   (c) Which of the two cellular arrangements would you choose and under what circumstances?

   (d) Assume that for the manufacturing cells determined in (a) two new parts have been introduced; part 9 with routes through machines 1, 3, and 4 and part 10 with routes 1, 2, 3. Formulate a model for optimal assignment of the two new parts to the existing cells.

10.5. Matrix (10.73) represents a manufacturing system with five machines and eight process plans corresponding to the eight parts shown. Determine two machine cells with a suitable algorithm. Which process plan (or process plans) needs to be modified to form two mutually exclusive machine cells? How would you modify this process plan (or process plans)?

Process plan

$$
\text{Machine}\quad
\begin{array}{c}
\\1\\2\\3\\4\\5
\end{array}
\begin{array}{c}
1\ \ \ 2\ \ \ 3\ \ \ 4\ \ \ 5\ \ \ 6\ \ \ 7\ \ \ 8\\
\left[
\begin{array}{cccccccc}
1 & & & 1 & & 1 & & \\
& 1 & & & & & & 1 \\
1 & & 1 & & 1 & & 1 & \\
& 1 & & 1 & & 1 & & 1 \\
& & 1 & & & & 1 &
\end{array}
\right]
\end{array}
\qquad (10.73)
$$

10.6. Consider the manufacturing system represented in the matrix (10.74)

Part

$$
\text{Machine}\quad
\begin{array}{c}
\\1\\2\\3\\4\\5
\end{array}
\begin{array}{c}
1\ \ \ 2\ \ \ 3\ \ \ 4\ \ \ 5\ \ \ 6\ \ \ 7\ \ \ 8\\
\left[
\begin{array}{cccccccc}
1 & & & & & 1 & & 1 \\
1 & & 1 & & & & 1 & 1 \\
& 1 & & 1 & & & & \\
& & 1 & & 1 & & 1 & \\
1 & 1 & & & & & &
\end{array}
\right]
\end{array}
\qquad (10.74)
$$

(a) Generate two machine cells and two part families by a suitable algorithm. Remove no more than two bottleneck parts.

(b) Two months after a cellular layout you recommended is implemented, the company receives new orders for four parts. Their machining sequences are shown next:

Part 9: (1, 2, 4)

Part 10: (2, 4, 6)

Part 11: (2, 3, 5, 6)

Part 12: (2, 5)

Write a model for assigning the parts to the two machine cells in a way that the flow of parts among the two cells is minimized. Not more than three new parts can be assigned to a cell.

CHAPTER 11

NEURAL NETWORKS

11.1 INTRODUCTION

Neural networks constitute a new approach to computation. The basic unit of a neural network is a neuron (illustrated in Figure 11.1). Neurons are organized into networks similar to those found in the brain. These networks have characteristics analogous to human intelligence and are used for solving problems that have proven difficult or impossible to solve using conventional computation approaches. One of many applications of neutral networks is in data mining (Groth, 1998). For detailed coverage of data mining approaches other than neural networks see Chapter 17. Input signals come into a neuron along weighted connections (synapses) from neighboring neurons and excite (in the case of positive links) or inhibit (for negative links) the neuron's "firing" activity (Matheus and Hohensee, 1987). The magnitude of a weight determines how strongly the output of one neuron influences another neuron's activity: The larger the absolute weight between two neurons, the greater is the efficacy of the connection and the stronger the influence. Typically, the weighted

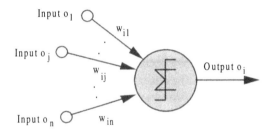

Figure 11.1. Artificial neuron (unit).

347

contributions from all inputs are summed to determine an activation level for that neuron:

$$a_i = \sum_{j=1}^{n} w_{ij} o_j$$

where

    $a_i =$ level of activation of neuron i

    $w_{ij} =$ weight of input j to neuron i

    $o_i =$ output signal from neuron i (input to subsequent neurons)

While a_i represents the neuron's "activation," it is not usually the same as the neuron's output signal o_i. Generally, o_i is calculated by applying a nonlinear output function, such as the simple step function:

$$o_i = \begin{cases} 1 & \text{if } a_i \geq \theta \\ 0 & \text{if } a_i < \theta \end{cases}$$

where θ is the threshold value at which the neuron turns "on." After its calculation, this output value is transmitted along the weighted outgoing connections to serve as input to subsequent neurons.

    Though this simple processing of input signals by an individual neuron is common to all neural networks, the details of implementation may vary between models. For example, a number of different output functions has been proposed. Figure 11.2 depicts the more common of these: a linear output function, a step function, and a sigmoidal output function. The range of output values assumed by the neurons may also vary with different models, for example, binary, graded, or continuous values. Some networks further incorporate the effect of activation decay and internal resistance, modeling functions similar to ones found in biological neurons. These and other details are, however, beyond the scope of this chapter and are presented in Rumelhart and McClelland (1986). It should be stressed that the basic operation of an individual neuron (unit) is simple:

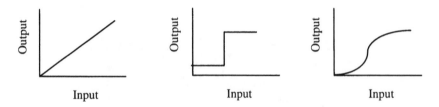

Figure 11.2. Output functions: (*a*) linear output function; (*b*) step function; (*c*) sigmoid output function.

1. A level of activity a_i is calculated from the weighted input signals from other neurons.
2. The neuron output signal o_i is calculated based on the neuron's activation.
3. The output o_i is transmitted along links to neighboring neurons for further processing.

A neural network is defined by specifying the connections between neurons. This specification usually takes the form of a connectivity or weight matrix. Connections are typically unidirectional, although bidirectional links are assumed in some models (e.g., Hopfield, 1982). In the most general case, each neuron in the network is connected to all other neurons, thus resulting in a totally connected network. In many neural networks, however, some connections are eliminated in order to improve their performance and simplify learning. Figure 11.3 illustrates a totally connected network and two networks with restricted architectures.

The architecture of a neural network defines the structure used for knowledge representation and processing. Different architectures impose different biases on the functions that can be represented within the network. As will become evident, knowledge representation in a neural network is quite different from knowledge-based systems. Two forms of knowledge representation can be identified within neural systems:

1. The state vector of unit activation
2. The representation of the mapping function encoded in the weights

Each of these is considered next as they play an important role in learning.

The *state vector* consists of the vector of output values (or, alternatively, activation values) taken across all neurons in the network. It represents the instantaneous piece of knowledge currently being processed by the network. For a given architecture, portions of this vector may be used to represent the input and output of the problem (see the input and output layers in Figures 3*b* and *c*).

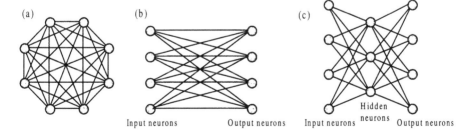

Figure 11.3. Network architectures: (*a*) completely connected network; (*b*) single-layer network; (*c*) multilayer network.

In some networks, such as the multilayered neural networks shown in Figure 11.3c, some neurons called *hidden neurons* do not serve either as input or as output. They are invisible to the external world. Hidden neurons are not initially assigned specific meaning but rather provide a network with an additional capability.

The second representational form is of longer duration, lasting beyond the short-term activation state vector. This persistent knowledge is encoded in the weights between the units, and it is through the modification of the distributed knowledge structure that learning occurs. Implicitly represented across these connections is the mapping function that defines how a given input state is translated into its associated output. Because all weights contribute to the representation, the acquired function is said to be "distributed" across the network, as opposed to being represented locally by an individual unit. The notion of a distributed representation is a matter of much debate and confusion. For a more thorough discussion of this topic the reader is referred to Rumelhart et al. (1986) and Hinton (1984).

11.2 NEURAL NETWORKS VERSUS OTHER INTELLIGENT APPROACHES

11.2.1 Knowledge-Based Systems

A traditional knowledge-based system is illustrated in Example 11.1 based on Zarefar and Goulding (1992).

Example 11.1. Consider the knowledge-based system for checking the parallelism of two lines A and B with three production rules R_1, R_2, and R_3 presented next.

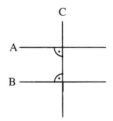

R_1: IF $A \perp C$
   THEN GOAL I
R_2: IF $B \perp C$
   THEN GOAL II
R_3: IF GOAL I AND GOAL II
   THEN $A//B$

The inference tree corresponding to the three rules is shown in Figure 11.4. The goal "line A is parallel to line B" can be easily inferenced given the facts A, B, and C. Note that the above knowledge-based system does not consider any tolerance between the

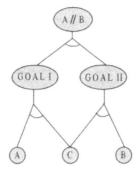

Figure 11.4. Inference tree for the rules R_1, R_2, and R_3.

angle of the lines; that is, it assumes that the lines are exactly perpendicular. The three production rules R_1', R_2', and R_3' presented next allow for a tolerance of $\pm 0.2°$.

R_1': IF $|\angle AC - 90°| \leq 0.2°$

    THEN GOAL I

R_2': IF $|\angle BC - 90°| \leq 0.2°$

    THEN GOAL II

R_3': IF GOAL I AND GOAL II

    THEN $A//B$

The nature of the decision made by the knowledge-based system with rules R_1', R_2', and R_3' is illustrated in Figure 11.5. The "Yes" solution is attained within the range $[-0.2°, +0.2°]$; otherwise, the "No" solution is generated.

One difficulty with this knowledge-based system is that the tolerance limit has to be specified in advance. Another problem is caused by the accumulation of errors during the inference process. A fuzzy-logic-based system discussed next allows for a continuous mapping of relationships between production rules.

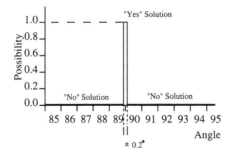

Figure 11.5. Binary interpretation of production rules R_1', R_2', R_3'.

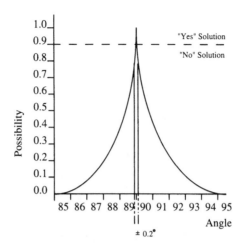

Figure 11.6. Possibility function.

11.2.2 Fuzzy-Logic-Based Systems

A fuzzy-logic-based system for checking of parallelism of two lines A and B from Example 11.1 is illustrated in Example 11.2.

Example 11.2. Rather than using a step function as in Figure 11.5, a continuous distribution of tolerances is considered in Figure 11.6 in the form of an exponential possibility function (see Zarefar and Goulding, 1992). The possibility function is used in the three production rules R_1'', R_2'', and R_3'' presented next.

$R_1'': P_{AC} = e^{|1.0536(AC-90°)|}$ FOR $AC = 90°$

$R_2'': P_{BC} = e^{|-1.0536(BC-90°)|}$ $P_{AC} = 1$

  FOR $BC = 90°$

  $P_{BC} = 1$

$R_3'':$ IF $(P_{AC} + P_{BC}) > 1.8$ $1 + 1 = 2 > 1.8$

  THEN $A//B$

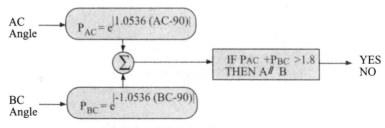

Figure 11.7. Logic-based system.

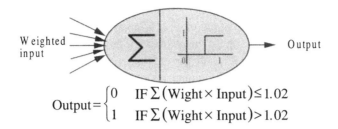

$$\text{Output} = \begin{cases} 0 & \text{IF } \Sigma\left(\text{Wight} \times \text{Input}\right) \le 1.02 \\ 1 & \text{IF } \Sigma\left(\text{Wight} \times \text{Input}\right) > 1.02 \end{cases}$$

Figure 11.8. Neural network concept.

As demonstrated above, if the angle $AC = 90°$ and $BC = 90°$, the value of the possibility function is $P_{AC} = 1$ and $P_{BC} = 1$, respectively. In this case, the system represented with the three rules R_1'', R_2'', and R_3'' is equivalent to the system represented with rules R_1, R_2, and R_3. Allowing for a certain tolerance of angles AC and BC, the production rules R_1', R_2', and R_3' can also be represented by the fuzzy-logic-based system. The fuzzy-logic-based system with the rules R_1'', R_2'', and R_3'' is more general than the two previous representations. The three rules R_1'', R_2'', and R_3'' are used to build a simple fuzzy-logic-based system, illustrated in Figure 11.7. If a neural network was to determine the parallelism of two lines, the weighted input would be added, and then a threshold value would determine the solution. Solving this problem with a neural network is conceptually illustrated in Figure 11.8. In the neural network in Figure 11.8, the threshold value of 1.02 is used to determine the parallelism of two lines.

11.3 LEARNING

Machine learning deals with computational methods for acquiring new knowledge, new skills, and new ways to organize existing knowledge (Carbonell and Langley, 1987).

Concept learning, a subdomain of machine learning (Michalski, 1987), has originated with studies of concept development in humans [see, e.g., Hoveland (1952) and Bruner et al. (1956)]. It subsequently continued to grow in the context of both artificial intelligence efforts to build machines with concept-learning capabilities and cognitive science studies to construct computational models of learning.

Two major orientations can be distinguished in concept learning: cognitive modeling and the engineering approach (Michalski, 1987). They parallel the orientation of efforts in cognitive science and artificial intelligence, respectively. Cognitive modeling attempts to develop computational theories of concept learning in humans or animals. It blends original cognitive psychology techniques with efforts to develop well-defined computational methods and computer programs embodying those methods. In contrast, the engineering approach attempts to explore and experiment with all possible learning mechanisms, irrespective of their occurrence in living organisms.

In any learning process, the learner (student) applies the knowledge possessed to information obtained from a source (e.g., a teacher) in order to derive new useful knowledge. This new knowledge is then stored for later use. Learning a new concept can occur in a number of ways, reflecting the type of inference the student performs on the information supplied. For example, one may learn the concept of an assembly system by being given a description of it, by generalizing examples of specific assembly systems, by constructing this concept in the process of observing and analyzing different types of manufacturing systems, or in another way. The type of inference performed by the student on the information supplied defines the strategy of concept learning and constitutes a useful criterion for classifying learning processes.

Several basic concept-learning strategies have been identified in machine learning (Michalski, 1987). These are presented next in the order of increasing complexity of inference performed by the learner. In any practical act of learning, more than one strategy is often simultaneously employed. It should also be noted that this classification of strategies applies not only to learning of concepts but also to any act of acquiring knowledge.

Direct Implanting of Knowledge.   This is an extreme case in which the learner does not have to perform any inference on the information provided. The learner directly accepts the knowledge supplied by the source. This strategy, also called *rote learning*, includes learning by direct memorization of given concept descriptions and learning by being programmed or constructed. For example, this strategy is employed when a specific algorithm for recognizing a concept is programmed into a computer or a database of facts about the concept is built.

Learning by Instruction.   The learner acquires concepts from a teacher or other organized source, such as a publication or textbook, but does not directly copy into memory the information supplied. The learning process may involve selecting the most relevant facts and/or transforming the source information to more useful forms.

Learning by Deduction.   The learner acquires a concept by deducing it from the knowledge given or possessed. In other words, this strategy includes any process in which knowledge learned is a result of a truth-preserving transformation of the knowledge given, including performing computation. An example of this strategy is determining that 5! is 120 by executing an already-known algorithm and having this fact for future use. This technique is also called *memo functions* (Michie, 1968). A form of learning by deduction is explanation-based learning, which transforms an abstract, not directly usable concept definition into an operational definition using a concept example for guidance (Mitchell et al., 1986). In general, deductive learning involves performing a sequence of deductions or computations on the information given or stored in background knowledge and memorizing the result.

More advanced deductive learning is exemplified by analytic or explanation-based learning methods (e.g., Mitchell et al., 1986). These methods start with the abstract concept definition and domain knowledge and by deduction derive an operational concept definition. A concept example is used to guide the deductive process. For instance, knowing that a vision camera is a device that has to be connected and fastened, an explanation-based method can produce an "operational" description of a vision camera. Such a description characterizes the vision camera in terms of lower level, more measurable features, such as the presence of lenses, cables, a tube, and a hole at the bottom. An attempt is being made to combine such analytical learning with inductive learning in order to learn concepts when the domain knowledge is incomplete, intractable, or inconsistent.

11.3.1 Learning Rules

Even if a particular neural network architecture has the capability to model the desired function, there remains the question of whether the learning rule that is used will be able to produce the appropriate representation. Since long-term representations are distributed across the weights of the interunit connections, learning new information is accomplished by modifying these weights in such a way as to improve the desired mapping between input and output vectors. This mapping function is learned by adjusting the weights according to some well-defined learning rule. It will be convenient to identify three general classes of rules. The first, which is called correlational rules, determine individual weight changes solely on the basis of levels of activity between connected neurons. The second class is the set of error-correcting rules, which rely on external feedback about the desired values of the output neurons. The third class comprises the unsupervised learning rules, in which learning occurs as self-adaptation to detected regularities in the input space without direct feedback from a teacher.

It is interesting to note the connection that can be drawn between these classes of neural network learning rules and traditional learning paradigms. The correlational rules are frequently found in autoassociative networks that perform rotelike learning of specific memory states (Hopfield, 1982; Kohonen, 1984). The error-correcting rules are predominantly employed for supervised learning from examples, since they use knowledge about how the system deviates from the teacher's examples to modify the connective weights (Rosenblatt, 1962; Rumelhart et al., 1986a; Sejnowski and Rosenberg, 1986). The unsupervised learning rules are used in situations where the network must learn to self-adapt to the environment without receiving specific feedback from a teacher (Kohonen, 1984; Rumelhart and Zipser, 1985; Klopf, 1982). These generalized relationships between learning rules and learning paradigms are only first approximations; there are, of course, exceptions to each of these claims. Viewing the learning rules from this perspective, however, can help to illustrate their general usage.

Next, each of the three classes of learning rules is examined.

Correlational Rules. The inspiration for many neural network learning rules can be traced to the early work of Hebb (1949). Hebb's research on learning in biological systems resulted in the postulate that the efficacy of synapses between two neurons increases when the firing activity between them is correlated. The implication is that the connection from neuron A to neuron B is strengthened whenever the firing of neuron A contributes to the firing of neuron B. This rule can be defined formally as

$$\Delta w_{ij} = o_i o_j$$

where

$\Delta w_{ij} =$ change in the connective weight from neuron j to neuron i

$o_i, o_j =$ output levels of the respective neurons

Notice that the only information affecting the weight change is "local"—it is derived solely from the levels of activity of the connected neurons and not from any knowledge of the global performance of the system. Most network learning rules use information about the correlated (or anticorrelated) activity of connected neurons, but what makes a rule fit into the class of correlational rules is that these activities are the only basis for weight changes.

In addition to Hebb's postulate, other correlational learning rules have been proposed for learning in neural networks. One such rule, sometimes termed anti-Hebbian, states that if a unit's firing is followed by a lack of activity at a second neuron to which the first unit's signal is being projected, then the connection strength between those two units is diminished (Levy and Desmond, 1985). There are only a small number of possible correlational rules of this sort. Since two connected units may assume two states of activation each (assuming binary-valued units), and their connection weight can be increased, decreased, or left unchanged, there are 3^4, or 81, possible correlational learning rules of this type. Many of these are uninteresting, such as the rules in which no modifications ever take place or in which they always occur. In the design of some learning rules, a temporal element is sometimes added to the basic Hebbian premise. Rather than taking into account the activities of the units at the same instant, the activities are sampled at some displaced time interval (Klopf, 1982; and Kosko, 1986).

Error-Correcting Rules. The second class of learning rules are those that employ error-correcting techniques. This class includes some of the most popular rules currently being used in actual implementations (Hinton et al., 1984; Rumelhart et al., 1986b; Sejnowski and Rosenberg, 1986). The general approach in these rules is to let the network produce its own output in response to some input stimulus, after which an external teacher presents the system with the correct or desired result. If the network's response is correct, no weights need be changed (although some systems use this information to strengthen the correct result). If there is an error in the network's output, then the difference between the desired and the achieved output can be used to guide the modification of weights appropriately. Since these methods strive to reach a global solution to the problem of representing the function by taking small steps in the direction of greatest local improvement, they are equivalent to a gradient descent search through the space of possible representations.

Unsupervised Learning. The third form of learning in neural networks involves self-organization in a completely unsupervised environment. In this category of learning rules, the focus is not on how actual neuron outputs match against externally determined desired outputs but rather on adapting weights to reflect the distribution of observed events. The "competitive learning" scheme proposed by Rumelhart and Zipser (1985) offers one approach toward this end. In this model, units of the network are arranged in predetermined "pools," in which the response by the units to input patterns is initially random. As patterns are presented, neurons within the pool are allowed to compete for the right to respond. That neuron that responds most strongly to the pattern is designated the winner. The weights of the network are then adjusted such that the response by the winner is reinforced, making it even more likely to identify with that particular quality of the input in the future. The result of this kind of competitive learning is that, over time, individual neurons evolve into distinct feature detectors that can be used to classify the set of input patterns. For an overview of learning in neural networks the reader may refer to Matheus and Hohensee (1987).

11.3.1.1 *Learning by Analogy.*

The learner acquires a new concept by modifying the definition of a similar known concept. In this case, rather than formulating a rule for a new concept from scratch, the student adapts an existing rule by modifying it appropriately to serve the new role. For example, if one knows the concept of a digital feedback control, learning the concept of a production control system can be accomplished by just observing the similarities and distinctions between the two. Another example is learning about electric circuits by drawing analogies from pipes conducting water.

Learning by analogy can be viewed as inductive and deductive learning combined, and for this reason it is placed between the two. Through inductive inference (see the following discussion) one determines general characteristics or transformation unifying concepts being compared. Then, by deductive inference, one derives from these characteristics the features expected of the concept being learned.

11.3.1.2 *Learning by Induction.*

In this strategy, the learner acquires a concept by drawing inductive inferences from the facts supplied or observations. Depending on what is provided and what the learner knows, two different forms of this strategy can be distinguished: learning from examples and learning by observation and discovery.

Learning from Examples. The learner induces a concept description by generalizing from teacher- or environment-provided examples and sometimes counterexamples of the concept. It is assumed that the concept already exists; the teacher knows it or there is some effective procedure for testing the concept's membership. The task for the learner is to determine a general concept description by analyzing individual concept examples.

An example of this strategy takes place when a senior quality engineer examines statistical process control charts in the presence of a newly appointed quality engineer, noting that "this batch of parts is defective," "this is another batch of defective parts, but notice that . . .," and so on.

Learning by Observation and Discovery. In this method, the learner analyzes given or observed entities and determines that some subsets of these entities can be grouped into certain classes (i.e., concepts). Because there is no teacher who knows the concepts beforehand, this strategy is also called unsupervised learning. Once a concept is formed, it is given a name. Concepts created in this way can then be used as terms in subsequent learning of other concepts.

An important form of this strategy is clustering (i.e., partitioning a collection of objects into classes) and the related process of constructing classifications. Classifications are typically organized into hierarchies of concepts. Such hierarchies exhibit an important property of inheritance. If an object is recognized as a member of some class, the properties associated specifically with this class, as well as with classes at the higher level of hierarchy, are (tentatively) assigned to the given object. This clustering concept is similar to one applied in group technology. Clustering in group technology is concerned with numerical data, whereas in conceptual learning mostly qualitative elements are considered.

11.4 BACK-PROPAGATION NEURAL NETWORK

The back-propagation neural network includes one or more layers of hidden neurons, that is, the neurons between the input and the output layers. While training the neural network, the target outputs should be taken into account as a portion of the training set in order to evaluate the error between each target output and actual output. The actual outputs are generalized during the learning process, and target outputs, which are regarded as a teacher, are provided as data in the training set. If the teacher determines that the error between itself and each actual output is larger than a tolerance, the learning process continues until the error is within the tolerance. From the point of this learning virtue, the back-propagation neural network belongs to the category of supervised learning networks since it uses target outputs (the teacher) to check whether the error is acceptable.

In contrast, a neural network learning process is called unsupervised if its training set does not include target outputs. This means that there is no teacher used to check the error of actual outputs of the neural network. The determination and comparison of outputs is performed by the neural network itself, not by the teacher (target output), to guide whether the learning process is accomplished. The unsupervised learning is discussed later in this chapter in the context of an adaptive resonance theory (ART) network.

Figure 11.9 shows a three-layer back-propagation neural network. In order to make the figure readable, only some of the connections are shown. Each circle represents a neuron, and each arrow represents a connection to be associated a weight. The neurons marked with the letter *b* are the bias neurons. They exist in each layer but the output layer. Each bias neuron serves as a threshold unit and has a constant activation value (i.e., output value) of 1.

11.4.1 Back-Propagation Learning

Learning is a process of acquiring a function that maps input vectors (stimuli) to the output vector (response). To learn this mapping function, a learning system receives a set of training events, from which it constructs an internal representation of the function. Considered at this level, learning in neural networks is no different from learning in traditional systems. In fact, neural networks can be effectively viewed within the context of traditional machine learning paradigms. The similarities between neural networks and traditional approaches, however, do not carry through to lower levels of analysis, where the details of representation and learning become significant.

The concept of back-propagation learning is illustrated in Figure 11.10. Assume that for a set of inputs and the corresponding set of outputs are known. The goal of learning is to determine a transformation function F that would produce the set of inputs for the known set of outputs. Assuming random values of the parameters (weights) of this function, the outputs produced based on the original inputs do not normally match the actual outputs. The difference between the actual and produced

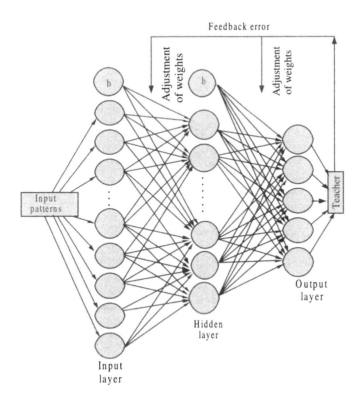

Figure 11.9. Three-layer back-propagation neural network.

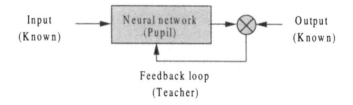

Figure 11.10. Back-propagation learning.

outputs is used as a basis for correcting the random weights, as illustrated by the feedback loop in Figure 11.10. The process of generating outputs for a known set of inputs and correcting the weights continues until the difference (error) between the actual outputs and the outputs produced for the latest set of network parameters (weights) is acceptable.

Back-propagation learning is illustrated in Example 11.3.

Example 11.3. Consider the neural network in Figure 11.11 with four neurons and the values $w_1, \ldots, w_5$ of the weights. The threshold values are provided for each of the three layers of neurons. The output value of each neuron is calculated for each pair of inputs (n_1, n_2) taking values from the XOR truth Table 11.1. For example, consider the hidden node n_3 in Figure 11.11 with its the threshold value 0.02. The activation value a_3 of neuron n_3 is computed based on its inputs and the corresponding weights:

$$a_3 = w_2 \cdot o_1 + w_3 \cdot o_2 \tag{11.1}$$

In this case the output value of neuron n_3 is computed as follows:

$$o_3 = \begin{cases} 0 & \text{if } |a_3| \le 0.02 \\ 1 & \text{if } |a_3| > 0.02 \end{cases} \tag{11.2}$$

Using the value of inputs $i_1 = 0$ and $i_2 = 0$ from Table 11.1, the output of nodes n_1 and n_2 can be easily determined as $o_1 = o_2 = 0$ using expressions (11.1) and (11.2). The

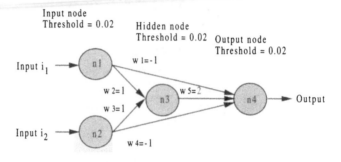

Figure 11.11. Simple neural network.

TABLE 11.1. XOR Truth Table

Input 1	Input 2	Output
i_1	i_2	o
0	0	0
0	1	1
1	0	1
1	1	0

two output values in turn become inputs to node n_3. The activation value of neuron n_3 is computed from (11.1) as

$$a_3 = 1 \cdot 0 + 1 \cdot 0 = 0$$

The output of neuron n_3 is determined from (11.2) as

$$o_3 = 0$$

Furthermore,

$$a_4 = w_1 \cdot o_1 + w_5 \cdot o_3 + w_4 \cdot o_2 = -1 \cdot 0 + 2 \cdot 0 + (-1 \cdot 0) = 0$$

and the output

$$o_4 = 0$$

which agrees with the output in Table 11.1.

Rather than using the step transfer function (threshold value) as in Example 11.3, one could apply a sigmoid transfer function shown next to reflect the fuzzy relationship:

$$o_i = \frac{1}{1 + e^{-\alpha(\Sigma\, \text{weight} \times \text{input} - \theta)}} \tag{11.3}$$

where

$\alpha =$ degree of fuzziness (constant during training)
$\theta =$ threshold level (its value changes)

11.4.1.1 *Back-Propagation Learning Algorithm.* In this section, a learning algorithm proposed by Rumelhart et al. (1986b) and known as the back-propagation algorithm is presented. The type of learning with the back-propagation algorithm belongs to a class learning with error-correcting rules.

Algorithm

Step 1. *Weight initialization.* Set all weights and node thresholds to small random numbers.

Step 2. *Calculation of output levels:*

(a) The output level of an input neuron is determined by the instance presented to the network.

(b) The output level o_j of a hidden neuron is determined as

$$o_j = f(\Sigma\, w_{ji} o_i - \theta_j) = \frac{1}{1 + e^{-\alpha(\Sigma\, w_{ij} o_i - \theta_j)}} \qquad (11.4)$$

where w_{ji} is the weight from neuron i to neuron j, α is a constant, θ_j is the node threshold, and f is a sigmoid function.

Step 3. *Weight training:*

(a) The error gradient is completed as follows: For the output neurons,

$$\delta_j = o_j(1 - o_j)(d_j - o_j) \qquad (11.5)$$

where d_j is the desired (target) output and o_j is the actual output of neuron j. For the hidden neurons,

$$\delta_j = o_j(1 - o_j) \sum_k \delta_k w_{kj} \qquad (11.6)$$

where δ_k is the error gradient at neuron k to which a connection points from hidden neuron j.

(b) The weight adjustment is computed as

$$\Delta w_{ji} = \eta \delta_j o_i \qquad (11.7)$$

where η is a trial-independent learning rate $(0 < \eta < 1)$ and δ_j is the error gradient at neuron j.

(c) Start with the output neuron and work backward to the hidden layers recursively. Adjust weights by

$$w_{ji}(t + 1) = w_{ji}(t) + \Delta w_{ji} \qquad (11.8)$$

where $w_{ji}(t)$ is the weight from neuron i to neuron j at iteration t and Δw_{ji} is the weight adjustment.

(d) Perform the next iteration (repeat Steps 2 and 3) until the error criterion is met, that is, the algorithm converges. Iteration includes presenting an instance, calculating output levels, and modifying weights.

The back-propagation learning algorithm is illustrated in Example 11.4.

Example 11.4. Consider the XOR (exclusive OR) problem represented with the back-propagation network in Figure 11.12, which has four neurons and five initial weights. The XOR truth table (see Table 11.1) is used to train the network. The data in this table is also used to validate the trained network; that is, for each known pair of inputs the network should produce an output (in this case is also known) within the error assumed.

The first iteration of the back-propagation learning algorithm is as follows:

Step 1. The weights are randomly initialized as follows: $w_{13} = 0.02$, $w_{14} = 0.03$, $w_{12} = -0.02$, $w_{23} = 0.01$, and $w_{24} = 0.02$

Step 2. *Calculation of output levels.* Consider a training instance (the fourth row from Table 11.1 with the input vector $(1, 1)$ and the desired output 0. From Figure 11.12,

$$o_3 = i_3 = 1 \qquad o_4 = i_4 = 1$$

From Equation (11.4) for $\alpha = 1$ and $\theta_j = 0$

$$o_2 = \frac{1}{1 + e^{-(1 \times 0.01 + 1 \times 0.02)}} = 0.678$$

$$o_1 = \frac{1}{1 + e^{-[0.678 \times (-0.02) + 1 \times 0.02 + 1 \times 0.03)]}} = 0.509$$

Step 3. *Weight training:*

Assume the learning rate $\eta = 0.3$.

$\delta_1 = 0.678(1 - 0.678)(0 - 0.678) = -0.148$ is computed from (11.5).

From (11.7), $\Delta w_{13} = 0.3(-0148) \times 1 = -0.044$.

From (11.8), $w_{13} = 0.02 - 0.044 = -0.024$.

From (11.6), $\delta_2 = 0.678(1 - 0.678)(-0.148)(-0.02) = 0.0006$.

From (11.7), $\Delta w_{23} = 0.3 \cdot 0.0006 \cdot 1 = 0.00018$.

From (11.8), $w_{23} = 0.01 + 0.00018 = 0.01018$.

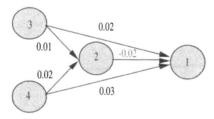

Figure 11.12. Back-propagation network for learning the XOR function with randomly generated weights.

The remaining iterations of the algorithm are omitted. The following set of final weights produces the output within the assumed mean square error of 0.01:

$$w_{13} = 4.98 \qquad w_{14} = 4.98 \qquad w_{12} = -11.30 \qquad w_{23} = 5.62 \qquad w_{24} = 5.60$$

For the new set of weights computed and an input from the XOR truth table (Table 11.1), the neural network in Figure 11.13 produces an output that, with the assumed error of 0.01, is the same as the corresponding output in the XOR table. For example, the output produced by the network in Figure 11.13 for the input vector (1, 0) from the XOR table is as follows:

$$o_3 = 1 \qquad o_4 = 0$$

$$o_2 = \frac{1}{1 + e^{-(1 \times 5.62 + 0 \times 5.62)}} = 0.9964$$

$$o_1 = \frac{1}{1 + e^{-(1 \times 4.98) + 0 \times 4.98 - 11.30 \times 0.9964}} = 0.9999$$

Of course, the value of $o_1 = 0.9999$ should be rounded off to 1.

Determining the number of hidden layers and the number of neurons in each hidden layer is a considerable task. Cybenko (1989) stated that one hidden layer is enough to classify input patterns into different groups. Experiments performed by Bounds and Lloyd (1988) showed that there was no obvious evidence that a network with two hidden layers performed better than with one layer only. However, Chester (1990) indicated that a two-hidden-layer network should perform better than a one-hidden-layer network. Here, the one-hidden-layer back propagation is selected.

Determining the number of neurons in a hidden layer is important. A conservative approach is to select a number between the number of input neurons and the number of output neurons. In Example 11.5 presented next, the number of neurons in the hidden layer is 15. Hecht-Nielson (1987) determined that $2N + 1$ hidden neurons, where N is the number of inputs, are required for a one-hidden-layer back-propagation network (in Example 11.5, $N = 11$).

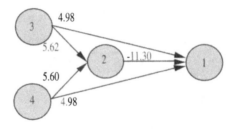

Figure 11.13. Network from Figure 11.12 with the new weights.

Mirchandani and Cao (1989) derived a relationship between the dimension d of input space, the maximum number of separable regions M, and the number of hidden neurons H, namely, $H = \log_2 M$. In Example 11.5, $H = 3$ and M is set to the number of training patterns equal to 5. Baum and Haussler (1989) presented some expressions related to training patterns, neurons, and weights under the consideration of various statistic confidence levels. Lippmann (1987) indicated that the maximum number of hidden neurons for a one-hidden-layer network is $O(N + 1)$, where O is the number of output neurons and N is the number of input neurons. In Example 11.5, the maximum number of hidden neurons for a one-hidden-layer network is $130 = 5(25 + 1)$. However, 10 hidden neurons have finally been selected (due to the limited memory of an IBM PC computer).

A back-propagation network is sensitive to initial values of weights (Kolen and Pollack, 1990). Properly selected initial weights can shorten learning time and result in stable weights. Initial weights that are too small increase the learning time, which may cause difficulties in converging to an optimal solution. If initial weights are too

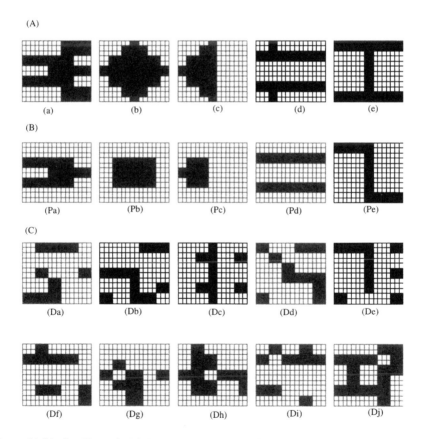

Figure 11.14. See Example 11.5 on next page. Binary image of twenty parts: (*A*) 5 training parts; (*B*) 5 testing parts with partial geometry; (*C*) 10 testing parts with distorted geometry.

TABLE 11.2. The Result of Part Classification

Training Part Geometry	Part Family	Partial Part Geometry	Part Family
a	1	Pa	1
b	2	Pb	2
c	3	Pc	3
d	4	Pd	4
e	5	Pe	5
Distorted Part Geometry	Part Family	Distorted Part Geometry	Part Family
Da	5	Df	4
Db	5	Dg	3
Dc	3	Dh	2
Dd	5	Di	4
De	5	Dj	1

large, the network may get unstable weights (Wasserman, 1989). In Example 11.5 presented next, the initial weights are randomly set between 0.3 and –0.3 (Chung and Kusiak, 1994).

Example 11.5.   In this example, a back-propagation neural network produces part families based on the part geometry images. Figure 11.14 (see page 365) shows binary images of 20 parts: 5 training parts, 5 testing parts with partial geometry, and 10 testing parts with distorted geometry.

The back-propagation neural network used to classify the parts in Figure 11.14 has a three-layer structure (similar to the network in Figure 11.9). If bias neurons are included in the input and hidden layers, there are 193 neurons in the input layer and 11 neurons in the hidden layer. In the output layer, one group is represented by a neuron with the output value 1 and the remaining 4 neurons have 0 output. This makes output patterns represent identification tags for a set of groups into which the input part shapes are to be classified. Each group includes a prototype (standard) part and all its variations. During the training phase, the system is presented only with standard (prototype) parts, each part corresponding to a different group. The result of classifying the 15 incomplete manufacturing parts from Figure 11.14 is shown in Table 11.2. All of the testing parts have been successfully classified into their corresponding part families on the basis of their geometry. The back-propagation learning involved an error tolerance of 0.001(for details see Chung and Kusiak, 1994).

11.5 SELF-LEARNING NEURAL NETWORK

In a back-propagation neural network, the input training set is sequentially presented to the network until the network finishes learning the entire training set. Learning a new pattern implies retraining of the network. As the result of retraining, the weights of the neural network may change. This raises the plasticity/stability dilemma. Plasticity is the ability to learn new patterns. The ability to generate appropriate patterns without changing the value of the weights is called stability. Allowing a neural network to learn new patterns and at the same time maintaining the stability of the existing weights is difficult. A back-propagation neural network cannot maintain plasticity and stability at the same time, which means the current weights are destroyed when new patterns are directly used to train the network without relearning the old patterns.

Besides, in a back-propagation neural network, one may never see the same training pattern more than once, or the network will often not learn anything; that is, it will not arrive at a stable state. This is another cause for getting unstable weights in the back-propagation neural network. The plasticity/stability dilemma and learning identical patterns, can be overcome with an ART (adaptive resonance theory) neural network. A learning algorithm in an ART neural network can learn new patterns (plasticity), including identical training patterns, and maintain weights stable (stability).

When meeting an input pattern (Step 2 in the ART learning algorithm presented in the next section), the ART neural network functions similar to a table, allowing us to identify an appropriate category in layer F2 (Figure 11.15) for the input pattern presented to layer F1. If the input pattern does not match any of the stored patterns, a new category is created (Step 5). Also, if an input pattern matches one of the patterns stored, it is classified into the same category as the stored pattern (Steps 4 and 5) and adapts the weights (Step 7) in order to produce a new stable state. In this manner, by creating a new category for an input pattern that is not similar to any of the patterns stored (Step 5) and keeping the existing weights unchanged (skip Steps 6–8), and by storing an input pattern that is similar to one of the stored patterns (Step 5) and modifying weights (Step 7), the plasticity/stability dilemma is resolved.

11.5.1 ART Neural Network

Figure 11.15 shows the structure of the ART neural network (Carpenter and Grossberg, 1987). The network consists of two subsystems: the attentional subsystem including two layers, an input representation field F1 and a category representation field F2, and the orienting subsystem. The two layers of neurons in the attentional system are fully interconnected by adaptive weights, which are updated by similar input patterns (Steps 5–7). The orienting system becomes active (Step 8) when a sufficiently distinct pattern is presented to the network (Step 5). In other words, it detects dissimilar input patterns (Step 5) and resets the attentional subsystem (Step 8) when it detects such a pattern.

There are control mechanisms in the ART network, gains 1 and 2, one for each layer of the attentional subsystem. The output of gain 1 is 1 if a value of the input pattern is 1; however, if a value of the stored pattern is 1, gain 1 becomes 0. The output of gain 2 is 1 if the input pattern has an element whose value is 1. There is no direct relationship between gains 1 and 2, but both are relevant to the input pattern. On the

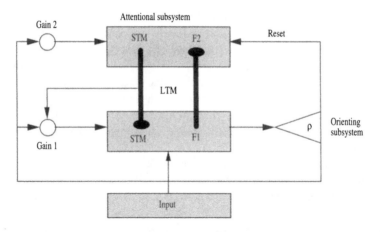

Figure 11.15. Structure of the ART network.

other hand, both gains modulate the overall weight. The two gains are involved in the derivation of the weight formula in the Steps 3, 5, and 7 of the algorithm presented in the next section.

The patterns of activity in layer F1 and F2 neurons are called short-term memory (STM), because their states may be quickly changed by applying a new input pattern after the processing of a previous input pattern is finished (Steps 2–4). Briefly speaking, STM is the temporary activity of neurons in either layer F1 or layer F2. The adaptive weights in the bottom-up and top-down pathways are called long-term memory (LTM), because they may be modified by input patterns and may persist for a long time after the training of the input patterns has been accomplished (Step 7). In other words, LTM saves patterns after training has been finished.

After an input pattern $\mathbf{x}$ (a vector) is presented to the neurons in layer F1 (Step 2), the pattern affects the neurons in layer F2 via bottom-up weights between the two layers. Each F2 neuron sums its inputs $b_{ij}x_i$ in Equation (11.9) of Step 3] to form its output, which responds accordingly to F1 (Step 3). After the neuron with the largest activity in layer 2 is determined (Step 4), the stored pattern is formed by calculating the sum of the product of the top-down weights connected to the neuron and the input pattern, that is, $\sum_{i=1}^{n} t_{ij}^{*}x_i$ in expression (11.11) of Step 5. In order to know whether this stored pattern is similar to the input pattern, its similarity to the input pattern is computed by their ratio; that is, $\sum_{i=1}^{n} t_{ij}^{*}x_i/\|\mathbf{x}\|$, where $\|\mathbf{x}\| = \sum_{i=1}^{n} x_i$. This similarity is used to compare with the vigilance that functions as a threshold. If the similarity is greater than the value of the vigilance, the two patterns are similar; otherwise they are dissimilar. This mutual matching between an input pattern in layer F1 and the stored pattern in layer F2 is called a resonance. In short, the input pattern and the stored pattern are said to resonate when they match each other (Stork, 1989).

An ART neural network classifies each input pattern in one category. The orienting subsystem is in charge of this task. Its main function is to judge, via a parameter called vigilance, ρ, whether the magnitude of a mismatch between F2 resonant activity and the input pattern is within the vigilance (Step 5) before it emits a reset signal (Step 8) to temporarily change the resonant activity that is the output of the active neuron with the largest output value (Step 4). In other words, the orienting subsystem evaluates the similarity (Step 5) between an input and the resonant activity. If they differ by more than the value of vigilance, a reset signal is sent to layer F2 (Step 8). The orienting subsystem calculates the similarity as the ratio of the number of elements 1 in the resonant activity [the numerator of the ratio in expression (11.11) of Step 5] to the number of elements 1 in the input pattern [the denominator of the ratio in expression (11.11) of Step 5]. If this ratio is below the vigilance, the reset signal is issued (Step 8).

Figure 11.16 illustrates the learning process in an ART network (Stork, 1989). Patterns are shown as a collection of dark squares on a 5 × 5 grid. The patterns on the left are inputs, and the output categories are on the top. The vigilance in Figure 11.16*a* is set to 0.7. First, the pattern A is input to the newly initialized ART neural network. Since initially there is no pattern stored, a new category is assigned to a neuron in layer F2. Next, pattern B is presented to the layer F1 of the ART neural network. Since this one fails in the orienting subsystem in charge of the vigilance test, another category is established. Then, pattern C is presented to the network. Since it still violates the vigilance test, another new category is created. Pattern D presented to the network is close enough to the stored pattern C and passes the vigilance test, so it is used to train the network. This training process is described in the next section. Finally, the ART neural network generates three categories: category 1 that includes pattern A, category 2 containing pattern B, and category 3 with patterns C and D. If the vigilance is

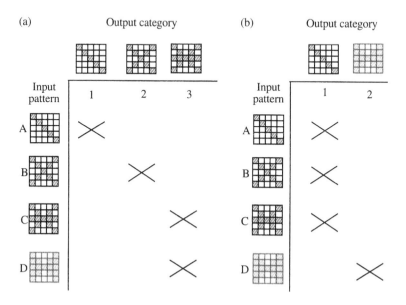

Figure 11.16. Learning process in ART network: (*a*) high vigilance = 0.7; (*b*) low vigilance = 0.3.

decreased to 0.3, as shown in Figure 11.16*b*, the total number of output categories becomes two: category 1 contains the first three patterns, category 2 with pattern D.

The above example illustrates the importance of setting of an appropriate value of the vigilance. So far, there are no guidelines for selecting the value of vigilance. If one regards an input pattern in Figure 11.16 as one machine row of a machine–part incidence matrix, there are four machines and 25 parts in the figure. The dark squares in the pattern in Figure 11.16 means that a part is processed on its corresponding machine; the blank square means that no processing is performed. In this example, three and two machine cells are generated for vigilance 0.7 and 0.3, respectively (for details see Kusiak and Chung, 1991).

After all, the ART neural network works as a pattern classifier. Each neuron in layer F2 represents one category. It classifies one machine row (an input pattern) in a particular category according to the similarity between the stored pattern and the input pattern. The ART learning algorithm searches neurons in layer F2 until the neuron with the largest output value, which satisfies the vigilance test, is found. The test is used to compare the similarity with the vigilance, which functions like a threshold in other algorithms for classification of machine rows.

11.5.1.1 Vigilance in ART Network.
In an ART neural network, as illustrated above, the selection of the value of vigilance, which is between 1 and 0, is paramount. The vigilance determines the maximum difference between two patterns in the same category. If the vigilance is too high, the ART neural network creates new categories for those patterns that originally were similar to the patterns stored but fail to fall into any of the existing category. This leads to a poor generalization, because slight variations of the same pattern become separate categories rather than one category. When the vigilance is too low, totally different patterns might be grouped together. This leads to classifying dissimilar patterns into the same category, which distorts the stored patterns. Unfortunately, there are no guidelines for setting the value of vigilance. One should make a decision regarding the difference between patterns that constitute distinct categories.

Carpenter and Grossberg (1987) proposed a feedback process to adjust the vigilance by imposing a penalty on an improper classification. Mekkaoui and Jespers (1990) discussed another way to adjust the value of vigilance. They used an ART neural network in two phases, a learning phase and a normal phase. The value of vigilance is set much higher in the learning phase and then reset to a desired level to perform the classification task in the normal phase. First, this approach allows an ART neural network to create distinct categories under the environment with a higher vigilance and to set up weights. Then, the input patterns are again presented to the network for classification. Entering the normal phase, where plasticity (the ability to learn new patterns) is not shut off, the vigilance is reset to a desired value, and the network relearns the input patterns. As a result of the above, the more precise output categories are obtained.

In order to determine a machine cell of best quality from alliterative solutions obtained for different values of vigilance, postprocessing is recommended. Various cost measures (e.g., suggested by Askin and Subramanian, 1989) or group efficiency (Chandrasekharan and Rajagopalan, 1989) may be considered.

11.5.2   Learning in ART Network

In this section, some notions inherent to the ART learning algorithm are discussed. The ART learning algorithm for classifying machine rows (input patterns) of a machine–part incidence matrix is also presented.

In order to find a stored pattern that is sufficiently close matching with an input machine row, GT/ART conducts a serial search (Steps 3–8, repeatedly). In other words, it looks up each machine cell (represented by the neurons in layer F2) through the stored patterns formed by the corresponding machine rows. One machine cell saves only one stored pattern, which is made of all machine rows belonging to the machine cell. For example, if there are three machine rows in a certain machine cell, the stored pattern in the machine cell is a mixed pattern of the three machine rows. If no stored pattern in an existing machine cell matches an input machine row (Steps 3, 4, 5 and 8), a new machine cell is created (Steps 4 and 5) that stores the input machine row. If there exists a stored pattern similar to an input machine row (Steps 3–7), the machine cell containing the stored pattern is selected for the input machine row (Steps 4 and 5).

Some advantages of the ART learning process are described next. First, its environment handles the learning process by adjusting the vigilance so that when all input rows are classified unsatisfactorily, an increment will automatically be added to the vigilance to ensure that the unsatisfactory machine cells do not reoccur. Second, the ART environment can generate stable weights after learning a finite number of machine rows of the machine–part incidence matrix. These stable weights increase the accuracy of forming machine cells. Third, ART is capable of learning a large number of machine rows. This is unlike a back-propagation neural network in which the number of training patterns is limited. Fourth, any order of arbitrary input patterns will produce a stable set of weights after a finite number of learning patterns; no repetitive sequence of training patterns will cause the weights of ART to cycle endlessly. Fifth, the ART algorithm cannot guarantee convergence to the global optimum. The ART learning algorithm is presented next.

ART Learning Algorithm

Step 1. Initialize bottom-up weights $t_{ij} = 1$ and top-down weights $b_{ij} = 1/(n + 1)$, $j = 1, 2, \ldots , m$, where n and m are the numbers of neurons in layers F1 and F2, respectively. Input vigilance ρ is between 0 and 1 and its increment is a small decimal positive real number.

Step 2. Present a new machine row $\mathbf{x} = \{ x_1, x_2, \ldots , x_n\}$ to the input layer.

Step 3. Compute the weighted sum,

$$y_j = \sum_{i=1}^{n} b_{ij} x_i \qquad j = 1, 2, \ldots , m \tag{11.9}$$

where y_j is the output of neuron j in layer F2, x_i is the element i of the input pattern, and b_{ij} is a bottom-up weight from neuron i in layer F1 to neuron j in layer F2.

Step 4. Find the neuron j^* with the largest value y_j in layer F2,

$$y_{j^*} = \max\{y_j\} \qquad j = 1, 2, \ldots, m \qquad (11.10)$$

In the case of a tie, select the neuron on the left.

Step 5. Compare the similarity between a stored pattern and input pattern with the vigilance ρ,

$$\sum_{i=1}^{n} \frac{t_{ij^*} x_i}{\|x\|} > \rho \qquad (11.11)$$

where t_{ij^*} is a top-down weight from neuron j^* in layer F2 to neuron i in layer F1 and $\| \, x \, \|$ is the norm of the vector x, that is, $\| \, x \, \| = \sum_{i=1}^{n} x_i$. If the above comparison holds, then x belongs to machine cell j^*, and go to the next Step; otherwise, go to Step 8.

Step 6. If all the input rows have been input, print out each machine cell, machines, parts, and bottleneck parts; then stop. Otherwise, go to the next step.

Step 7. Update b_{ij^*} and t_{ij^*},

$$t_{ij^*}(k + 1) = t_{ij^*}(k)x_i \qquad i = 1, 2, \ldots, n$$

$$b_{ij^*}(k + 1) = \frac{t_{ij^*}(k)x_i}{0.5 + \left(\sum_{i=1}^{n} t_{ij^*}(k)x_i\right)} \qquad i = 1, 2, \ldots, n \qquad (11.12)$$

where k is a time step. Go to Step 2.

Step 8. Since x does not belong to the machine cell considered in Step 4, temporarily reset the output of neuron j^* to zero, so that it is never considered in Step 4. Go to Step 3.

The ART learning algorithm is illustrated with the next example.

Example 11.6.   Given three input patterns (parts), (1 0 1), (1 1 0), and (0 0 1), group them into categories (part families). The iterations of the ART algorithm for the three input patterns are presented next.

Step 1. Set $\rho = 0.8$, $n = 3$, and $m = 3$,

$$t_{ij} = 1, \qquad b_{ij} = \frac{1}{(1 + 3)} = 0.25 \qquad \text{for } i, j = 1, 2, 3$$

Step 2. Input (1 0 1).
Step 3. Compute

$$y_1 = b_{11}x_1 + b_{21}x_2 + b_{31}x_3 = 0.25 \cdot 1 + 0.25 \cdot 0 + 0.25 \cdot 1 = 0.5$$

$$y_2 = b_{12}x_1 + b_{22}x_2 + b_{32}x_3 = 0.25 \cdot 1 + 0.25 \cdot 0 + 0.25 \cdot 1 = 0.5$$

$$y_3 = b_{13}x_1 + b_{23}x_2 + b_{33}x_3 = 0.25 \cdot 1 + 0.25 \cdot 0 + 0.25 \cdot 1 = 0.5.$$

Step 4. Since $y_1 = y_2 = y_3 = 0.5$, neuron 1 in layer F2 is selected. Set $j^* = 1$.
Step 5. The following values are computed:

$$\| \mathbf{x} \| = 1 + 0 + 1 = 2$$

$$t_{11}x_1 + t_{21}x_2 + t_{31}x_3 = 1 \cdot 1 + 1 \cdot 0 + 1 \cdot 1 = 2$$

Since $\frac{2}{2} = 1 > 0.8$, the pattern (1 0 1) is classified as a machine cell 1.
Step 6. There are other input rows to be considered for grouping.
Step 7. The weights $t_{11} = 1 \cdot 1 = 1$, $t_{21} = 1 \cdot 0 = 0$, and $t_{31} = 1 \cdot 1 = 1$ are updated.
The values of the remaining t_{ij}'s are not changed.
The weights

$$b_{11} = \frac{t_{11}x_1}{0.5 + (t_{11}x_1 + t_{21}x_2 + t_{31}x_3)} = \frac{1}{0.5 + 1 \cdot 1 + 1 \cdot 0 + 1 \cdot 1} = 0.4$$

$$b_{21} = \frac{t_{21}x_2}{0.5 + (t_{11}x_1 + t_{21}x_2 + t_{31}x_3)} = \frac{0}{0.5 + 1 \cdot 1 + 1 \cdot 0 + 1 \cdot 1} = 0$$

$$b_{31} = \frac{t_{31}x_3}{0.5 + (t_{11}x_1 + t_{21}x_2 + t_{31}x_3)} = \frac{1}{0.5 + 1 \cdot 1 + 1 \cdot 0 + 1 \cdot 1} = 0.4$$

are updated. The values of the remaining b_{ij}'s are unchanged. Go back to Step 2.
Step 2. Input (1 1 0).
Step 3. Compute

$$y_1 = b_{11}x_1 + b_{21}x_2 + b_{31}x_3 = 0.4 \cdot 1 + 0 \cdot 1 + 0.4 \cdot 0 = 0.4$$

$$y_2 = b_{12}x_1 + b_{22}x_2 + b_{32}x_3 = 0.25 \cdot 1 + 0.25 \cdot 1 + 0.25 \cdot 0 = 0.5$$

$$y_3 = b_{13}x_1 + b_{23}x_2 + b_{33}x_3 = 0.25 \cdot 1 + 0.25 \cdot 1 + 0.25 \cdot 0 = 0.5$$

Step 4. $j^* = 2$.
Step 5. $\| \mathbf{x} \| = 1 + 1 + 0 = 2$ and $t_{12}x_1 + t_{22}x_2 + t_{32}x_3 = 1 \cdot 1 + 1 \cdot 1 + 1 \cdot 0 = 2$. Since $\frac{2}{2} = 1 > 0.8$, the pattern (1 1 0) is classified as a machine cell 2.

Step 6. There are input rows available for grouping.

Step 7. The weights $t_{12} = 1 \cdot 1 = 1$, $t_{22} = 1 \cdot 1 = 1$, and $t_{32} = 1 \cdot 0 = 0$ are updated. The values of the remaining t_{ij}'s are unchanged. The weights

$$b_{12} = \frac{t_{12}x_1}{0.5 + (t_{12}x_1 + t_{22}x_2 + t_{32}x_3)} = \frac{1}{0.5 + 1 \cdot 1 + 1 \cdot 1 + 1 \cdot 0} = 0.4$$

$$b_{22} = \frac{t_{22}x_2}{0.5 + (t_{12}x_1 + t_{22}x_2 + t_{32}x_3)} = \frac{1}{0.5 + 1 \cdot 1 + 1 \cdot 1 + 1 \cdot 0} = 0.4$$

$$b_{32} = \frac{t_{32}x_3}{0.5 + (t_{12}x_1 + t_{22}x_2 + t_{32}x_3)} = \frac{1}{0.5 + 1 \cdot 1 + 1 \cdot 1 + 1 \cdot 0} = 0$$

are updated. The values of the remaining b_{ij}'s are unchanged. Go back to Step 2.

Step 2. Input (0 0 1).

Step 3. Compute

$$y_1 = b_{11}x_1 + b_{21}x_2 + b_{31}x_3 = 0.4 \cdot 0 + 0 \cdot 0 + 0.4 \cdot 1 = 0.4$$

$$y_2 = b_{12}x_1 + b_{22}x_2 + b_{32}x_3 = 0.4 \cdot 0 + 0.4 \cdot 0 + 0 \cdot 1 = 0$$

$$y_3 = b_{13}x_1 + b_{23}x_2 + b_{33}x_3 = 0.25 \cdot 0 + 0.25 \cdot 0 + 0.25 \cdot 1 = 0.25$$

Step 4. $j^* = 1$.

Step 5. $\| \mathbf{x} \| = 0 + 0 + 1 = 1$ and $t_{11}x_1 + t_{21}x_2 + t_{31}x_3 = 1 \cdot 0 + 0 \cdot 0 + 1 \cdot 1 = 1$. Since $\frac{1}{1} = 1 > 0.8$, the pattern (1 0 1) is classified as machine cell 1.

Step 6. There are no other inputs to be grouped. The algorithm generates two machine cells. Both vectors (1 0 1) and (0 0 1) are classified as machine cell 1, (1 1 0) as machine cell 2. Machine cell 1 processes parts 1 and 3, and machine cell 2 processes parts 1 and 2. It is easy to note that part 1 is a bottleneck part as it visits the two cells. Stop.

If the vigilance test in Step 5 is not satisfied, the neuron obtained in Step 4 is removed from the set of neurons representing machine cells in F2. If any neuron remains in the set, the ART algorithm reiterates from Step 3; otherwise it regenerates an uncommitted "machine cell neuron" and encodes a stored pattern onto this cell neuron's reference pattern.

11.5.3 Computational Experience

Consider the machine–part incidence matrix shown in the Figure 11.17 (Kusiak and Cheng, 1990). The input data to ART involve vigilance and its increment and the incidence matrix. The input patterns are the rows of the incidence matrix.

To run the ART algorithm, the initial values of input vigilance and its increment are set to 0.5 and 0.02, respectively. The ART clusters the seven rows of the incidence matrix into one machine cell. Usually, this is not an acceptable solution of the GT problem. In order to improve this solution, the value of vigilance should be changed. Increasing the value of vigilance by 0.02 results in two part families PF-1 = {1, 3, 4, 5, 7} and PF-2 = {1, 2, 6, 8} and two machine cells MC-1 = {1, 2, 4, 6} and MC-2 = {3, 5, 7} are produced. This solution includes one bottleneck part, part 1. The solution does not alter until the value of vigilance is reset to 0.58, which leads to three machine cells. The summary of the results of test problem in Figure 11.17 for values of vigilance from 0.50 to 1.00 is shown in Table 11.3. Obviously, the higher the value of vigilance, the larger the number of machine cells. Depending on the user's preference, the best solution among the alternatives in Table 11.3 is the one with the minimum cost, better cell quality, or higher cell utilization.

Figure 11.18 presents an output generated by the ART algorithm for the test machine–part matrix in Figure 11.17. The first three machine rows were classified into three different machine cells, because they violated inequality (11.11) in the ART learning algorithm, where the value of vigilance was 0.60. However, row 4 was classified as machine cell 1 because ART regarded the stored pattern (row 1) in the machine cell as its closest pattern. Similarly, machine rows 5 and 6 were classified as machine cells 3 and 2, respectively, because the stored pattern (row 3) in the machine cell 3 was similar with row 5, and the stored pattern (row 2) in the machine cell 2 matched row 6. As for the last machine row, row 7, because ART regarded it as the pattern similar to the stored pattern (the mixed pattern of the machine row 3 and the machine row 5) in the machine cell 3, it was assigned to machine cell 3.

The comprehensive testing results included in Kusiak and Chung (1991) indicate that some of the solutions generated by the ART algorithm are of lesser quality (in terms of the number of machine cells and bottleneck parts) than those generated by traditional algorithms. However, it is expected that the developments in neural networks will improve the quality of solutions generated.

11.6 SUMMARY

In this chapter, the basic concepts of neural networks were introduced. Neural networks were compared with knowledge-based systems and fuzzy-logic-based systems. Numerous learning concepts were outlined. Two of these concepts, back-propagation (supervised) learning and unsupervised learning, were explored in

```
                           Part
                  1 2 3 4 5 6 7 8
               1 ⎡ 0 0 0 0 0 0 1 0 ⎤
               2 ⎢ 1 0 0 1 0 0 0 0 ⎥
               3 ⎢ 0 1 0 0 0 1 0 1 ⎥
     Machine   4 ⎢ 0 0 0 0 1 0 1 0 ⎥
               5 ⎢ 0 1 0 0 0 1 0 0 ⎥
               6 ⎢ 0 0 1 1 0 0 0 0 ⎥
               7 ⎣ 1 1 0 0 0 0 0 1 ⎦
```

Figure 11.17. Input matrix for test problem.

TABLE 11.3. Results for Test Problem in Figure 11.17

Vigilance	Number of Cells	Parts	Machines	Bottleneck Parts
0.50	1	1, 2, 3, 4, 5, 6, 7	1, 2, 3, 4, 5, 6, 7	None
0.52–0.56	2	1, 3, 4, 5, 7	1, 2, 4, 6	1
		1, 2, 6, 8	3, 5, 7	
0.58–0.66	3	5, 7	1, 4	1
		1, 3, 4	2, 6	
		1, 2, 6, 8	3, 5, 7	
0.68–0.70	4	5, 7	1, 4	1, 2, 8
		1, 3, 4	2, 6	
		2, 6, 8	3, 5	
		1, 2, 8	7	
0.72–1.00	5	5, 7	1, 4	1, 2, 4, 8
		1, 4	2	
		2, 6, 8	3, 5	
		3, 4	6	
		1, 2, 8	7	

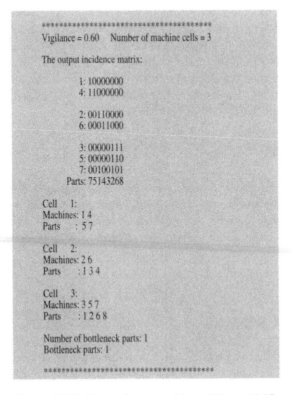

Figure 11.18. Output for test problem of Figure 11.17.

the context of two neural networks. The back-propagation and ART neural networks were illustrated with examples. In particular, the ART network was used to solve two types of group technology problems, forming part families based on part geometry and machine cells and part families using the machine–part incidence matrix. The latter formulation was extensively studied in Chapter 10.

REFERENCES

Askin, R., and S. Subramanian (1989), A cost-based heuristic for group technology configuration, *International Journal of Production Research*, Vol. 25, No. 1, pp. 101–114.

Bounds, D. G., and P. J. Lloyd (1988), A multilayer perceptron network for the diagnosis of low back pain, *Proceedings of the 2nd IEEE Annual International Conference on Neural Networks*, San Diego, CA, IEEE Press, Piscataway, NJ. June 21–24, pp. II.481–II.489.

Bruner, J. S., J. J. Goodnow, and G. A. Austin (1956), *A Study of Thinking*, John Wiley, New York.

Carbonell, J., and P. Langley (1987), Machine learning, in S. C. Shapiro (Ed.), *Encyclopedia of Artificial Intelligence*, Vol. 1, John Wiley, New York, pp. 464–488.

Carpenter, G., and S. Grossberg (1987), A massively parallel architecture for a self-organizing neural pattern recognition machine, *Computer Vision, Graphics and Image Understanding*, Vol. 37, No. 2, pp. 54–115.

Chandrasekharan, M. P., and R. Rajagopalan (1989), Groupability—an analysis for group technology, *International Journal of Production Research*, Vol. 27, No. 6, pp. 1035–1052.

Chester, D. L. (1990), Why two hidden layers are better than one? *Proceedings of the 4th IEEE Annual International Conference on Neural Networks*, Washington, DC, IEEE Press, Piscataway NJ. January 15–19, pp. I.265–I.268.

Chung, Y. K., and A. Kusiak (1994), Grouping parts with a neural network, *SME Journal of Manufacturing Systems*, Vol. 13, No. 4, pp. 262–275.

Cybenko, G. (1989), Approximation by superpositions of a sigmoidal function, *Mathematics of Control, Signals and Systems*, Vol. 2, No. 4, pp. 303–314.

Fu, L.-M. (1994), *Neural Networks in Computer Intelligence*, McGraw-Hill, New York.

Groth, R. (1998), *Data Mining: A Hands-On Approach for Business Professionals*, Prentice-Hall, Upper Saddle River, NJ.

Hebb, D. O. (1949), *The Organization of Behavior*, John Wiley, New York.

Hecht-Nielsen, R. (1987), Kolmogorov's Mapping Neural Network Existence Theorem, *Proceedings of the 1st IEEE Annual International Conference on Neural Networks*, San Diego, CA, June 21–24, pp. III.11–III.14.

Hinton, G. E. (1984), Distributed representations, Technical Report CMU-CS-84-157, Carnegie-Mellon University, Pittsburgh, PA.

Hinton, G. E., T. J. Sejnowski, and D. H. Ackley (1984), Boltzmann machines: Constraint satisfaction networks that learn, Technical Report CMU-CS-84-119, Carnegie-Mellon University, Pittsburgh, PA.

Hopfield, J. J. (1982), Neural networks and physical systems with emergent collective computational properties, *Proceedings of the National Academy of Science USA*, Vol. 79, pp. 2554–2558.

Hoveland, C. I. (1952), A communication analysis of concept learning, *Psychological Review*, Vol. 59, No. 6, pp. 461–472.

Klopf, A. H. (1982), *The Hedonistic Neuron: A Theory of Memory, Learning, and Intelligence*, Hemisphere, Washington, DC.

Kohonen, T. (1984), *Self-Organizing and Associative Memory*, Springer-Verlag, Berlin.

Kolen, J. F., and J. B. Pollack (1990), Backpropagation is sensitive to initial conditions, *Complex Systems*, Vol. 4, No. 3, pp. 269–280.

Kosko, B. (1986), Differential Hebbian learning, *AIP Conference Proceedings 151, Neural Networks for Computing*, pp. 227–282.

Kusiak, A., and C. H. Cheng (1990), A branch-and-bound algorithm for solving the group technology problem, *Annals of Operations Research*, Vol. 26, pp. 415–431.

Kusiak A., and Y.-K. Chung (1991), GT/ART: Using neural networks to form machine cells, *ASME Manufacturing Review*, Vol. 4, No. 4, pp. 293–301.

Levy, W. B., and N. L. Desmond (1985), The rules of elemental synaptic plasticity, in J. A. Anderson, W. B. Levy, and S. Lehmkuhle (Eds.), *Synaptic Modification, Neuron Selectivity, and Nervous System Organization*, Lawrence Erlbaum, Hillsdale, NJ, pp. 105–121.

Lippmann, R. P. (1987), An introduction to computing with neural nets, *IEEE Transactions on Acoustics, Speech and Signal Processing*, April, pp. 4–22.

Matheus, C. J., and W. E. Hohensee (1987), Learning in artificial neural systems, *Computational Intelligence*, Vol. 3, No. 4, pp. 283–294.

Mekkaoui, A., and P. Jespers (1990), An optimal self-organizing pattern classifier, *Proceedings of the 4th IEEE Annual International Conference on Neural Networks*, IEEE Press, Piscataway, NJ. pp. I.447–I.450.

Michalski, R. S. (1987), Concept learning, in S. C. Shapiro (Ed.), *Encyclopedia of Artificial Intelligence*, Vol. 1, John Wiley, New York, pp. 185–194.

Michie, D. (1968), Memo functions and machine learning, *Nature*, Vol. 218, pp. 19–22.

Mirchandani, G., and W. Cao (1989), On hidden nodes for neural nets, *IEEE Transactions on Circuits and Systems*, Vol. 36, No. 5, pp. 661–664.

Mitchell, T. M., R. M. Keller, and S. T. Kedar-Cebelli (1986), Explanation based generalization: A unifying view, *Machine Learning*, Vol. 1, No. 1, pp. 47–80.

Rosenblatt, F. (1962), *Principles of Neurodynamics*, Spartan, Washington, DC.

Rumelhart, D. E., and J. L. McClelland (1986), *Parallel Distributed Processing: Explorations in the Microstructures of Cognition*, Vol. 1, MIT Press, Cambridge, MA.

Rumelhart, D. E., and D. Zipser (1985), Feature discovery by competitive learning, *Cognitive Science*, Vol. 9, pp. 75–112.

Rumelhart, D. E., G. E. Hinton, and R. J. Williams (1986a), Learning representations by back-propagation errors, *Nature*, Vol. 323, pp. 533–536.

Rumelhart, D. E., G. E. Hinton, and R. J. Williams (1986b), Learning internal representations by error propagation, in D. E. Rumelhart and J. L. McClelland (Eds.), *Parallel Distributed Processing: Explorations in the Microstructure of Cognition*, MIT Press, Cambridge, MA, pp. 318–362.

Sejnowski T. J., and C. R. Rosenberg (1986), NETtalk: A parallel network that learns to read aloud, Electrical Engineering and Computer Science Technical Report JHU/EECS-86/01, Johns Hopkins University, Baltimore, MD.

Stork, D. G. (1989), Self-organization, pattern recognition, and adaptive resonance theory, *Journal of Neural Network Computing*, Vol. 1, No. 1, pp. 26–42.

Wasserman, P. D. (1989), *Neural Computing: Theory and Practice*, Van Nostrand Reinhold, New York.

Zarefar, H., and J. R. Goulding (1992), Neural networks in design of products: A case study, in A. Kusiak (Ed.), *Intelligent Design and Manufacturing*, John Wiley, New York, pp. 179–201.

QUESTIONS

11.1. What is a neuron?

11.2. What are the basic functions of a neuron?

11.3. What is a neuron transfer function?

11.4. What are three different types of learning? Give an example of each.

11.5. What are two basic types of neural networks?

11.6. Which of the two neural network implementations is more widely used: hardware or software?

11.7. What are the basic steps in designing a neural network?

11.8. What is vigilance in neural networks?

11.9. What is the ART network?

11.10. What are the main differences between the self-learning neural network and the back-propagation neural network?

PROBLEMS

11.1. Given the input vector $[i_1, i_2] = [1.1, 2.5]$, weights w_1, w_2, and w_3, and threshold values for the neural network in Figure 11.19, calculate the output.

11.2. Determine the value of output O for the neural network in Figure 11.20 that diagnoses tool breakage on a machining center. Each neuron uses a step transfer function with the value of the thresholds shown in the network. The value of each weight is shown on the corresponding connection:

(a) Is the tool operational or broken?

(b) What should be the value of threshold T_O to get a response opposite to the one in (a)?

11.3. The back-propagation network has been applied to machine diagnosis concerned with identifying the most likely machine failure given a set of failure symptoms. To solve this problem, we can use a back-propagation network in which the input units encode failure symptoms and the output units encode possible machine failures and then train the network on a set of examples. Once the network is well trained, it can be used to predict the type of machine failure based on failure symptoms. Suppose the domain concerned involves eight

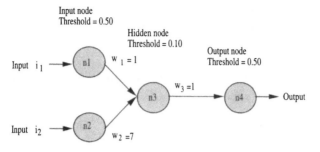

Figure 11.19. Simple neural network.

binary attributes to describe failure symptoms, such as lack of motion or noise, and three possible failure types. The training examples are listed in Table 11.4 (Fu, 1994).

Use a neural network with one hidden layer to model the machine diagnosis problem. The network is fully connected and the number of hidden units is determined in the following way. Start with one hidden unit, then two, then three, and so on, until you have the best performance (i.e., the network cannot be improved by adding more hidden units). To evaluate performance, consider the mean-square error (MSE) of the training data, the training convergence (number of epochs), and the test error. (*Hint:* Your final determination of the number of hidden units will be somewhat arbitrary, but it should not be necessary to exceed four or five hidden units). Use the first eight examples in Table 11.4 for training and the remaining four for testing. Once you have

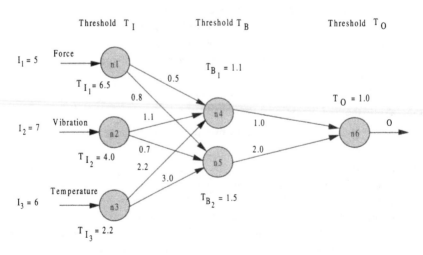

Figure 11.20. Neural network for tool breakage diagnosis.

**TABLE 11.4. Training Examples
for the Machine Diagnosis Problem**

Input Vector	Output Vector
(1,1,1,0,0,0,0,0)	(1,0,0)
(0,1,0,0,1,0,0,0)	(0,1,0)
(1,0,0,0,0,1,0,1)	(0,0,1)
(1,0,1,1,0,0,0,0)	(1,0,0)
(0,1,0,1,1,0,1,0)	(0,1,0)
(0,0,0,0,0,1,1,1)	(0,0,1)
(0,0,0,1,0,1,1,1)	(1,0,0)
(0,1,0,0,1,1,1,0)	(0,1,0)
(0,0,1,0,0,0,1,1)	(0,0,1)
(0,1,1,0,0,0,0,1)	(1,0,0)
(0,0,0,0,1,1,0,0)	(0,1,0)
(1,0,0,1,0,1,0,1)	(0,0,1)

designed an acceptable network and trained the weights, answer the following questions:

(a) What are the trained weights in the network? You can print or redraw your network and label the weights accordingly.

(b) Label the three machine-type failures corresponding to the output vectors (1,0,0), (0,1,0), and (0,0,1) as A, B, and C, respectively. What are the failure types for the following two vectors of machine failure symptoms:

Failure symptom vector 1: (1,1,0,0,1,1,0,0)

Failure symptom vector 2: (0,0,1,1,0,0,1,1)

(c) Based on the test data vectors (i.e., the last four vectors in Table 11.4), does it appear that the network is not trained to recognize any machine-type failure? If so, how do you suggest improving the performance of the network?

(d) With the current weights, what change can be made to produce the exact output vectors for the training data? (*Hint*: Assuming you used the default activation function for the output nodes, what different activation functions might be used?)

CHAPTER 12

LAYOUT OF MACHINES AND FACILITIES

12.1 INTRODUCTION

The layout of manufacturing facilities involves machines, material handling equipment, and all support areas. The layout pattern of machines is to a large degree determined by the type of material handling equipment used. The three most commonly used material handling carriers and their corresponding machine layout are:

- Handling robot—circular machine layout
- Automated guided vehicle (AGV)—linear single-row or double-row machine layout
- Gantry robot—cluster machine layout

In the next two sections, the layout of machine cells served by AGVs is considered. An AGV is most efficient while moving along a straight line (Muller, 1983). This technical limitation has forced designers of manufacturing systems to arrange

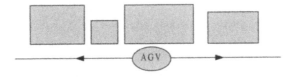

Figure 12.1. Single-row machine layout.

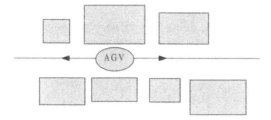

Figure 12.2. Double-row machine layout.

machines along a straight line. Two types of machine layout considered in the next two sections are the linear single-row layout (see Figure 12.1) and the linear double-row layout (see Figure 12.2). These two types of machine layouts are frequently used in cellular manufacturing systems.

Readers interested in the background information on the facility and machine layout problem may refer to Tompkins and White (1984), Kouvelis and Kiran (1989), Kusiak (1990), and Singh (1996).

12.2   SINGLE-ROW MACHINE LAYOUT

Two major decision factors considered in arranging machines on a factory floor are:

- Frequency of trips of an AGV between each pair of machines
- Travel time between each pair of machines

The frequencies of trips between the machines are entered as a flow matrix $[f_{ij}]$. The elements of the flow matrix indicate the frequency of trips to be made by the material handling carrier between each pair of machines in a given time horizon.

For example, the frequency of trips f_{ij} between machines i and j in a manufacturing system could be computed using the following formula (Kusiak, 1990):

$$f_{ij} = \sum_{k=1}^{n_{ij}} \left\lceil \frac{v_{ij}^k}{u_k} \right\rceil$$

where

v_{ij}^k = volume of part type k to be moved from machine i to machine j in a given horizon (e.g., 1 year)

n_{ij} = number of different parts to be moved from machine i to machine j in a given time horizon

u_k = number of parts to be moved in a single trip of the material handling carrier

$\lceil \bullet \rceil$ = smallest integer greater than or equal to $\bullet$

The travel time between machines is another important data. As the AGV velocity is a nonlinear function of time, rather than the distance between sites, the AGV travel time should be considered. The AGV travel time between any pair of machines consists of five different components:

- Loading time
- Acceleration time
- Travel time
- Deceleration time
- Unloading time

The loading and unloading times are constant for a given type of AGV. The AGV travel time between machines i and j, t_{ij}, is a nonlinear function of the uni-axis distance r_{ij} between two machines, that is, $t_{ij} = g(r_{ij})$. Since an AGV travels along its path only (back and forth), the uni-axis distance r_{ij} is measured along that path (one coordinate) only. The uni-axis distance between any two machines can be computed based on the matrix of adjacent uni-axis distances computed in advance and remaining constant throughout the entire computational process.

The distance between two adjacent machines depends on the sizes of the two machines and the clearance between them. Using the information about the dimensions and the orientation of all the machines, an adjacent uni-axis distance matrix $[d_{ij}]$ is constructed. A value d_{ij} in the matrix indicates the distance between machines i and j when they are located adjacent to each other. The uni-axis distance r_{pv} between machines p and v when machine q is placed between machines p and v is $d_{pq} + d_{qv}$. Then, the AGV travel time t_{pv} between machines p and v is $g(d_{pq} + d_{qv})$.

In this chapter, in addition to the above-mentioned adjacent uni-axis distance, uni-axis and rectilinear distances are used. The uni-axis distance is a distance between any two machines when they are located anywhere in the layout measured along one direction only (e.g., along an AGV path). The uni-axis distance is further discussed and illustrated in Section 12.3. The rectilinear distance is defined and illustrated in Section 12.4. It should be stressed that in the algorithms discussed in this chapter, time should be used rather than any of the three distances above.

In this chapter, it is assumed that the costs of locating a machine to any site are identical. This is a realistic assumption because the site preparation and machine location costs are independent of the sites in an automated manufacturing system. It is also assumed that a machine can be oriented in only one particular direction, irrespective of its location. A longer side of each machine is equidistant from the AGV path.

To solve the single-row layout problem with m machines, a heuristic algorithm (Kusiak and Park, 1994) is presented. The algorithm presented next first selects two machines and connects them. Then the next machine is selected and located either on the right-hand side or the left-hand side of the previously located machines. This process continues until all machines are assigned. The AGV travel time t_{ij} between machines i and j is computed using the uni-axis distance r_{ij}. The solution generated by the algorithm does not produce the layout but produces only the sequence in which the machines are placed in the layout. The final layout can be drawn using a CAD system.

Algorithm 12.1

Step 0.   Set iteration number $k = 1$. From the flow matrix $[f_{ij}]$ compute

$$f_{i^*j^*} = \max\{f_{ij}: i, j = 1, 2, \ldots, m\}.$$

If there is a tie, $f_{i^*j^*} = \max\{f_{ij} \cdot t_{ij}: i, j = 1, 2, \ldots, m\}$. Connect i^*, j^* and include them in the solution set. Set the solution set $U = \{i^*, j^*\}$.

Step 1.   Compute

$$f_{p^*q^*} = \max\{f_{i^*k}, f_{j^*l}: k \in \{\{1, 2, \ldots, m\} - U\}\}$$

Set $s^* = q^*$. Consider two alternatives:
   (a)   Place machine s^* left of machine i^*.
   (b)   Place machine s^* right of machine j^*.
Compute

$$q_{s^*i^*} = \Sigma \{f_{s^*r} \cdot t_{s^*r}: r \in U \text{ and machine } s^* \text{ is placed left of machine } i^*\}$$

$$q_{j^*s^*} = \Sigma \{f_{s^*r} \cdot t_{s^*r}: r \in U \text{ and machine } s^* \text{ is placed right of machine } j^*\}.$$

If $q_{s^*i^*} \le q_{j^*s^*}$:
   1.   Select alternative (a) above.
   2.   Set $i^* = s^*$.

If $q_{s^*i^*} > q_{j^*s^*}$:
   1.   Select alternative (b) above.
   2.   Set $j^* = s^*$.

Then $U = U + s^*$.

Step 2.   Set iteration number $k = k + 1$. Repeat Step 1 until the final solution is obtained (i.e., until all the machines are included in the solution set U).

Algorithm 12.1 builds a layout by selecting adjacent machines with the maximum flow between them in an attempt to minimize the total product of flow and travel time or distance. Rather than selecting $\max\{f_{ij}\}$, one can modify Steps 0 and 1 of the above algorithm and consider selecting pairs of machines based on the ratio $\max\{f_{ij}/t_{ij}\}$ or

$\max\{f_{ij}/d_{ij}\}$. The quality of the solutions generated using these ratios appear to be better than those generated by computing $\max\{f_{ij}\}$ as presented in Algorithm 12.1. Algorithm 12.1 is illustrated in Example 12.1 (Kusiak and Park, 1994).

Example 12.1.   Determine the layout of four machines. The matrix of frequencies $[f_{ij}]$ is provided in (12.1). Machine dimensions are given in Table 12.1. The clearance between each pair of machines is assumed to be one unit:

$$[f_{ij}] = \text{Machine} \begin{array}{c} \\ 1 \\ 2 \\ 3 \\ 4 \end{array} \begin{array}{cccc} \overset{\text{Machine}}{\overset{\displaystyle 1 \quad 2 \quad 3 \quad 4}{}} \\ \begin{bmatrix} 0 & 10 & 15 & 15 \\ 10 & 0 & 0 & 5 \\ 15 & 0 & 0 & 40 \\ 15 & 5 & 40 & 0 \end{bmatrix} \end{array} \qquad (12.1)$$

Based on the machine dimensions in Table 12.1 and the unit clearance between the machines, the distance matrix between any adjacent pair of machines located in adjacent positions is computed as

$$[d_{ij}] = \text{Machine} \begin{array}{c} \\ 1 \\ 2 \\ 3 \\ 4 \end{array} \begin{array}{cccc} \overset{\text{Machine}}{\overset{\displaystyle 1 \quad 2 \quad 3 \quad 4}{}} \\ \begin{bmatrix} 0 & 4 & 5 & 3 \\ 4 & 0 & 6 & 4 \\ 5 & 6 & 0 & 5 \\ 3 & 4 & 5 & 0 \end{bmatrix} \end{array} \qquad (12.2)$$

For simplicity, assume that the AGV travel time is identical to the uni-axis distance between any two machines i and j, that is, $t_{ij} = r_{ij}$.

Iteration 1

Step 0. From the flow matrix $[f_{ij}]$, $\max\{f_{ij}: i,j = 1, 2, 3, 4\} = f_{34} = 40$ is obtained. Thus $i^* = 3, j^* = 4$ is determined.

TABLE 12.1. Dimensions of Machines in Example 12.1

Machine Number	Machine Dimensions
1	2×2
2	4×4
3	6×6
4	2×2

Machines 3 and 4 are connected and included in the solution, symbolically denoted as

The solution set is updated to $U = \{3, 4\}$ and columns 3 and 4 of the matrix $[f_{ij}]$ are removed from further consideration as indicated with asterisks:

$$\begin{array}{cc} & \text{Machine} \\ & \begin{array}{cccc} 1 & 2 & 3^* & 4^* \end{array} \\ [f_{ij}] = \text{Machine} \begin{array}{c} 1 \\ 2 \\ 3 \\ 4 \end{array} & \left[\begin{array}{cccc} 0 & 10 & 15 & 15 \\ 10 & 0 & 0 & 5 \\ 15 & 0 & 0 & 40 \\ 15 & 5 & 40 & 0 \end{array}\right] \end{array} \qquad (12.3)$$

Step 1. Compute

$$\max\{f_{3k}, f_{4l;} \, k, l = 1, 2\} = f_{31} = f_{41} = 15$$

Set $s^* = 1$. Consider two alternatives:

(a) Place machine 1 left of machine 3:

(b) Place machine 1 right of machine 4:

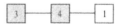

Compute

$$q_{s^*i^*} = q_{13} = f_{13} \cdot t_{13} + f_{14} \cdot t_{14}$$

$$= f_{13} \cdot r_{13} + f_{14} \cdot r_{14}$$

$$= f_{13} \cdot d_{13} + f_{14} \cdot (d_{13} + d_{34})$$

$$= 15 \cdot 5 + 15 \cdot (5 + 5)$$

$$= 225$$

$$q_{j^*s^*} = q_{41} = f_{14} \cdot t_{14} + f_{13} \cdot t_{13} = f_{14} \cdot r_{14} + f_{13} \cdot r_{13}$$

$$= f_{14} \cdot d_{14} + f_{13} \cdot (d_{14} + d_{43})$$

$$= 15 \cdot 3 + 15 \cdot (3 + 5)$$

$$= 165$$

Since $q_{13} > q_{41}$:

1. Alternative (b) is selected; machine 1 is placed right of machine 4:

2. Set $j^* = 1$.

3. The set U is updated to $U + s^* = \{1, 3, 4\}$ and column 1 is excluded from further consideration, as shown in the matrix.

$$
[f_{ij}] = \text{Machine} \quad
\begin{array}{c}
1 \\ 2 \\ 3 \\ 4
\end{array}
\begin{array}{cccc}
\overset{\text{Machine}}{} \\
1^* \quad 2 \quad 3^* \quad 4^* \\
\left[\begin{array}{cccc}
0 & 10 & 15 & 15 \\
10 & 0 & 0 & 5 \\
15 & 0 & 0 & 40 \\
15 & 5 & 40 & 0
\end{array}\right]
\end{array}
\qquad (12.4)
$$

Step 2. Since machine 2 is not included in the solution set U, go to Step 1.

Iteration 2

Step 1. Compute $\max\{f_{3k}, f_{1l}: k, l = 2\} = f_{12} = 10$. Set $s^* = 2$. Consider two alternatives:

(a) Place machine 2 left of machine 3:

(b) Place machine 2 right of machine 1:

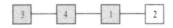

Compute

$$q_{s^* i^*} = q_{23} = f_{23} \cdot t_{23} + f_{24} \cdot t_{24} + f_{21} \cdot t_{21}$$

$$= f_{23} \cdot r_{23} + f_{24} \cdot r_{24} + f_{21} \cdot r_{21}$$

$$= f_{23} \cdot d_{23} + f_{24} \cdot (d_{23} + d_{34}) + f_{21} \cdot (d_{23} + d_{34} + d_{41})$$

$$= 0 \cdot 6 + 5 \cdot (6 + 5) + 10 \cdot (6 + 5 + 3)$$

$$= 195$$

$$q_{j^*s^*} = q_{12} = f_{21} \cdot t_{21} + f_{24} \cdot t_{24} + f_{23} \cdot t_{23}$$

$$= f_{21} \cdot r_{21} + f_{24} \cdot r_{24} + f_{23} \cdot r_{23}$$

$$= f_{21} \cdot d_{21} + f_{24} \cdot (d_{21} + d_{14}) + f_{23} \cdot (d_{21} + d_{14} + d_{43})$$

$$= 10 \cdot 4 + 5 \cdot (4 + 3) + 0 \cdot (4 + 3 + 5)$$

$$= 75.$$

Since $q_{23} > q_{12}$:

1. Alternative (b) is selected:

2. Set $j^* = 2$.
3. The solution set U is updated to $U + s^* = \{1, 2, 3, 4\}$.

Step 2. Since all machines have been included in the solution set U, stop.
The total cost of the solution is 440. The final layout for the data in Example 12.1 is shown in Figure 12.3.

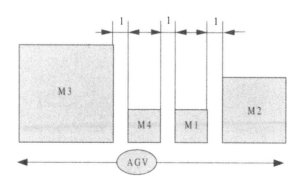

Figure 12.3. Machine layout for the problem in Example 12.1.

12.3 DOUBLE-ROW MACHINE LAYOUT

In this section, a heuristic algorithm (Kusiak and Park, 1994) for generating a double-row machine layout problem is presented. The machine dimensions are not equal. This model along with the previous model are of importance to industry as almost any machine layout on a factory floor reduces to double or a combination of a double-row and single-row machine layout.

The algorithm presented next places a pair of machines at a time. The machines are always being located on two opposite sides of the AGV track as the travel time is typically the shortest. Computational experience has shown that arranging two machines at a time produces the best solution quality in a reasonable computation time. The centers of the first pair of machines selected are aligned with a straight line, used as a reference line for locating machines in the final layout and computing uni-axis distances between the machines. Then the next pair of machines is selected and located, as closely as possible, either right (used in this chapter) or left to the sites of the first two machines. Then the algorithm locates the remaining machines, two at a time, adjacent to the previously located machines. Figure 12.4 presents an example of the layout with seven machines.

To compute the AGV travel time, the uni-axis distance between machines should be computed, which is not simple. It is difficult to identify the location of a machine because the machine dimensions are not equal and the machines are arranged in two rows. In this section, in order to compute the distance between machines, the line crossing the centers of the first two machines selected in the layout serves as a reference line. Thus the uni-axis distance between the first two machines is always 0.

The computation of the distances between any pair of machines is illustrated in Figure 12.5. The broken vertical line crossing the geometric centers of the first two machines (M_1, M_2) is the reference line. For example, the uni-axis distance between machines 3 and 4 is $r_{34} = |\, r_{13} - r_{24}\,|$.

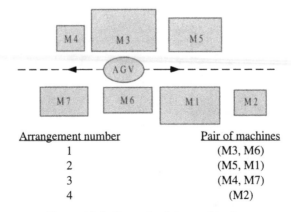

Arrangement number	Pair of machines
1	(M3, M6)
2	(M5, M1)
3	(M4, M7)
4	(M2)

Figure 12.4. Example of the machine layout.

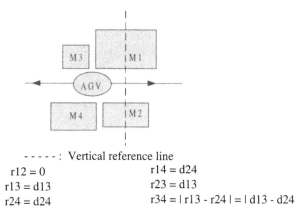

- - - - - : Vertical reference line

r12 = 0   r14 = d24
r13 = d13   r23 = d13
r24 = d24   r34 = | r13 - r24 | = | d13 - d24

Figure 12.5. Computation of uni-axis distances between pairs of machines.

The algorithm for solving the double-row layout problem with m machines is presented next (Kusiak and Park, 1994).

Algorithm 12.2

Step 0. Set iteration number $k = 1$. From the flow matrix $[f_{ij}]$, compute

$$f_{i^*j^*} = \max\{f_{ij}: i, j = 1, 2, \ldots, m\}$$

If there is a tie, $f_{i^*j^*} = \max\{f_{ij} \cdot t_{ij}: i, j = 1, 2, \ldots, m\}$.
Place i^*, j^* on the opposite sites of the AGV path and include them in the solution.
Set the solution set $U = \{i^*, j^*\}$. Remove columns i^* and j^* of matrix $[f_{ij}]$ from further consideration.

Step 1. Compute

$$f_{p^*q^*} = \max\{f_{i^*k}, f_{j^*l}: k, l \in \{1, 2, \ldots, m\} - U\}.$$

Set $s^* = q^*$ and remove q^* from further consideration. Compute

$$f_{x^*y^*} = \max\{f_{i^*k}, f_{j^*l}, f_{s^*v}: k, l, v \in \{1, 2, \ldots, m\} - U - q^*\}$$

Set $t^* = y^*$ and remove y^* from further consideration. Consider two alternatives:
(a) Place s^* right of i^* and t^* right of j^*.
(b) Place t^* right of i^* and s^* right of j^*.
Compute

$$q_{i^*s^*j^*t^*} = \sum\{f_{s^*r} \cdot t_{s^*r} + f_{t^*r} \cdot t_{t^*r}: r \in U\} + f_{s^*t^*} \cdot t_{s^*t^*}$$

where s^* is placed right of i^* and t^* is placed right of j^*, and

$$q_{i^*t^*j^*s^*} = \sum \{f_{s^*r} \cdot t_{s^*r} + f_{t^*r} \cdot t_{t^*r} : r \in U\} + f_{s^*t^*} \cdot t_{s^*t^*}$$

where t^* is placed right of i^* and s^* is placed right of j^*.
If $q_{i^*s^*j^*t^*} < q_{i^*t^*j^*s^*}$, select alternative (a). Otherwise, select alternative (b). Set $U = U + \{s^*, t^*\}$.

Step 2. If only one machine has not been assigned, go to Step 4; otherwise go to Step 3.

Step 3. Compute

$$f_{p^*q^*} = \max\{f_{i^*k}, f_{j^*l}, f_{s^*v}, f_{t^*w} : k, l, v, w \in \{1, 2, \ldots, m\} - U\}$$

Set $c^* = q^*$ and remove q^* from further consideration. Compute

$$f_{x^*y^*} = \max \{f_{i^*k}, f_{j^*l}, f_{s^*v}, f_{t^*w}, f_{c^*z} : k, l, v, w, z \in \{\{1, 2, \ldots, m\} - U - q^*\}\}$$

Set $d^* = y^*$ and remove y^* from further consideration.
Consider four alternatives:

    (a)   Place c^* left of i^* and d^* left of j^*.
    (b)   Place d^* left of i^* and c^* left of j^*.
    (c)   Place c^* right of s^* and d^* right of t^*.
    (d)   Place d^* right of s^* and c^* right of t^*.

Compute

$$q_{c^*i^*d^*j^*} = \sum \{f_{c^*r} \cdot t_{c^*r} + f_{d^*r} \cdot t_{d^*r} : r \in U\} + f_{c^*d^*} \cdot t_{c^*d^*}$$

where c^* is placed left of i^* and d^* is placed left of j^*,

$$q_{d^*i^*c^*j^*} = \sum \{f_{c^*r} \cdot t_{c^*r} + f_{d^*r} \cdot t_{d^*r} : r \in U\} + f_{c^*d^*} \cdot t_{c^*d^*}$$

where d^* is placed left of i^* and c^* is placed left of j^*,

$$q_{s^*c^*t^*d^*} = \sum \{f_{c^*r} \cdot t_{c^*r} + f_{d^*r} \cdot t_{d^*r} : r \in U\} + f_{c^*d^*} \cdot t_{c^*d^*}$$

where c^* is placed right of s^* and d^* is placed right of t^*, and

$$q_{s^*d^*t^*c^*} = \sum \{f_{c^*r} \cdot t_{c^*r} + f_{d^*r} \cdot t_{d^*r} : r \in U\} + f_{c^*d^*} \cdot t_{c^*d^*}$$

where d^* is placed right of s^* and c^* is placed right of t^*. If
$$q_{c^*i^*d^*j^*} = \min\{q_{c^*i^*d^*j^*}, q_{d^*i^*c^*j^*}, q_{s^*c^*t^*d^*}, q_{s^*d^*t^*c^*}\}:$$

1. Select alternative (a).
2. Set $i^* = c^*$ and $j^* = d^*$.

If $q_{d^*i^*c^*j^*} = \min\{q_{c^*i^*d^*j^*}, q_{d^*i^*c^*j^*}, q_{s^*c^*t^*d^*}, q_{s^*d^*t^*c^*}\}$:
1. Select alternative (b).
2. Set $i^* = d^*$ and $j^* = c^*$.

If $q_{s^*c^*t^*d^*} = \min\{q_{c^*i^*d^*j^*}, q_{d^*i^*c^*j^*}, q_{s^*c^*t^*d^*}, q_{s^*d^*t^*c^*}\}$:
1. Select alternative (c).
2. Set $s^* = c^*$ and $t^* = d^*$.

If $q_{s^*d^*t^*c^*} = \min\{q_{c^*i^*d^*j^*}, q_{d^*i^*c^*j^*}, q_{s^*c^*t^*d^*}, q_{s^*d^*t^*c^*}\}$:
1. Select alternative (d).
2. Set $s^* = d^*$ and $t^* = c^*$.

Set $U = U + \{c^*, d^*\}$. Go to Step 5.

Step 4. Set c^* to the last machine that has not been assigned. Consider four alternatives:

(a)   Place c^* left of i^*.
(b)   Place c^* left of j^*.
(c)   Place c^* right of s^*.
(d)   Place c^* right of t^*.

Compute

$$q_{c^*i^*} = \sum \{f_{c^*r} \cdot t_{c^*r} : r \in U\} \quad \text{where } c^* \text{ is placed left of } i^*$$

$$q_{c^*j^*} = \sum \{f_{c^*r} \cdot t_{c^*r} : r \in U\} \quad \text{where } c^* \text{ is placed left of } j^*$$

$$q_{s^*c^*} = \sum \{f_{c^*r} \cdot t_{c^*r} : r \in U\} \quad \text{where } c^* \text{ is placed right of } s^*$$

$$q_{t^*c^*} = \sum \{f_{c^*r} \cdot t_{c^*r} : r \in U\} \quad \text{where } c^* \text{ is placed right of } t^*$$

Then, if $q_{c^*i^*} = \min\{q_{c^*i^*}, q_{c^*j^*}, q_{s^*c^*}, q_{t^*c^*}\}$, select alternative (a). If $q_{c^*j^*} = \min\{q_{c^*i^*}, q_{c^*j^*}, q_{s^*c^*}, q_{t^*c^*}\}$, select alternative (b). If $q_{s^*c^*} = \min\{q_{c^*i^*}, q_{c^*j^*}, q_{s^*c^*}, q_{t^*c^*}\}$, select alternative (c). If $q_{t^*c^*} = \min\{q_{c^*i^*}, q_{c^*j^*}, q_{s^*c^*}, q_{t^*c^*}\}$, select alternative(d). Set $U = U + \{c^*\}$. Go to Step 5.

Step 5.   Set iteration number $k = k + 1$. Repeat Steps 2–4 until the final solution is obtained (i.e., all the machines are included in the solution set U).

Analogous to Algorithm 12.1, in Steps 0, 1, and 3 of Algorithm 12.2, rather than using the ratio $\max\{f_{ij}\}$, the pairs of machines can be selected based on the ratios $\max\{f_{ij}/t_{ij}\}$, $\max\{f_{ij}/d_{ij}\}$, or other ratios. These ratios may improve the quality of solutions generated by Algorithm 12.2.

Algorithm 12.2 is illustrated in Example 12.2 (Kusiak and Park, 1994).

Example 12.2. Design a layout for five machines served by an AGV. The matrix of frequencies $[f_{ij}]$ is given in (12.5). Based on the clearance matrix (12.6) and machine dimensions provided in Table 12.2, the matrix of adjacent uni-axis distances $[d_{ij}]$ is computed in (12.7). For simplicity, assume that the AGV travel time is identical to the uni-axis distance, that is, $t_{ij} = r_{ij}$, for all i and j:

$$
[f_{ij}] = \text{Machine} \quad
\begin{array}{c}
1 \\ 2 \\ 3 \\ 4 \\ 5
\end{array}
\begin{bmatrix}
0 & 5 & 1 & 4 & 1 \\
5 & 0 & 3 & 0 & 2 \\
1 & 3 & 0 & 0 & 0 \\
4 & 0 & 0 & 0 & 5 \\
1 & 2 & 0 & 5 & 0
\end{bmatrix}
\quad (12.5)
$$

Machine
1 2 3 4 5

$$
[c_{ij}] = \text{Machine} \quad
\begin{array}{c}
1 \\ 2 \\ 3 \\ 4 \\ 5
\end{array}
\begin{bmatrix}
0 & 1 & 1 & 1 & 2 \\
1 & 0 & 1 & 1 & 1 \\
1 & 1 & 0 & 1 & 1 \\
1 & 1 & 1 & 0 & 3 \\
2 & 1 & 1 & 3 & 0
\end{bmatrix}
\quad (12.6)
$$

Machine
1 2 3 4 5

$$
[d_{ij}] = \text{Machine} \quad
\begin{array}{c}
1 \\ 2 \\ 3 \\ 4 \\ 5
\end{array}
\begin{bmatrix}
0 & 36 & 38.5 & 56 & 42 \\
36 & 0 & 23.5 & 41 & 26 \\
38.5 & 23.5 & 0 & 43.5 & 28.5 \\
56 & 41 & 43.5 & 0 & 48 \\
42 & 1 & 28.5 & 48 & 0
\end{bmatrix}
\quad (12.7)
$$

TABLE 12.2. Dimensions of Machines in Example 12.2

Machine Number	Machine Dimensions
1	50×30
2	20×20
3	25×20
4	60×35
5	30×15

Iteration 1

Step 0. For the flow matrix (12.5), $\max\{f_{ij}: i, j = 1, 2, \ldots, 5\} = f_{12} = f_{45} = 5$ is determined. The flow value f_{45} is selected because $f_{45} \cdot t_{45} = f_{45} \cdot d_{45} = 5 \cdot 48 = 240 > f_{12} \cdot t_{12} = f_{12} * d_{12} = 5 \cdot 36 = 180$. Thus $i^* = 4$ and $j^* = 5$. Machines 4 and 5 are assigned to the opposite sites of the AGV path and are included in the solution:

$$
\begin{array}{c}
\boxed{4} \\[-2pt]
-\; -\; -\; - \\[-2pt]
\boxed{5}
\end{array}
$$

The solution set is updated to $U = \{4, 5\}$. Columns 4 and 5 of matrix (12.7) are removed from further consideration, as indicated with asterisks:

$$
[f_{ij}] = \text{Machine} \quad
\begin{array}{c}
1 \\ 2 \\ 3 \\ 4 \\ 5
\end{array}
\begin{array}{ccccc}
\text{Machine} & & & & \\
1 & 2 & 3 & 4^* & 5^* \\
\left[\begin{array}{ccccc}
0 & 5 & 1 & 4 & 1 \\
5 & 0 & 3 & 0 & 2 \\
1 & 3 & 0 & 0 & 0 \\
4 & 0 & 0 & 0 & 5 \\
1 & 2 & 0 & 5 & 0
\end{array}\right]
\end{array}
\qquad (12.8)
$$

Step 1. Compute

$$
\max\{f_{4k}, f_{5l}: k, l = 1, 2, 3\} = f_{41}
$$

Set $s^* = 1$ and delete column 1 from further consideration, as indicated in

$$
[f_{ij}] = \text{Machine} \qu
\begin{array}{c}
1 \\ 2 \\ 3 \\ 4 \\ 5
\end{array}
\begin{array}{ccccc}
\text{Machine} & & & & \\
1^* & 2 & 3 & 4^* & 5^* \\
\left[\begin{array}{ccccc}
0 & 5 & 1 & 4 & 1 \\
5 & 0 & 3 & 0 & 2 \\
1 & 3 & 0 & 0 & 0 \\
4 & 0 & 0 & 0 & 5 \\
1 & 2 & 0 & 5 & 0
\end{array}\right]
\end{array}
\qquad (12.9)
$$

From (12.9) compute

$$
\max\{f_{5l}, f_{1v}: l, v = 2, 3\} = f_{12}
$$

Set $t^* = 2$ and column 2 is removed:

$$[f_{ij}] = \text{Machine} \quad \begin{array}{c} \\ 1 \\ 2 \\ 3 \\ 4 \\ 5 \end{array} \begin{array}{c} \text{Machine} \\ \begin{array}{ccccc} 1^* & 2^* & 3 & 4^* & 5^* \end{array} \\ \begin{bmatrix} 0 & 5 & 1 & 4 & 1 \\ 5 & 0 & 3 & 0 & 2 \\ 1 & 3 & 0 & 0 & 0 \\ 4 & 0 & 0 & 0 & 5 \\ 1 & 2 & 0 & 5 & 0 \end{bmatrix} \end{array} \qquad (12.10)$$

Consider two alternatives:

   (a) Place machine 1 right of machine 4 and machine 2 right of machine 5:

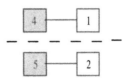

   (b) Place machine 2 right of machine 4 and machine 1 right of machine 5:

Compute:

$$q_{s^*i^*t^*j^*} = q_{4152}$$

$$= f_{14} \cdot t_{14} + f_{15} \cdot t_{15} + f_{24} \cdot t_{24} + f_{25} \cdot t_{25} + f_{12} \cdot t_{12}$$

$$= f_{14} \cdot r_{14} + f_{15} \cdot r_{15} + f_{24} \cdot r_{25} + f_{25} \cdot r_{25} + f_{12} \cdot r_{12}$$

$$= f_{14} \cdot d_{14} + f_{15} \cdot d_{14} + f_{24} \cdot d_{25} + f_{25} \cdot d_{25} + f_{12} \cdot |d_{14} - d_{25}|$$

$$= 4 \cdot 56 + 1 \cdot 56 + 0 \cdot 26 + 2 \cdot 26 + 5 \cdot 30$$

$$= 482$$

$$q_{s^*j^*t^*i^*} = q_{4251}$$

$$= f_{24} \cdot t_{24} + f_{25} \cdot t_{25} + f_{14} \cdot t_{14} + f_{15} \cdot t_{15} + f_{12} \cdot t_{12}$$

$$= f_{24} \cdot r_{24} + f_{25} \cdot r_{25} + f_{14} \cdot r_{14} + f_{15} \cdot r_{15} + f_{12} \cdot r_{12}$$

$$= f_{24} \cdot d_{24} + f_{25} \cdot d_{24} + f_{14} \cdot d_{15} + f_{15} \cdot d_{15} + f_{12} \cdot |d_{24} - d_{15}|$$

$$= 0 \cdot 41 + 2 \cdot 41 + 4 \cdot 42 + 1 \cdot 42 + 5 \cdot |41 - 42|$$

$$= 297$$

Since $q_{4251} < q_{4152}$, alternative (b) is selected. Set $U = U + \{1, 2\}$.

Step 2. Since only machine 3 has not been assigned, go to Step 4.

Step 4. Set $c^* = 3$. Consider four alternatives:

(a) Place machine 3 left of machine 4:

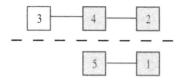

(b) Place machine 3 left of machine 5:

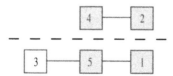

(c) Place machine 3 right of machine 2:

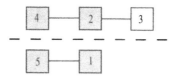

(d) Place machine 3 right of machine 1:

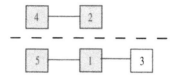

Compute:

$$q_{c^*i} = q_{34} = f_{34} \cdot t_{34} + f_{35} \cdot t_{35} + f_{32} \cdot t_{32} + f_{31} \cdot t_{31}$$

$$= f_{34} \cdot r_{34} + f_{35} \cdot r_{35} + f_{32} \cdot r_{32} + f_{31} \cdot r_{31}$$

$$= f_{34} \cdot d_{34} + f_{35} \cdot d_{34} + f_{32} \cdot (d_{34} + d_{42}) + f_{31} \cdot (d_{34} + d_{51})$$

$$= 0 \cdot 43.5 + 0 \cdot 43.5 + 3 \cdot (43.5 + 41) + 1 \cdot (43.5 + 42)$$

$$= 109.5 \cdot 1 + 67 \cdot 3$$

$$= 339$$

$$q_{c^*j} = q_{35} = f_{35} \cdot t_{35} + f_{34} \cdot t_{34} + f_{32} \cdot t_{32} + f_{31} \cdot t_{31}$$

$$= f_{35} \cdot r_{35} + f_{34} \cdot r_{34} + f_{32} \cdot r_{32} + f_{31} \cdot r_{31}$$

$$= f_{35} \cdot d_{35} + f_{34} \cdot d_{35} + f_{32} \cdot (d_{35} + d_{42}) + f_{31} \cdot (d_{35} + d_{51})$$

$$= 0 \cdot 28.5 + 0 \cdot 28.5 + 3 \cdot (28.5 + 41) + 1 \cdot (28.5 + 42)$$

$$= 279$$

$$q_{s^*c^*} = q_{23} = f_{32} \cdot t_{32} + f_{31} \cdot t_{31} + f_{34} \cdot t_{34} + f_{35} \cdot t_{35}$$

$$= f_{32} \cdot r_{32} + f_{31} \cdot r_{31} + f_{34} \cdot r_{34} + f_{35} \cdot r_{35}$$

$$= f_{32} \cdot d_{32} + f_{31} \cdot d_{32} + f_{34} \cdot (d_{32} + d_{24}) + f_{35} \cdot (d_{32} + d_{24})$$

$$= 3 \cdot 23.5 + 1 \cdot 23.5 + 0 \cdot (23.5 + 41) + 0 \cdot (23.5 + 41)$$

$$= 94$$

$$q_{t^*c^*} = q_{13} = f_{31} \cdot t_{31} + f_{32} \cdot t_{32} + f_{34} \cdot t_{34} + f_{35} \cdot t_{35}$$

$$= f_{31} \cdot r_{31} + f_{32} \cdot r_{32} + f_{34} \cdot r_{34} + f_{35} \cdot r_{35}$$

$$= f_{31} \cdot d_{31} + f_{32} \cdot d_{31} + f_{34} \cdot (d_{31} + d_{15}) + f_{35} \cdot (d_{31} + d_{15})$$

$$= 1 \cdot 38.5 + 3 \cdot 38.5 + 0 \cdot (38.5 + 42) + 0 \cdot (38.5 + 42)$$

$$= 154$$

Since $q_{23} = \min\{q_{34}, q_{35}, q_{23}, q_{13}\}$, alternative (c) is selected:

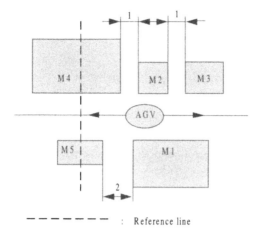

Figure 12.6. Layout of machines for the problem in Example 12.2.

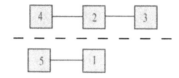

Set $U = U + \{3\} = \{1, 2, 3, 4, 5\}$.

Step 5.   Since all machines have been included in the set U, stop.

The total cost of the final solution shown in Figure 12.6 is 391.

The two algorithms have low computational complexity and provide solutions of good quality. For details of computational studies see Kusiak and Park (1994).

12.4   MULTIROW FACILITY AND MACHINE LAYOUT

The facility layout problem has been broadly discussed in the literature. To model the facility layout problem, numerous approaches have been applied (e.g., graph theory, integer programming, and mathematical programming).

12.4.1   Quadratic Assignment Model

Koopmans and Beckmann (1957) developed a quadratic assignment model for locating plants. To present the quadratic assignment model, the following need to be defined:

$n =$ number of facilities and their locations

$f_{ik} =$ flow of material between facility i and facility k

$c_{jl} =$ cost per unit flow of material between location j and location l

$a_{ij} =$ cost of locating and operating facility i at location j

$$x_{ij} = \begin{cases} 1 & \text{if facility } i \text{ is at location } j \\ 0 & \text{otherwise} \end{cases}$$

Koopmans and Beckmann (1957) assumed that:

- a_{ij} includes gross revenue minus cost of primary input but does not include the transportation cost of material between plants,
- f_{ik} is independent of the locations of the plants, and
- c_{jl} is independent of the facilities and that it is cheaper to transport material directly from facility i to facility k than through a third location.

The objective function (12.11) minimizes the total of the product flow and unit transportation cost and the cost of locating and operating a facility:

$$\min \sum_{i=1}^{n} \sum_{j=1}^{n} \sum_{k=1}^{n} \sum_{l=1}^{n} f_{ik} c_{jl} x_{ij} x_{kl} + \sum_{i=1}^{n} \sum_{j=1}^{n} a_{ij} x_{ij} \tag{12.11}$$

subject to

$$\sum_{j=1}^{n} x_{ij} = 1 \qquad i = 1, \ldots, n \tag{12.12}$$

$$\sum_{i=1}^{n} x_{ij} = 1 \qquad j = 1, \ldots, n \tag{12.13}$$

$$x_{ij} = 0, 1 \qquad i = 1, \ldots, n, \; j = 1, \ldots, n \tag{12.14}$$

The constraints of this model are the same as in the assignment problem, that is, constraint (12.12) makes sure that each facility is assigned to one location and constraint (12.13) assigns one location to one facility. The integrality of decision variables is imposed by constraint (12.14).

In the model (12.11)–(12.14) it is assumed that the number of facilities is equal to the number of locations. When the number of facilities is less than the number of locations, dummy facilities can be introduced and the flow values from these dummy facilities to all other facilities need to be set to zero.

If all a_{ij} are equal to zero or are identical, then the objective function (12.11) transforms to

$$\min \sum_{i=1}^{n} \sum_{j=1}^{n} \sum_{k=1}^{n} \sum_{l=1}^{n} f_{ik} c_{jl} x_{ij} x_{kl} \tag{12.15}$$

The formulation involving objective function (12.15) and constraints (12.12)–(12.14) is most often referred to as the quadratic assignment model.

Not all facility layout problems can be formulated with model (12.11)–(12.14), for example, when the facility locations are not known in advance. In the latter case the distance between the locations cannot be computed. The distance between two locations j and i depends on the arrangement of the machines in the layout.

This situation does not arise in layout problems with facilities of equal area.

The quadratic assignment model is NP complete (Sahni and Gonzalez, 1976). Heuristic approaches are likely to provide solutions of acceptable quality. Some earlier heuristic methods used flowcharts and the experience of the facility analyst to determine layouts. Other methods used the relationship chart to determine the layout. The relationship chart shows the closeness desired between pairs of facilities. This concept was first introduced by Muther (1955). The closeness desired between pairs of facilities is represented in the relationship chart by letters A, E, O, U, X, and I. For any pair of facilities (i, j), the letters A, E, I, O, U, and X indicate that the closeness between facilities i and j is absolutely necessary (A), especially important (E), important (I), ordinary (O), unimportant (U), and undesirable (X), respectively. The relationship chart formed the basis for the development of a popular method called systematic layout planning. Wimmert (1958) presented an algorithm for the facility layout problem that minimizes the product of flow values and distances between all combinations of facilities. Buffa (1955) proposed another method called sequence analysis, which is based on the analysis of the sequence of operations of parts in a plant. In addition, there were other methods developed in the late 1950s and early 1960s that did not provide solutions of good quality. These methods are discussed in Foulds (1983) as schematic methods and systematic methods.

CRAFT (computerized relative allocation of facilities technique) is one of the first and most widely used heuristic for solving the facility layout problems. It performs pairwise exchanges of facilities in order to improve an initial solution specified by the user. Pairwise exchanging cannot be generally regarded as a "greedy" method because each variable is not changed optimally at each iteration of the solution method. A restriction is imposed at each move by specifying that if it is cost-attractive to move facility i from site A to site B, for instance, CRAFT also moves the facility presently in site B to site A. This is done even when a better strategy may be to move this latter facility elsewhere. For this reason and others, CRAFT may not obtain optimal solutions. Of course, there are many ways to modify the basic approach of CRAFT to obtain improved versions. To illustrate CRAFT, consider the following example, discussed in Love et al. (1988).

Example 12.3. Determine the layout of four facilities with sites A, B, C, and D located as shown in Figure 12.7.

For simplicity, unit square locations and rectilinear distances are assumed. The rectilinear distance d_{ij} between two points (x_i, y_i) and (x_j, y_j) in the Euclidean space

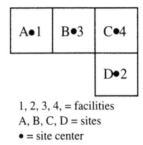

1, 2, 3, 4, = facilities
A, B, C, D = sites
• = site center

Figure 12.7. Location of four facilities at four sites.

equals $d_{ij} = |x_j - x_i| + |y_j - y_i|$. Note that the distances in the CRAFT algorithm are established between sites as opposed to the machines (facilities) in the previous two examples. In the facility layout model, as the facilities get exchanged (moved), the values of distances between sites do not change. While this is acceptable in the design of facilities where a facility might be allowed to change shape, the footprint of a machine is fixed. A solution that would not be able to consider actual distances (or travel times) between machines would likely not be acceptable.

The following data are given:

- Flow matrix:

$$
[f_{ij}] = \begin{array}{c} \\ 1 \\ 2 \\ 3 \\ 4 \end{array} \begin{array}{cccc} 1 & 2 & 3 & 4 \\ \left[\begin{array}{cccc} - & 50 & 20 & 100 \\ 50 & - & 30 & 10 \\ 20 & 30 & - & 70 \\ 100 & 10 & 70 & - \end{array} \right] \end{array}
$$

- Matrix of rectilinear distances (obtained from Figure 12.7):

$$
[d_{ij}] = \begin{array}{c} \\ A \\ B \\ C \\ D \end{array} \begin{array}{cccc} A & B & C & D \\ \left[\begin{array}{cccc} - & 1 & 2 & 3 \\ 1 & - & 1 & 2 \\ 2 & 1 & - & 1 \\ 3 & 2 & 1 & - \end{array} \right] \end{array}
$$

12.4.2 CRAFT Algorithm

The CRAFT algorithm is presented by solving the problem in Example 12.3. Consider the initial solution presented in Figure 12.7, which is denoted symbolically as

$$\text{Cost } 510 \quad \text{A} \bullet 1 \; \text{B} \bullet 3 \; \text{C} \bullet 4$$
$$\text{D} \bullet 2$$

and has a corresponding cost 510. Note that the value of cost is calculated from expression (12.15) for $c_{jl} = d_{jl}$ as follows:

$$\text{Cost} = f_{12}d_{AD} + f_{13}d_{AB} + f_{14}d_{AC} + f_{23}d_{DB} + f_{24}d_{DC} + f_{34}d_{BC}$$

$$= 50 \cdot 3 + 20 \cdot 1 + 100 \cdot 2 + 30 \cdot 2 + 10 \cdot 1 + 70 \cdot 1$$

$$= 510$$

Pairwise exchanges of facilities result in the following cost changes:

Exchange pair	(1, 2)	(1, 3)	(1, 4)	(2, 3)	(2, 4)	(3, 4)
Cost change	−100	−50	−80	−60	90	−100

For example, the cost change for pair (1, 2) above is computed as follows. Exchanging facilities 1 and 2 (in Figure 12.7 facility 1 is replaced with 2 and facility 1 with 2) results in the following cost:

$$\text{Cost} = f_{12}d_{AD} + f_{13}d_{BD} + f_{14}d_{CD} + f_{23}d_{AB} + f_{24}d_{AC} + f_{34}d_{BC}$$

$$= 50 \cdot 3 + 20 \cdot 2 + 100 \cdot 1 + 30 \cdot 1 + 10 \cdot 2 + 70 \cdot 1$$

$$= 410$$

The cost change is then $410 - 510 = -100$.

Of the six possible pairwise exchanges, either pair (1, 2) or pair (3, 4) could be exchanged to obtain a cost reduction of 100. Arbitrarily choose to exchange facilities 1 and 2, which becomes a new initial solution. Based on the new initial solution, the second solution with the corresponding cost and the six possible pairwise exchanges is shown next:

Cost 410 A $\bullet$ 2 B $\bullet$ 3 C $\bullet$ 4
          D $\bullet$ 1

Exchange pair	(1, 2)	(1, 3)	(1, 4)	(2, 3)	(2, 4)	(3, 4)
Cost change	100	−40	10	30	100	100

Facilities 1 and 3 are now exchanged. The corresponding solution with the cost 370 and pairwise exchanges is:

Cost 370 A $\bullet$ 2 B $\bullet$ 1 C $\bullet$ 4
          D $\bullet$ 3

Exchange pair	(1, 2)	(1, 3)	(1, 4)	(2, 3)	(2, 4)	(3, 4)
Cost change	80	40	90	90	40	60

The CRAFT algorithm terminates owing to the positive cost changes for each exchange pair possibility. As this is one of the two optimal solutions, the CRAFT heuristic has performed well for the data provided in the example.

12.5 SUMMARY

In this chapter, issues in modeling layout of facilities were discussed. One of the tasks in the design of facilities is machine layout. Two heuristic algorithms for generating layouts of machines were presented. Two types of machine layout considered were the linear single-row machine layout and linear double-row machine layout. The two algorithms can be applied to generate layouts for machines arranged in any pattern. For a general layout case, the distances between machines and travel time have to be defined appropriately. In fact, the two-row machine layout model covers most of the layout scenarios that might be encountered on the factory floor. The algorithms solve the layout cases with equal and unequal machine dimensions. The algorithms are simple, easy to implement, and produce solutions of good quality. The CRAFT algorithm for generating facility layouts was also presented.

REFERENCES

Buffa, E. S. (1955), Sequence analysis for functional layouts, *Journal of Industrial Engineering*, Vol. 6, No. 1, pp. 12–13.

Foulds, L. R. (1983), Techniques for facilities layout: Deciding which pairs of activities should be adjacent, *Management Science*, Vol. 29, No. 12, pp. 1414–1426.

Koopmans, T. C., and M. Beckmann (1957), Assignment problems and the location of economic activities, *Econometrica*, Vol. 25, No. 1, pp. 53–76.

Kouvelis, P., and K. S. Kiran (1989), Layout problem in flexible manufacturing systems: Recent research results and further research directions, in K. E. Stecke and R. Suri (eds.), *Proceedings of the Third ORSA/TIMS Conference on Flexible Manufacturing Systems*, Elsevier, New York, pp. 147–152.

Kusiak, A. (1990), *Intelligent Manufacturing Systems*, Prentice-Hall, Englewood Cliffs, NJ.

Kusiak, A., and K. Park (1994), Layout of machines in cellular manufacturing systems, *International Journal of Manufacturing Systems Design*, Vol. 1, No. 1, pp. 31–39.

Love, R. F., J. G. Morris, and G. O. Wesolowsky (1988), *Facilities Location: Models and Methods*, Elsevier, New York.

Muller, T. (1983), *Automated Guided Vehicles*, IFS Publications, Kempston, Bedford, United Kingdom.

Muther, R. (1955), *Practical Plant Layout*, McGraw-Hill, New York.

Sahni, S., and T. Gonzalez (1976), P-complete approximation problem, *Journal of Association for Computing Machinery*, Vol. 23, No. 3, pp. 555–565.

Singh, N. (1996), *Computer-Integrated Design and Manufacturing*, John Wiley, New York.

Tompkins, J. A., and J. A. White (1984), *Facilities Planning and Design*, John Wiley, New York.

Wimmert, R. J. (1958), A mathematical method for equipment location, *Journal of Industrial Engineering*, Vol. 9, No. 3, pp. 498–505.

Zoller, K., and K. Adendorff (1972), Layout planning by computer simulation, *AIIE Transactions*, Vol. 4, No. 2, pp. 116–125.

QUESTIONS

12.1. What are the basic types of layout patterns in manufacturing systems?

12.2. Which type of layout is most often used in manufacturing systems?

12.3. Can a single- or double-row layout model be used to determine the U shape layout of a manufacturing system?

12.4. Is machine layout equivalent to faculty layout? Describe the differences, if any.

12.5. What is the frequently used objective function used in determining the layout of machines?

12.6. Define the Euclidean distance.

12.7. Define the rectilinear distance.

12.8. What is meant by "minimum" operand in the objective function of the layout problem and the "maximum" operand in Step 1 of Algorithm 12.2? Are the two in conflict?

12.9. What are the limitations of the quadratic assignment model in modeling layout problems?

12.10. In your own words, what are the steps of the CRAFT heuristic algorithm?

PROBLEMS

12.1. Three machines (2, 3, and 4) and two pallet stands (1 and 5) have been arranged as shown in Figure 12.8. The number of visits per hour of each machine–machine pair (or machine–pallet stand pair) by a robot arm is given by the matrix

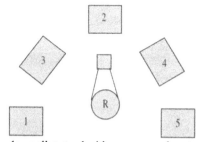

1: pallet stand with unprocessed parts
2: lathe - 1
3: lathe - 2
4: milling machine
5: pallet stand with processed parts
R: robot

Figure 12.8. Manufacturing cell layout.

$$[f_{ij}] = \begin{array}{c} \\ 1 \\ 2 \\ 3 \\ 4 \\ 5 \end{array} \begin{array}{ccccc} 1 & 2 & 3 & 4 & 5 \\ \left[\begin{array}{ccccc} 0 & 2 & 6 & 1 & 5 \\ & 0 & 1 & 5 & 7 \\ & & 0 & 2 & 4 \\ & & & 0 & 3 \\ & & & & 0 \end{array}\right] \end{array}$$

The travel time between any machine–machine pair or machine–pallet stand pair is 2 minutes, while the cost of one trip per minute is 1 cent.

(a) Determine whether the layout in Figure 12.2 is the "best" in terms of the total cost of the robot travel.

(b) Write an expression for determining the total material handling cost per hour. How would you modify the data for this layout problem if it would be required that machines 3 and 4 be located in adjacent positions?

(c) Write an expression for determining the total material handling cost per hour.

(d) How would you modify the data for this layout problem if it would be required that machines 3 and 4 be located in adjacent positions?

12.2. An industrial engineer is to determine the layout of a manufacturing cell with five machines. The data in the three matrices (a), (b), and (c) are given, where l_i denotes the length of machine i and w_i denotes the width of machine i:

(a) Matrix of Frequency of Trips

$$[f_{ij}] = \begin{array}{c} \\ 1 \\ 2 \\ 3 \\ 4 \\ 5 \end{array} \begin{array}{ccccc} 1 & 2 & 3 & 4 & 5 \\ \left[\begin{array}{ccccc} 0 & 60 & 50 & 90 & 50 \\ 60 & 0 & 20 & 10 & 30 \\ 50 & 20 & 0 & 15 & 50 \\ 90 & 10 & 15 & 0 & 20 \\ 50 & 30 & 50 & 20 & 0 \end{array}\right] \end{array}$$

(b) Clearance Matrix

$$[c_{ij}] = \begin{array}{c} \\ 1 \\ 2 \\ 3 \\ 4 \\ 5 \end{array} \begin{array}{ccccc} 1 & 2 & 3 & 4 & 5 \\ \left[\begin{array}{ccccc} 0 & 2 & 2 & 2 & 1 \\ 2 & 0 & 1 & 1 & 1 \\ 2 & 1 & 0 & 4 & 1 \\ 2 & 1 & 4 & 0 & 2 \\ 1 & 1 & 1 & 2 & 0 \end{array}\right] \end{array}$$

(c) Machine Dimensions

Machine Number	Dimension $l_i \times w_i$
1	5×2
2	2×2
3	3×2
4	5×4
5	3×1

In addition to the above data it is known that:

- The travel time between any pair of machines is proportional to the corresponding Euclidean distance.

- The machines are arranged lengthwise along the isle.

  (i) Determine the single-row machine layout.

  (ii) Draw the layout.

12.3. Given the data below:

(a) Matrix of Frequency of Trips   (b) Matrix of Handling Times

$$
[f_{ij}] =
\begin{array}{c}
 \\ 1 \\ 2 \\ 3 \\ 4 \\ 5
\end{array}
\begin{array}{ccccc}
1 & 2 & 3 & 4 & 5 \\
\left[\begin{array}{ccccc}
0 & 50 & 25 & 45 & 60 \\
40 & 0 & 72 & 12 & 34 \\
55 & 72 & 0 & 24 & 50 \\
40 & 12 & 14 & 0 & 21 \\
50 & 34 & 50 & 21 & 0
\end{array}\right]
\end{array}
\qquad
[t_{ij}] =
\begin{array}{c}
 \\ 1 \\ 2 \\ 3 \\ 4 \\ 5
\end{array}
\begin{array}{ccccc}
1 & 2 & 3 & 4 & 5 \\
\left[\begin{array}{ccccc}
0 & 9 & 2 & 6 & 4 \\
9 & 0 & 2 & 6 & 3 \\
2 & 2 & 0 & 5 & 3 \\
6 & 6 & 5 & 0 & 5 \\
4 & 3 & 3 & 5 & 0
\end{array}\right]
\end{array}
$$

(c) Clearance Matrix

$$
[c_{ij}] =
\begin{array}{c}
 \\ 1 \\ 2 \\ 3 \\ 4 \\ 5
\end{array}
\begin{array}{ccccc}
1 & 2 & 3 & 4 & 5 \\
\left[\begin{array}{ccccc}
0 & 2 & 2 & 1 & 2 \\
2 & 0 & 1 & 1 & 1 \\
2 & 1 & 0 & 3 & 1 \\
1 & 1 & 3 & 0 & 2 \\
2 & 1 & 1 & 2 & 0
\end{array}\right]
\end{array}
$$

(d) Matrix Dimensions

Machine Number	Dimension $l_i \times b_i$
1	40×20
2	30×20
3	30×20
4	50×40
5	30×15

Knowing that the handling time does not depend on the orientation of any two adjacent machines, determine the best single-row machine layout.

12.4. Determine the double-row layout of machines, given the data below for five machines:

(a) Matrix of Frequency of Trips

$$
[f_{ij}] =
\begin{array}{c}
 \\ 1 \\ 2 \\ 3 \\ 4 \\ 5
\end{array}
\begin{array}{ccccc}
1 & 2 & 3 & 4 & 5 \\
\left[\begin{array}{ccccc}
0 & 50 & 40 & 20 & 60 \\
 & 0 & 80 & 10 & 20 \\
 & & 0 & 15 & 60 \\
 & & & 0 & 15 \\
 & & & & 0
\end{array}\right]
\end{array}
$$

(b) Machine Dimensions

Machine Number	Dimension $l_i \times w_i$
1	6×2
2	3×2
3	3×2
4	7×3
5	4×2

(c) Clearance Matrix

$$[c_{ij}] = \begin{array}{c} \\ 1 \\ 2 \\ 3 \\ 4 \\ 5 \end{array} \begin{array}{ccccc} 1 & 2 & 3 & 4 & 5 \\ \left[\begin{array}{ccccc} 0 & 3 & 1 & 2 & 1 \\ & 0 & 4 & 1 & 2 \\ & & 0 & 3 & 1 \\ & & & 0 & 2 \\ & & & & 0 \end{array}\right] \end{array}$$

Assume that:

- The travel time between any pair of machines is proportional to the corresponding rectilinear distance.
- The width of the isle is 2 units.
- The machines are arranged lengthwise along the isle.

In the table above, l_i denotes the length of machine i and w_i denotes the width of machine i.

12.5. Given the data below:

(a) Matrix of Frequency of Trips

$$[f_{ij}] = \begin{array}{c} \\ 1 \\ 2 \\ 3 \\ 4 \\ 5 \end{array} \begin{array}{ccccc} 1 & 2 & 3 & 4 & 5 \\ \left[\begin{array}{ccccc} 0 & 40 & 30 & 23 & 50 \\ 40 & 0 & 40 & 12 & 40 \\ 30 & 40 & 0 & 14 & 50 \\ 23 & 12 & 14 & 0 & 20 \\ 50 & 40 & 50 & 20 & 0 \end{array}\right] \end{array}$$

(b) Matrix of Handling Times

$$[t_{ij}] = \begin{array}{c} \\ 1 \\ 2 \\ 3 \\ 4 \\ 5 \end{array} \begin{array}{ccccc} 1 & 2 & 3 & 4 & 5 \\ \left[\begin{array}{ccccc} 0 & 5 & 4 & 2 & 4 \\ 5 & 0 & 2 & 2 & 2 \\ 4 & 2 & 0 & 5 & 1 \\ 2 & 2 & 5 & 0 & 5 \\ 4 & 2 & 1 & 5 & 0 \end{array}\right] \end{array}$$

(c) Clearance Matrix

$$[d_{ij}] = \begin{array}{c} \\ 1 \\ 2 \\ 3 \\ 4 \\ 5 \end{array} \begin{array}{ccccc} 1 & 2 & 3 & 4 & 5 \\ \left[\begin{array}{ccccc} 0 & 2 & 2 & 3 & 1 \\ 2 & 0 & 1 & 1 & 1 \\ 2 & 1 & 0 & 3 & 1 \\ 3 & 1 & 3 & 0 & 2 \\ 1 & 1 & 1 & 2 & 0 \end{array}\right] \end{array}$$

(d) Machine Dimensions

Machine Number	Dimension $l_i \times b_i$
1	30×20
2	20×20
3	30×20
4	30×25
5	20×15

Solve the single-row machine layout problem with the heuristic algorithm.

12.6. Corporation AA is redesigning the layout of a manufacturing cell located at Building XL-20. The manufacturing cell will include five machines. Facility planners have analyzed the production plan for the next two years and estimated the frequency of trips (trips per shift) to be made by the forklift as shown in matrix (a). The handling time (minutes) between any pair of machines that include unloading, travel, and loading was also estimated [see matrix (b)]. Based on the processing and maintenance specifications, the minimum clearance (feet) between machines was specified in matrix (c). Machine dimensions (feet) are provided in matrix (d).

(a) Matrix of Frequency of Trips

$$[f_{ij}] = \begin{array}{c} \\ 1 \\ 2 \\ 3 \\ 4 \\ 5 \end{array} \begin{array}{ccccc} 1 & 2 & 3 & 4 & 5 \\ \begin{bmatrix} 0 & 40 & 30 & 25 & 60 \\ 40 & 0 & 70 & 15 & 40 \\ 30 & 70 & 0 & 14 & 50 \\ 25 & 15 & 14 & 0 & 20 \\ 60 & 40 & 50 & 20 & 0 \end{bmatrix} \end{array}$$

(b) Matrix of Handling Times

$$[t_{ij}] = \begin{array}{c} \\ 1 \\ 2 \\ 3 \\ 4 \\ 5 \end{array} \begin{array}{ccccc} 1 & 2 & 3 & 4 & 5 \\ \begin{bmatrix} 0 & 8 & 4 & 2 & 5 \\ 8 & 0 & 2 & 2 & 2 \\ 4 & 2 & 0 & 5 & 1 \\ 2 & 2 & 5 & 0 & 2 \\ 5 & 2 & 1 & 2 & 0 \end{bmatrix} \end{array}$$

(c) Clearance Matrix

$$[c_{ij}] = \begin{array}{c} \\ 1 \\ 2 \\ 3 \\ 4 \\ 5 \end{array} \begin{array}{ccccc} 1 & 2 & 3 & 4 & 5 \\ \begin{bmatrix} 0 & 2 & 1 & 2 & 1 \\ 2 & 0 & 1 & 1 & 1 \\ 1 & 1 & 0 & 2 & 1 \\ 2 & 1 & 2 & 0 & 2 \\ 1 & 1 & 1 & 2 & 0 \end{bmatrix} \end{array}$$

(d) Machine Dimensions

Machine Number	Dimension $l_i \times w_i$
1	4×2
2	3×2
3	6×3
4	4×3
5	6×2

(e) Closeness Constraints

$$R = \begin{array}{c} \\ 1 \\ 2 \\ 3 \\ 4 \\ 5 \end{array} \begin{array}{ccccc} 1 & 2 & 3 & 4 & 5 \\ \left[\begin{array}{ccccc} - & U & U & A & U \\ & - & X & U & O \\ & & - & U & U \\ & & & - & U \\ & & & & - \end{array}\right] \end{array}$$

The location constraints imposed on the five machines are specified in matrix (e) based on the rating scale provided in the Table 12.3. The machines are to be arranged in two rows along an isle that is 9 feet wide due to the fire hazard requirements. For your information, the operator-run forklift is 2 feet wide and 3 feet long.

(i) Determine the best layout of machines.

(ii) Draw the layout.

12.7. Given the data below:

(a) Matrix of Frequency of Trips

$$[f_{ij}] = \begin{array}{c} \\ 1 \\ 2 \\ 3 \\ 4 \\ 5 \\ 6 \end{array} \begin{array}{cccccc} 1 & 2 & 3 & 4 & 5 & 6 \\ \left[\begin{array}{cccccc} 0 & 60 & 40 & 20 & 60 & 50 \\ & 0 & 30 & 50 & 80 & 40 \\ & & 0 & 70 & 90 & 10 \\ & & & 0 & 20 & 90 \\ & & & & 0 & 40 \\ & & & & & 0 \end{array}\right] \end{array}$$

TABLE 12.3 Closeness Rating

Rating	Closeness
A	Absolutely necessary
E	Especially important
I	Important
O	Ordinary closeness
U	Unimportant
X	Not desirable

(b) Machine Dimensions

Machine Number	Dimension $l_i \times w_i$
1	4×2
2	4×2
3	4×2
4	4×3
5	4×2
6	2×2

(c)   Clearance Matrix

$$[d_{ij}] = \begin{array}{c} \\ 1 \\ 2 \\ 3 \\ 4 \\ 5 \\ 6 \end{array} \begin{array}{cccccc} 1 & 2 & 3 & 4 & 5 & 6 \\ \left[\begin{array}{cccccc} 0 & 3 & 1 & 6 & 2 & 3 \\ & 0 & 4 & 1 & 2 & 1 \\ & & 0 & 3 & 1 & 4 \\ & & & 0 & 3 & 1 \\ & & & & 0 & 2 \\ & & & & & 0 \end{array}\right] \end{array}$$

- The isle width is 6.
- If needed, make any other assumptions.

Solve the double-row machine layout problem with the heuristic algorithm. *Hint:* Make the transportation time proportional to the rectilinear distance traveled.

CHAPTER 13

INVENTORY SPACE ALLOCATION

13.1 INTRODUCTION

This chapter discusses the allocation of storage space to inventory. The space allocation problem is modeled as a generalized transportation problem that minimizes the cost of material handling. The basic formulation of the model considers material type, material flow transitions, and distance.

Allegri (1984) suggested minimizing distance traveled and moving material by line of flight as two approaches to reducing material handling efforts. Factors such as material flow, required storage space, available storage space, and type of material and its average inventory levels must be considered. For example, different types of material require different material flow routes through a facility. The material transfer routes depend on the manufacturing operations required as well as on the location of related in-process storage areas.

13.2 RELATED MODELS

Much of the recent inventory management literature has covered topics related to space allocation. However, the problem is seldom approached directly. Inman (1993) presented inventory as the result of a problem, rather than the problem itself. Crandall and Burwell (1993) studied the effect of work-in-process inventory levels on throughput and lead times. Both papers recognized the reality of inventory buffers, even in JIT (just-in-time) systems. With the existence of in-process inventory, it is necessary to implement a strategy for allocating storage space and managing inventory levels. MacMillan (1993) discussed point-of-use storage as a technique for supporting the implementation of the JIT approach. The procedure uses dedicated storage locations

412

rather than a centralized stockroom. The concept of uniform plant loading was discussed in Park (1993). Level production stabilizes inventory levels, thus justifying the use of dedicated storage zones with fixed capacity for space allocation.

Most related literature in the area of modeling material flow focuses on two classical problems: the transportation problem and the vehicle routing problem. The transportation problem was introduced by Hitchcock (1941) and has received extensive coverage in the literature. Applications have been proposed in areas such as waste disposal (Rautman et al., 1993), education (Beheshtian-Ardekani and Mahmood, 1986), public safety (Taylor, 1985), employee scheduling (Strevell and Chong, 1985), and agriculture (Harrison and Wills, 1983). Dantzig (1947) developed efficient methods for solving the transportation problem [see also Koopmans (1949) and Cooper and Charnes (1954)]. One extension, the transshipment problem, was introduced by Orden (1956) to describe the transport of material from source to destination via an intermediate point. Most applications of the transshipment problem to modeling material flow consider the multiple-facility problem, rather than the space allocation formulation considered in this chapter.

The vehicle routing problem (VRP) is concerned with routing vehicles from a central depot to several delivery or pickup points. The objective is to minimize the total delivery cost (i.e., distance traveled) subject to limited vehicle capacity. Chien et al. (1989) and Federgruen and Zipkin (1984) have integrated the VRP with inventory allocation to multiple customers. Once again, these applications are multiple-facility routing and distribution problems, rather than the local space allocation and material handling problem addressed here.

The problem discussed in this chapter is similar to the transshipment model; however, the formulation is that of the generalized transportation problem. The model focuses on material transitions via intermediate storage points. Inventory is allocated a storage zone as to minimize material handling within the facility. Elements of the VRP apply to the space allocation problem. The model presented in Section 13.3 minimizes the distance material is transported. However, it does not provide information on routing and scheduling of material handling devices. Integrating the VRP with the model presented in Section 13.3 is of practical importance and, thus, is recognized as a topic of future research.

13.3   SPACE ALLOCATION MODEL

This section presents a model known in the operations research literature as the generalized transportation problem. The model is applied to the allocation of storage space to inventory in a manner that minimizes material handling. A basic formulation of the model is presented in the next section.

13.3.1   Basic Model

In order to minimize material handling, one has to minimize not only the distance material is moved but also the amount of material that is moved. For example, it is more costly to move 2000 lb over 500 ft than it is to move 20 lb over 500 ft. The model considers all possible transitions of material (i.e., from one operation to the next operation), the amount and type of material making the transition, and all possible storage locations, in order to minimize the total material handling cost (Larson and Kusiak, 1995). The following notation is used in the model:

$n =$ number of storage locations
$m =$ number of transitions
$x_{ij} =$ amount of material making transition i at storage location j
$d_{ij} =$ distance from source to destination in transition i via storage location j
$S_j =$ total capacity of storage location j
$T_i =$ total amount of inventory making transition i
$\tau_i =$ conversion factor for type of material making transition i

$$\min z = \sum_{i=1}^{m} \sum_{j=1}^{n} d_{ij} x_{ij} \tag{13.1}$$

such that

$$\sum_{i=1}^{m} \tau_i x_{ij} \le S_j \qquad j = 1, 2, \ldots, n \tag{13.2}$$

$$\sum_{j=1}^{n} x_{ij} = T_i \qquad i = 1, 2, \ldots, m \tag{13.3}$$

$$x_{ij} \ge 0 \qquad j = 1, \ldots, n; i = 1, \ldots, m \tag{13.4}$$

Model (13.1)–(13.4) is known as the generalized transportation problem. The objective function (13.1) minimizes the amount of material handling, that is, the amount of material that must be transferred over a distance for each transition. Inequality (13.2) represents the capacity constraint for each storage location. Constraint (13.3) represents the amount of storage that is required for material in each transition. Inequality (13.4) is a nonnegativity constraint that is necessary to prevent a negative solution.

Model (13.1)–(13.4) considers transitions, rather than supply-and-demand points. Transitions are easily determined by constructing a material flow chart of the facility. It is also necessary to include a constant τ_i in constraint (13.2). The constant is used to convert the amount of material x_{ij} in transition i at storage location j into common units for storage capacity constraints. Since there are m transitions, there can be m

different types of material. In practice, there are usually fewer than m different types of material. Each transition i has an associated conversion factor τ_i.

To apply model (13.1)–(13.4) in practice, it is necessary to make several assumptions. First, it is necessary to assume that all material in transition must be routed to storage, possibly as a result of busy machines or production scheduling constraints. Second, the only cost factor of interest is the distance material is transported. Costs related to material handling resources and load/unload time are not captured in the objective function. Finally, to determine T_i it may be necessary to aggregate production routing and production volume information for many different products. Although determining the optimal production volume based on material handling costs and capabilities is an important consideration, it is not accomplished with this formulation of the problem. For an extended formulation of model (13.1)–(13.4), see Larson and Kusiak (1995).

13.4 MODEL FORMULATION

The following case study illustrates model (13.1)–(13.4). Material handling adds no value to a product and, in many cases, adds significant cost. Therefore, engineers, managers, and production planners must seek to minimize material handling. Material flow within the facility is the target of many efforts to reduce handling time and distance. Management of in-process inventory is a component of material flow that must not be neglected.

Most production facilities rely on in-process storage areas to buffer the flow of material between successive operations. Consider two scenarios: (1) a production facility routes all in-process inventory to the same storage area; (2) a production facility routes material to one of several in-process storage areas. The first approach to managing in-process inventory provides a simple method of routing and tracking material; however, material handling costs may increase. For example, if two successive machines in the production cycle are located adjacent to each other in the facility, routing material through the in-process storage area will increase material handling time and distance. The second approach provides several storage areas for in-process material. Routing material to the nearest available storage area decreases material handling; however, tracking material becomes more complicated in large facilities.

Facilities that make use of many in-process storage areas must employ a strategy for managing inventory that minimizes material handling. This requires dedicated storage areas for different types of material. Material type is determined by the machines visited during the production cycle, or the material flow.

13.4.1 Definitions

Before presenting the case study, several terms used throughout the section are defined. A *storage zone* is an area of floor space that is available for storage of inventory. Storage zones may be used for *pallet storage* or *rack storage*. Pallet storage holds material on pallets, which cannot be stacked. Rack storage holds material in racks that are stacked three high. Material is classified as *paper* or *film* (plastic). Material is stored in *rolls* that are wound on to cardboard cores.

A name and a number label each machine in the facility. Material flow between machines is described in the next section. All material leaves the facility through the *dock*. Material handling between machines and the dock is done with *fork lifts*.

13.4.2   System Description

In this chapter, a paper and plastic film coating facility is modeled to determine an inventory management strategy that minimizes material handling. The facility takes paper and film raw stock, applies a specified coating to the surface of the raw stock, and slits and rolls the product to the customer requirements. A company logo is also printed on the surface of the product if required. The machines that perform the operations are quite large and considered "fixed" in the facility. Storage areas are spread throughout the facility and are available for holding any type of inventory. Figure 13.1 presents a simplified floor plan of the facility.

A strategy for managing work-in-process (WIP) and finished inventory is the focus of the study. The WIP is stored in racks that can be moved throughout the facility. Finished inventory is stored on pallets that are set directly on the floor and cannot be stacked. As shown in Figure 13.1, there are 15 different storage zones. Table 13.1 lists the area and capacity of each storage zone for each type of material. The following assumptions, which are based on data collected, are used to calculate the capacity of each storage zone.

Many different customers requiring a number of different products are served by the facility. Based on the required operations for each product and the machines used in the facility, the material flow chart in Figure 13.2 is developed. Each machine in the facility is represented by a corresponding node on the flow chart. There are also nodes representing the raw stock warehouse, where material enters the system, and the dock, where material exits the system. Each arc on the flow chart represents a possible material transition. If an arc does not connect two nodes, no material flow occurs between the two corresponding machines.

All material must pass through at least one of the slitter stations and the roll wrap machine. There are eight different slitter machines arranged in two groups in the facility. Therefore, the system is simplified by considering two slitter groups (1 and 2), rather than each machine independently.

A transition is defined as the movement of material from a source node to a destination node in the flow chart. Due to scheduling requirements, all material in transition is routed to a storage zone.

13.4.3   Case Study Objective

At ABC Corporation material was stored according to the rule *store it wherever space permits*. The lack of a fundamental strategy for managing in-process and finished inventory resulted in inefficient use of space and increased material handling. The

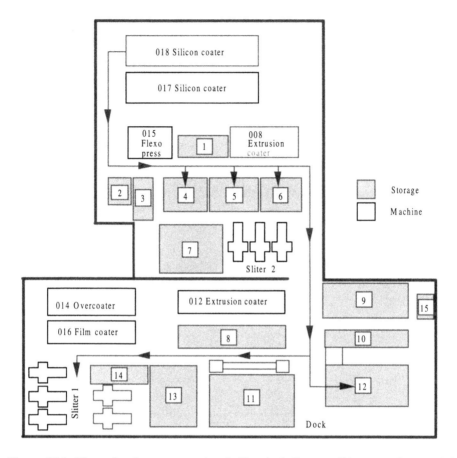

Figure 13.1. Floor plan for paper coating facility, including possible routes for material making transition 23.

purpose of this case study was to determine the optimal storage strategy based on the following objectives:

- Minimize material handling (i.e., pounds of material transported over a given distance).
- Maximize storage capacity of the facility for WIP and finished goods.
- Provide justification for dedicated storage zones.
- Illustrate advantages of decreasing finished product inventory and increasing the amount of WIP.

Material handling is measured by the distance traveled as well as the amount of material transported over the given distance. The units used to express material handling are pound-feet. For each transition, storage space is allocated according to

TABLE 13.1. Storage Zone Capacities

		Capacity (lb)	
Storage Zone	Area (ft^2)	WIP (rolls)	Finished (pallets)
1	1,124	232,476	98,080
2	736	152,226	64,223
3	625	129,268	54,538
4	1,919	396,905	167,452
5	3,630	750,789	316,754
6	1,214	251,090	105,934
7	4,877	1,008,705	425,567
8	3,046	630,001	265,794
9	3,639	752,651	317,539
10	2,075	429,170	181,065
11	6,858	1,418,433	598,429
12	6,379	1,319,362	556,632
13	5,000	1,034,145	436,300
14	1,588	328,444	138,569
15	534	110,447	46,597
Total	43,244	8,944,113	3,773,471

Note: WIP (rolls) = 206.83 lb/ft^2; finished (pallets) = 87.26 lb/ft^2.

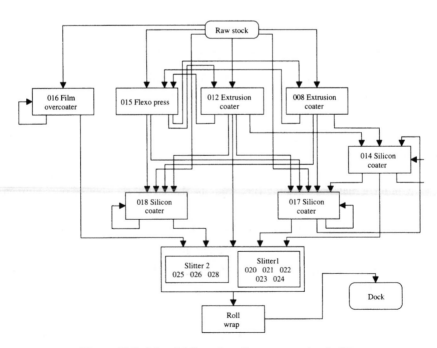

Figure 13.2. Material flow chart for paper coating facility.

the average amount of inventory making the transition. Optimal storage zones are selected to minimize the total pound-feet of material handling.

Referring to Table 13.1, it is seen that the capacity of the facility is considerably higher for WIP than for finished inventory. Therefore, to maximize total capacity, the level of WIP must increase and the level of finished inventory must decrease. This is an obvious assertion if the methods of holding the two types of material are considered. In-process inventory is stored in stacked racks and finished inventory is stored on pallets that cannot be stacked.

A problem that currently exists in the facility is misplaced material. Dedicated storage zones will improve tracking of material. Material in transition is routed to preassigned storage zones that are selected to minimize material handling and meet storage requirements.

13.4.4 Constructing the Space Allocation Model

This section presents the data required to construct the model. It is first necessary to identify all material flow transitions from Figure 13.2 that are relevant to the study. The transitions are summarized in Table 13.2.

Sources and *destinations* in the table refer to the machine numbers, the dock (DK), the roll wrap machine (RW), slitter group 1 (SL1), or slitter group 2 (SL2). The *storage* column identifies possible storage zones for material in the transition; all locations are available for material in each transition. The *type* column identifies the material type for the transition, which is used to determine the material type conversion factor τ.

The sources and destinations for material flow transitions and the 15 storage zones are shown in Figure 13.1. Each transition in Table 13.2 has 15 possible routes, one for each storage zone. A CAD drawing of the facility is used to determine the distances needed to construct the distance matrix, shown in Figure 13.3.

The distances are measured in feet traveled by a fork truck transporting material. The 31 transitions are listed vertically and the 15 storage zones are listed horizontally. For example, the entry in row 22, column 8, is the distance in feet traveled by a fork truck making transition 22 (from machine 017 to machine 014) via storage zone 8. It follows from Figure 13.1 that storage zones located on the most direct path between source and destination of a given transition will have the lowest values in the row corresponding to that transition.

Since Figure 13.1 shows a simplified drawing of the facility, distances that appear to be the same in Figure 13.1 vary slightly in the distance matrix. Several possible routes for material making transition 23 are also shown in Figure 13.1. Although routes via storage zones 4, 5, and 6 appear to be the same distance, the distance matrix gives values of 979 ft, 983 ft, and 971 ft, respectively. Distances to and from storage locations are measured to the center of the storage zone. The CAD drawing is more precise and, therefore, yields slightly different values for each route.

The objective statement of the model contains 465 decision variables x_{ij} for $i = 1, 2, \ldots, 31$ and $j = 1, 2, \ldots, 15$. Each decision variable represents the amount of material

TABLE 13.2. Data for Material Flow Transitions

Transition	Source	Destination	Storage	Type	Average Level (lb)	τ
01	016	016	All	WIP	0	1
02	016	SL1	All	WIP	46,809	1
03	016	SL2	All	WIP	2,464	1
04	015	018	All	WIP	24,970	1
05	015	017	All	WIP	8,323	1
06	015	012	All	WIP	41,617	1
07	015	008	All	WIP	8,323	1
08	012	015	All	WIP	3,432	1
09	012	018	All	WIP	3,432	1
10	012	017	All	WIP	89,227	1
11	012	SL1	All	WIP	3,432	1
12	012	SL2	All	WIP	3,432	1
13	012	014	All	WIP	168,159	1
14	008	015	All	WIP	13,669	1
15	008	018	All	WIP	2,734	1
16	008	014	All	WIP	182,076	1
17	008	014	All	WIP	54,677	1
18	018	018	All	WIP	0	1
19	018	SL1	All	WIP	2,260	1
20	018	SL2	All	WIP	201,382	1
21	017	017	All	WIP	3,739	1
22	017	014	All	WIP	18,696	1
23	017	SL1	All	WIP	142,087	1
24	017	SL2	All	WIP	195,370	1
25	014	014	All	WIP	143,466	1
26	014	017	All	WIP	3,587	1
27	014	SL1	All	WIP	303,432	1
28	014	SL2	All	WIP	17,933	1
29	SL1	RW	All	Finished	35,500	2.37
30	SL2	RW	All	Finished	14,500	2.37
31	RW	DK	All	Finished	3,000,000	2.37

making transition i in storage location j. In the objective statement, each decision variable x_{ij} is multiplied by the corresponding distance d_{ij} from the distance matrix. The objective statement for the model is developed using a program written in C. The user is asked to supply a file containing the distances from the distance matrix and enter the number of transitions and storage areas.

The information provided in Table 13.2 is used to determine the material type conversion factor τ_i for each transition and the storage zone capacity constraints. Values for the average level of inventory and τ_i are listed in Table 13.2. The C program generates the constraints of the model by asking the user to enter the information found in Tables 13.1 and 13.2. A file containing model (13.1)–(13.4) is loaded into the LINDO software, which solves the model with the simplex algorithm (Schrage, 1987).

Trans.	Storage 1	2	3	4	5	6	7	8	9	10	11	12	13	14	15
01	1308	1534	1488	1420	1266	1128	1236	450	814	778	898	822	298	230	852
02	1447	1673	1627	1559	1405	1267	1375	589	953	917	1037	891	431	233	991
03	980	1206	1164	1092	938	800	818	636	778	742	942	796	678	628	816
04	495	523	363	531	535	523	899	1119	1045	1009	1289	1143	1361	1459	1083
05	450	478	478	486	490	478	854	1074	1000	964	1244	1098	1316	1414	1035
06	517	545	545	553	557	545	825	785	711	675	955	809	1027	1125	749
07	358	386	386	394	398	386	762	982	908	872	1152	1006	1224	1322	946
08	698	848	802	730	738	726	1006	712	860	824	1024	878	782	880	898
09	923	1149	943	1031	881	743	963	669	817	781	981	835	739	837	855
10	878	1104	1058	986	836	698	918	624	772	736	936	790	694	792	807
11	1322	1548	1502	1430	1280	1142	1250	458	828	792	912	766	328	278	866
12	855	1081	1039	963	813	675	693	505	653	617	817	671	575	673	691
13	1135	1361	1315	1243	1093	955	1079	362	631	595	805	659	260	352	669
14	160	330	284	216	378	492	928	1148	1074	1038	1318	1172	1390	1488	1112
15	385	631	425	517	521	509	885	1105	1031	995	1275	1129	1347	1445	1069
16	340	586	540	472	476	464	840	1060	986	950	1230	1084	1302	1400	1021
17	597	843	797	729	733	721	1001	798	845	809	1099	953	868	960	883
18	593	621	461	629	633	621	997	1217	1143	1107	1387	1241	1459	1557	1181
19	992	1020	1020	1028	1032	1020	1284	1006	1154	1118	1318	1172	1048	998	1192
20	525	553	557	561	565	553	727	1053	979	943	1223	1077	1295	1393	1017
21	499	527	527	535	539	527	883	1123	1049	1013	1293	1147	1365	1463	1084
22	756	784	784	792	796	784	1044	861	908	872	1162	1016	931	1023	946
23	943	971	971	979	983	971	1215	957	1105	1069	1269	1123	999	949	1143
24	476	504	508	512	516	504	658	1004	930	894	1174	1028	1246	1344	968
25	1300	1526	1480	1412	1258	1120	1244	533	796	760	970	824	403	347	834
26	1043	1269	1223	1155	1001	863	1083	795	937	901	1101	955	837	787	972
27	1487	1713	1667	1599	1445	1307	1415	629	993	957	1077	931	471	273	1031
28	1020	1246	1204	1132	978	840	858	676	818	782	982	836	718	668	856
29	1252	1478	1432	1364	1210	1072	1196	394	758	722	842	696	298	248	796
30	855	1081	1035	967	804	675	799	359	507	471	671	525	463	561	581
31	935	1161	1115	1047	893	755	879	355	441	405	215	273	597	695	479

Figure 13.3. Distance matrix for 31 transitions and 15 storage zones.

13.4.5 Solving the Space Allocation Model

The optimal values for the decision variables minimize the objective function, which represents the cost of material handling. Therefore, the recommended strategy for inventory management is based on the optimal values for the decision variables. The decision variables define the amount of material that is to be stored in each storage zone for each transition.

The decision variables suggest which storage zones should be rack storage, which storage zones should be pallet storage, and which storage zones should be both. Figure 13.4 shows how floor space should be used based on the model results.

Table 13.3 defines the inventory management strategy for the facility. When the fork truck at the source of the transition picks up material, the fork truck driver determines the transition based on the next machine in the production schedule, or the destination. Once the transition is identified, the driver consults Table 13.3 to determine the proper storage zone for the material. The material is then transported to the allocated storage zone.

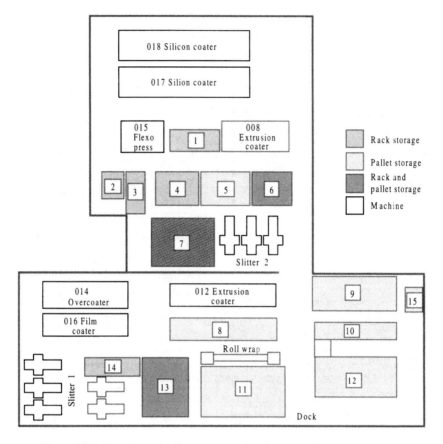

Figure 13.4. Location of pallet storage and rack storage for current inventory mix.

From the model input data, it is determined that approximately 70% of the inventory is finished product. To increase total capacity of the facility, it is necessary to decrease the level of finished inventory and increase the level of WIP. The effect of finished inventory on total storage capacity is illustrated in Figure 13.5.

The model is modified to analyze the effects of several alternative inventory mixes as the inventory management strategies. The values in Table 13.2 for average level of inventory are changed for each transition to increase WIP and decrease finished product. As shown in Figure 13.5, the total capacity of the facility increases as finished inventory decreases because it is more efficient to hold in-process inventory in racks than finished inventory on pallets. Solving the model for each alternative inventory mix illustrates the effect on storage capacity of the facility as well as material handling effort. Table 13.4 shows the total storage capacity of the facility and the material handling effort for each alternative inventory mix.

The optimal value of the objective function for the original inventory mix (70% finished product and 30% WIP) is 2,216,118,000 ft-lb of material handling to maintain the required levels of inventory. The optimal value of the objective function

TABLE 13.3. Storage Zones for Material in Transition

Transition	Storage Zone(s)
01	—
02	13, 14
03	13
04	3
05	3
06	3
07	3
08	4
09	6
10	6
11	13
12	7
13	13
14	4
15	3
16	1
17	1, 4
18	—
19	4
20	2, 4
21	4
22	3
23	3, 4
24	4
25	13
26	6
27	14
28	13
29	13
30	13
31	5–13, 15

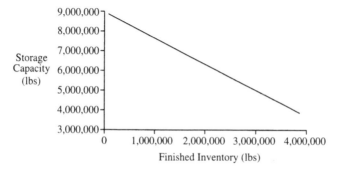

Figure 13.5. Total storage capacity vs. finished inventory.

TABLE 13.4. Total Storage Capacity, Total Material Handling Effort, and Average Distance per Transition for Several Inventory Mixes

Inventory Mix	Total Capacity	Total Material Handling Effort	Average Distance per Transition
30% WIP and 70% finished	4,561,000	2,216,118,000	486
40% WIP and 60% finished	4,904,000	2,320,728,250	473
50% WIP and 50% finished	5,303,000	2,465,970,500	465
60% WIP and 40% finished	5,772,000	2,675,434,750	464
70% WIP and 30% finished	6,332,000	3,038,259,250	480

for the 70% WIP and 30% finished product inventory mix is 3,038,259,250 ft-lb of material handling. Changing the inventory mix results in a 37.1% increase in material handling effort. However, the total capacity of the facility is increased from 4,561,000 lb 6,332,000 lb, or 38.8%. This suggests a slight reduction in material handling effort. Furthermore, a 60% WIP and 40% finished product inventory mix results in a 26.6% increase in total capacity and only a 20.7% increase in material handling effort.

To better illustrate the reduction in material handling effort, consider the objective function of model (13.1)–(13.4). The value of the objective function is a static measure of the total material handling effort for material in transition. Assuming the storage capacity constraints are tight (i.e., 100% utilization of storage space), the average distance per transition is obtained by dividing the optimal value for the objective function by the total capacity of the facility for a given inventory mix. These values are listed in Table 13.4 and plotted in Figure 13.6. According to this information, the 60% WIP and 40% finished product inventory mix minimizes material handling for the five alternatives tested. It should be noted that the average values for material in transition were determined for 100% storage space utilization.

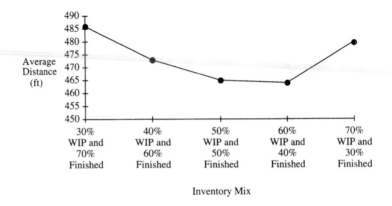

Figure 13.6. Average distance per transition for several different inventory mixes.

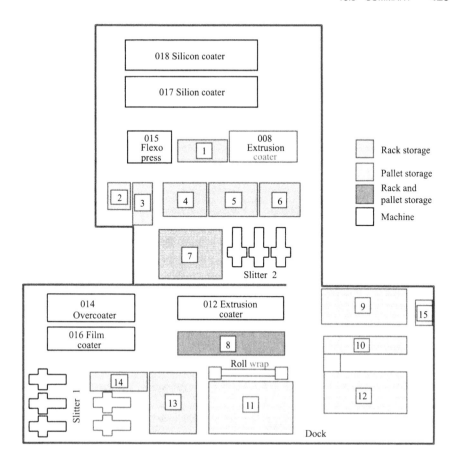

Figure 13.7. Location of pallet and rack storage for 60% WIP and 40% finished product.

If this is not true, it is necessary to subtract any slack in the storage zone capacity constraints from the total storage capacity of the facility before dividing.

The 60% WIP and 40% finished inventory mix also results in a more desirable arrangement of storage zones. Comparing Figures 13.4 and 13.7, it is seen that the latter places finished inventory closer to the dock and WIP closer to the machines. Furthermore, storage zones containing both types of storage, pallets and racks, are centrally located in the facility.

13.5 SUMMARY

The development of the model presented in this chapter was motivated by the need to allocate storage space so that material handling, a non-value-added operation, is minimized. The space allocation problem is modeled as a generalized transportation problem; however, transporting material within the facility is the focus. The model considers material type, material flow transitions, and distance transported. The model is constructed by first developing a material flow chart of the facility. The arcs on the flow chart represent material flow transitions. Storage areas are identified and a distance matrix is constructed. Constraints of the model consider storage capacity and storage requirements.

A C code was used to construct a linear programming model. The code required the user to enter the number of transitions, the number of storage zones, the capacity of each storage zone, the average level of inventory for material in each transition, and the material type conversion factor τ_i for each transition. The file containing the model was read by LINDO.

The optimal values for decision variables suggested an inventory management strategy that minimized material handling for the current inventory mix. However, it was recognized that the total storage capacity of the facility increased by decreasing the level of finished inventory and increasing the level of in-process inventory. The model was solved for several alternative inventory mixes, thus resulting in an increase of total storage capacity, a decrease in material handling effort, and a more desirable arrangement of storage zones.

REFERENCES

Allegri, T. H. (1984), *Materials Handling Principles and Practice,* Van Norstrand, New York.

Beheshtian-Ardekani, M., and M. A. Mahmood (1986), Development and validation of a tool for assigning students to groups for class projects, *Decision Sciences,* Vol. 17, pp. 92–113.

Chien, T. W., A. Balakrishnan, and R. T. Wong (1989), An integrated inventory allocation and vehicle routing problem, *Transportation Science,* Vol. 23, No. 2, pp. 67–76.

Cooper, W. W., and A. Charnes (1954), The stepping stone method of explaining linear programming calculation in transportation problems, *Management Science,* Vol. 1, No. 1, pp. 17–30.

Crandall, R. E., and T. H. Burwell (1993), The effect of work-in-process inventory levels on throughput and lead times, *Production and Inventory Management Journal,* Vol. 34, No. 1, pp. 6–12.

Dantzig, G. B. (1947), *Linear Programming and Extensions,* Princeton University Press, Princeton, NJ.

Federgruen A., and P. Zipkin (1984), A combined vehicle routing and inventory allocation problem, *Operations Research,* Vol. 32, pp. 1019–1037.

Harrison, H., and D. R. Wills (1983), Product assembly and distribution optimization in an agribusiness cooperative, *Interfaces,* Vol. 13, pp. 1–9.

Hitchcock, F. L. (1941), The distribution of a product from several sources to numerous locations, *Journal of Mathematics and Physics,* Vol. 20, pp. 224–230.

Inman, R. R. (1993), Inventory is the flower of all evil, *Production and Inventory Management Journal,* Vol. 34, No. 4, pp. 41–45.

Koopmans, T. C. (1949), Optimum utilization of the transportation system, *Econometrica,* Vol. 17, pp. 136–146.

Larson, N., and A. Kusiak (1995), Work-in-process space allocation: A model and an industrial application, *IIE Transactions: Design and Manufacturing*, Vol. 27, No. 4, pp. 497–506.

MacMillan, J. M. (1993), Principles of point-of-use storage, *Production and Inventory Management Journal*, Vol. 34, No. 4, pp. 53–55.

Nahmias, S. (1993), *Production and Operations Analysis*, Irwin, Homewood, IL.

Orden, A. (1956), The transshipment problem, *Management Science*, Vol. 2, pp. 276–285.

Park, P. S. (1993), Uniform plant loading through level production, *Production and Inventory Management Journal*, Vol. 34, No. 2, pp. 12–17.

Rautman, C. A., R. A. Reid, and E. E. Ryder (1993), Scheduling the disposal of nuclear waste material in a geologic repository using the transportation model, *Operations Research,* Vol. 41, pp. 459–469.

Schrage, L. E. (1987), *User's Manual for Linear, Integer, and Quadratic Programming with LINDO*, Scientific Press, Redwood City, CA.

Strevell, M. W., and P. S. Chong (1985), Gambling on vacation, *Interfaces*, Vol. 15, pp. 63–67.

QUESTIONS

13.1. What is the usual objective function of the vehicle routing problem?

13.2. What is the objective function of the generalized transportation problem?

13.3. What is a storage zone?

13.4. What is a transition?

13.5. What are the objectives of the case study discussed in this chapter?

13.6. What is the justification for the function presented in Figure 13.6?

PROBLEMS

13.1. Design a supply chain network and apply model (13.1)–(13.4).

13.2. Modify model (13.1)–(13.4) to consider inventory holding cost and transportation cost.

13.3. Apply the assignment model to the problem discussed in the case study. Discuss and compare the assignment solution with the generalized transportation problem solution.

CHAPTER 14

LAYOUT OF A WAREHOUSE

14.1 INTRODUCTION

Warehousing involves three functions: (1) receiving goods from a source, (2) storing goods until they are needed by a customer (internal or external), and (3) retrieving the goods when requested. Storing material for an internal customer implies the need for work-in-process (WIP) storage, while storing goods for an external customer may imply the need for finished product storage. However, the functions of warehousing remain the same; thus, successful warehouse layouts must accomplish the following objectives, regardless of the material being stored (Tompkins, 1988):

- Maximize the use of space
- Maximize the use of equipment
- Maximize the use of labor
- Maximize accessibility to all items
- Maximize protection of all items

Warehouse layout is often complicated by large varieties of products needing storage, varying areas of required storage space, and drastic fluctuations in product demand. Optimal approaches to warehouse layout often consider a single objective (i.e., maximize floor space utilization) and/or provide a solution to a static problem. Some recent research in warehouse layout is discussed in Section 14.2.

Alternative storage methods and equipment further complicate warehouse design problems. For example, material may be handled by pallet, case, or bulk items and the system may use lift trucks, conveyors, or automated storage and retrieval (AS/R) systems. To narrow the scope of problems encountered in practice, this chapter

428

focuses on single-command lift truck storage and retrieval of pallets (i.e., one storage or one retrieval of a pallet is performed per trip between storage and the input/output point). This system is typical of many industries observed by the authors and should be of interest to a large audience. Furthermore, a clear definition of a particular type of warehousing system allows us to present a sensible heuristic procedure and illustrate it with a meaningful case study. As a result, we intend to provide a method for warehouse layout and operation that is applicable to a variety of industries.

To achieve effective use of floor space and equipment, the layout must maximize floor space utilization and minimize travel time (and/or distance). The layout must also be robust to fluctuations in inventory level and demand.

14.2 RELATED LITERATURE

The literature on warehouse layout and operations is extensive. Three areas of research are particularly applicable to the problem discussed in this chapter: (1) stock location policy selection, (2) travel time and/or distance considerations, and (3) floor space allocation. This section provides a brief overview of some related research.

Several stock location policies have been discussed in the literature, including dedicated storage, randomized storage, and class-based storage. Francis et al. (1992) provided an overview of these methods as well as the shared storage location policy. An early paper by Ballou (1967) discussed the role of storage location policies in warehouse layout. Several authors have provided comparisons of storage location policies, including Harmatuck (1976), Roll and Rosenblatt (1983), and Wilson (1977). Eynan and Rosenblatt (1994) applied class-based stock location to the study of AS/R systems. Rosenwein (1994) used cluster analysis and order picking data to establish class-based storage. The procedure presented in this chapter establishes classes based on floor space utilization, frequency of single-command storage or retrieval, and average inventory level.

Much of the research in the area of travel time and/or distance reduction has been motivated by the need to improve the efficiency of the order picking operations. The problem has frequently been formulated as a traveling salesman model (Ratliff and Rosenthal, 1983). However, the problem considered in their paper deals with single-command operations, and there is no need for the development of a formal model. The heuristic presented in Section 14.3 considers the rectilinear (or rectangular) distance from warehouse input point, to storage location, to output point. Tompkins and White (1984) address a variety of topics related to movement of material handling devices.

Floor space allocation in storage systems has received less attention in the literature. Larson and Kusiak (1995) presented a generalized transportation problem formulation for WIP space allocation and illustrated the model with a case study. Space allocation within the facility is more generally treated as the facility layout problem in which departments are arranged according to some cost objective and qualitative material flow and/or administrative constraints. In the procedure presented in this chapter, storage regions are determined for each class and prioritized for the facility layout. The regions are sized in a way that reduces "honeycombing" (Smith and Peters, 1988) and, as a result, the storage regions assume varying areas. A recent paper by Lacksonen (1994) addressed facility layout problems with varying areas.

In addition to the research cited in the areas of stock location policies, travel time considerations, and floor space allocation, the following publications are of interest. Lanker (1988) provided a detailed description of lift trucks as material handling devices. Ross (1993a,b) discussed the basics of pallet storage systems in a two-part series.

14.3 PROCEDURE FOR WAREHOUSE LAYOUT

This section presents a heuristic for determining warehouse layout in a lift truck pallet storage and retrieval environment (Larson et al., 1997). The procedure is motivated by (1) the utilization of lift truck pallet storage in many industries and (2) the need for a practical method that provides an alternative to the ad hoc layouts that are often implemented in the absence of formal analysis and design.

14.3.1 Class-Based Storage Rationale

Warehouse operations are usually governed by one of three storage policies: randomized storage, dedicated storage, or class-based storage. In randomized storage, inventory is allocated a storage location based on the available space at the time of the storage job. The location is recorded for future retrieval jobs. The dedicated storage policy assigns material to a predetermined location based on throughput and storage requirements. All storage jobs of a given material are routed to the dedicated storage location. Class-based storage is a compromise between randomized and dedicated storage. Inventory is assigned a class based on some criteria (i.e., demand, product type, size) and each class is assigned a block of storage locations. Within each block of storage locations, material is stored randomly.

A cost trade-off is often realized between dedicated and randomized storage. Although dedicated storage typically reduces the material handling cost, more storage space is required for each product to prevent overflow. Alternatively, randomized storage usually requires less storage space; however, the material handling cost is often greater because there is no effort to store fast-moving material in the most desirable storage locations.

This chapter proposes class-based storage for lift truck pallet storage and retrieval warehouses. The objective is to group material with similar characteristics and allocate floor space based on group priority. The classification procedure seeks to increase floor space utilization by limiting the effects of "honeycombing" (Ross, 1993a; Smith and Peters, 1988) and decrease material handling by ranking items based on throughput and storage space requirements. Each class is assigned to a storage region; however, within the storage region material is stored randomly. Random storage within the region provides flexibility to accommodate variations in inventory level for material assigned to the class. Thus, the proposed class-based storage policy increases the floor space utilization, decreases the material handling cost, and increases flexibility.

14.3.2 Computational Procedure

Prior to presenting the class-based procedure for warehouse layout, it is necessary to establish several assumptions. Although these assumptions may appear to be limiting, the method can be generalized to accommodate most practical scenarios, as illustrated in Section 14.4. Furthermore, enhancements of the procedure are an expected result of the future research issues discussed in Section 14.5. The assumptions are summarized below:

1. The facility is rectangular and the dimensions are known.
2. Material enters the facility via a single input point (i.e., a receiving dock).
3. Material exits the facility via a single output point (i.e., a shipping dock).
4. Each storage region consists of only one storage medium (i.e., rack or floor stack).
5. All primary aisles are parallel and the same length.

The procedure can accommodate more complex facilities by decomposing the layout problem into rectangular components with single input and output points. In fact, the physical characteristics of the facility are only of interest in determining the direction and length of an initial primary aisle, which is the basis for subsequent primary aisles. Classes are formed and storage regions are sized according to throughput and storage requirements. The computational procedure in this chapter includes the following three phases:

Phase 1: determination of aisle layout and dimensions
Phase 2: assignment of material to a storage medium
Phase 3: allocation of floor space

In phase 1, the direction and length of the initial aisle are determined as well as the other necessary dimensions. Each item is assigned a storage medium in phase 2. Classes are allocated floor space in phase 3 and subsequent primary aisles are added to the layout.

14.3.2.1 The Procedure

Phase 1: Determine Aisle Layout and Dimensions. The direction of aisles must first be determined; however, it is not necessary to determine the frequency (or spacing) of aisles a priori. The objective of the heuristic layout procedure is to determine classes that define different storage regions in the facility. It is assumed that each aisle serves the two storage regions on either side of the aisle. A region may also be referred to as a zone of some storage medium, such as racks or floor stacking. The rack dimensions or row depth determines the size of the storage zone. Aisles and zones of storage are added to the layout until all material that requires storage has been assigned a class and storage zone in the warehouse.

It is assumed that material enters the warehouse at a single input point and exits the warehouse at a single output point. Therefore, the most desirable path of travel covers the shortest rectilinear distance from input point to output point. The shortest travel time path is determined so that (1) the rectilinear distance from input point to output point is minimized, (2) the number of corners on the path is minimized, and (3) the length of the longest arc on the path is maximized. Based on these objectives, the most desirable path is easily determined by examination. The direction and length of the longest arc on the path define the direction and length of the initial primary aisle. Figure 14.1 illustrates three possible paths through a facility with a single input point and a single output point. According to the objectives stated above, path (b) is obviously the best solution.

When the input point and output point of the warehouse are at the same location, selection of an initial aisle is somewhat arbitrary. In general, if many classes are desired and it is expected that the corresponding storage zones be of varying sizes (i.e., varying row depths), aisles should be shorter and more frequent. Alternatively, if fewer classes are desired, a lesser number of longer aisles are desired. Figure 14.2 shows primary aisle layouts for various input/output point configurations.

To simplify the classification procedure, it is useful to represent dimensions in units of floor space, or slots, rather than some known unit of measurement (i.e., feet and meters). Let a slot represent the smallest unit of assignable floor space. Typically, the slot length S_l and slot depth S_d are equal to the pallet dimensions (plus the necessary clearance allowance). Recall that a zone defines an area that can be allocated for storage (i.e., racks or floor stacking). The procedure establishes dimensions for zone length L and zone depth D_k, where k denotes the kth storage zone. Although the zone depth D_k can vary among zones, it is assumed that the zone length L is the same for all storage zones. The aisle length is used to determine the zone length L. The zone length should exclude any allowances for secondary (or service) aisles at either end of the zone (and possibly even throughout the zone). Thus, storage zone k is said to

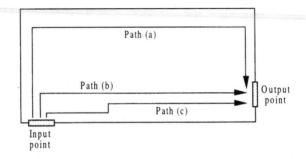

Figure 14.1. Alternative paths through a rectangular warehouse.

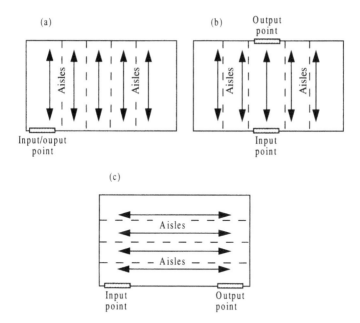

Figure 14.2. Primary aisle layouts for various input/output point configurations.

be L slots long and D_k slots deep. Figure 14.3 illustrates storage zone and slot dimensions.

The zone depth D_k is a function of the storage medium. For each item, the objective is to determine the storage medium (and corresponding zone depth) that minimizes the effects of "honeycombing." Honeycombing occurs when storage locations are partially filled by items or lots that cannot be mixed or blocked from access. Two types

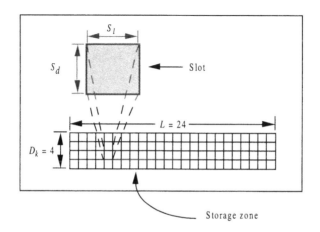

Figure 14.3. Storage zone and slot dimensions.

of storage are typical of lift truck pallet storage and retrieval warehouses, rack and floor stacking. A combination of both types of storage may be used in the warehouse and, furthermore, there may be several different types of racks and/or varying row depths for floor stacking. Prior to assigning items to classes, it is necessary to identify the types of storage to consider. For example, the warehouse may use four-tier one-deep rack storage for items that cannot be stacked and three-deep, four-deep, and five-deep floor stacking for items that can be stacked. Each item (or lot number) is assigned to a storage location in the warehouse. However, since several different types of storage are used, the size and capacity of the storage location vary. Table 14.1 lists the floor space consumed and capacity of a storage location for each storage medium. The capacity of a storage location for floor stacking depends on the stack height h of the item stored. Phase 2 of the procedure determines the storage medium for each item such that storage location utilization increases.

The following steps summarize the activities in phase 1:

1. Identify a path from input point to output point such that:
    (a) The rectilinear distance from input point to output point is minimized
    (b) The number of turns is minimized
    (c) The length of the longest arc on the path is maximized

    Note: If the input/output points are at the same location, consider the number and size of desired classes and arbitrarily choose an initial primary aisle.

2. Establish an initial primary aisle along the longest arc of the path identified in 1.
3. Establish slot dimensions based on pallet dimensions, rack dimensions, and the necessary clearance allowances.
4. Determine the storage zone length L (in slots) based on the length of the initial primary aisle and any secondary aisles.
5. Identify storage mediums for consideration in phase 2 (i.e., rack and/or n-deep floor stacking).

Phase 2: Determine Storage Medium. The row-depth algorithm presented next determines the storage medium for an item (or a lot) based on average inventory level, honeycombing, fluctuations in inventory level, and stack height. It is suitable for

TABLE 14.1. Floor Space Consumed and Capacity of a Storage Location for Various Storage Mediums

Storage Medium	Floor Space Consumed per Storage Location (slots)	Capacity per Storage Location (pallets)
Four-tier one-deep rack	0.25	1
Three-deep floor stack	3	$3h$
Four-deep floor stack	4	$4h$
Five-deep flour stack	5	$5h$

implementation using a spreadsheet and standard database functions. The following data are needed for n items requiring storage:

- Average inventory level s_j for $j = 1, \ldots, n$
- Stack height h_j for $j = 1, \ldots, n$
- Maximum row depth $d_{\max}$ for floor stacking
- Percent allowance for honeycombing and inventory fluctuation α
- Minimum stack height $h_{\min}$ for floor stacking
- Minimum row depth $d_{\min}$ for floor stacking

The algorithm is specifically designed to determine the row depth that maximizes storage location utilization for floor stacking. If rack storage is available, it is necessary to consider classification criteria for assigning an item to that medium. Two possible criteria are (1) a minimum stack height for floor stacking (i.e., an item with a stack height h_j less than an established minimum is assigned to rack storage) and (2) a minimum row depth for floor stacking (i.e., an item with a calculated row depth less than an established minimum is assigned to rack storage). The following notation is used in the row-depth algorithm:

$$n = \text{number of items requiring storage}$$
$$j = \text{index for item } j = 1, \ldots, n$$
$$h_j = \text{stack height for item } j = 1, \ldots, n$$
$$s_j = \text{average inventory level for item } j = 1, \ldots, n$$
$$\alpha = \text{percentage allowance for honeycombing}$$
$$h_{\min} = \text{minimum stack height for floor stacking}$$
$$d_{\max} = \text{maximum row depth for floor stacking}$$
$$d_{\min} = \text{minimum row depth for floor stacking}$$
$$N_j = \text{number of storage locations required for item } j = 1, \ldots, n$$
$$d_j = \text{optimal row depth for item } j = 1, \ldots, n$$
$$N^* = \text{current number of storage locations required}$$
$$F(s_j) = \text{a function of the rack design that provides the number of storage locations needed to store } s_j \text{ pallets}$$

Row Depth Algorithm

Input: n, h_j ($j = 1, \ldots, n$), s_j ($j = 1, \ldots, n$), α, $h_{\min}$, $d_{\max}$, $d_{\min}$
Output: N_j, d_j
Step 0. Let $j = 1$.
Step 1. IF $h_j < h_{\min}$, go to Step 5. ELSE go to Step 2.

Step 2. Let $N^* = \left\lceil \dfrac{(1 + \alpha)s_j}{d_{\max}h_j} \right\rceil$

$$d_j = \left\lceil \frac{(1+\alpha)s_j}{N^*h_j} \right\rceil$$

Step 3. IF $d_j \geq d_{min}$, go to Step 4. ELSE go to Step 5.

Step 4. Let $N_j = N^*$. Go to Step 6.

Step 5. Let $d_j = 0$

$$N_j = F(s_j)$$

Step 6. IF $j < n$, let $j = j + 1$. Go to Step 1. ELSE stop.

For each item, the *row-depth algorithm* determines the minimum number of storage locations needed for floor stacking and then decreases the row depth until it is necessary to add a storage location. The minimum row depth that does not cause an increase in the number of storage locations is assigned to the item. This heuristic algorithm accomplishes two objectives. First, minimizing the number of storage locations needed to store items maximizes the number of different items that can be stored in a single zone. Thus, the most desirable storage zones can hold larger varieties of items. Second, minimizing the number of storage locations per item increases row depths. As a result, fewer aisles are needed and floor space utilization is increased.

The following steps summarize the activities in phase 2:

1. Collect the necessary data for n items requiring storage.
2. Apply the *row depth algorithm* to obtain:
   (a) Optimal row depth for each item d_j
   (b) Number of storage locations needed to store each item N_j

Phase 3: Allocate Floor Space.   Allocate Floor Space. The third phase of the procedure is to allocate floor space to classes based on storage medium, required number of storage locations, and throughput. The throughput t_j for $j = 1, \ldots, n$ items is measured by the number of pallets stored and retrieved per a given time period. For each storage medium, items (or lot numbers) are ranked according to the ratio of throughput t_j to storage requirement N_j, denoted as R_j. The items with the highest R_j's for a given storage medium are assigned to the most desirable storage zones such that the storage zone capacity L is not exceeded. It is assumed that the length of the storage zone L, measured in storage slots (locations), determines the capacity of the zone.

The above procedure provides the same solution that would be obtained by solving the linear programming relaxation of the knapsack problem and rounding the fractional assignment to zero. The knapsack can only hold a given weight; thus, the problem is to maximize the value of material in the knapsack while not exceeding the weight constraint. For the problem considered here, the value of storing a particular item is equal to its throughput t_j and the weight is equal to the storage space required N_j. For each storage medium, the algorithm solves successive knapsack problems until all items requiring that storage medium are assigned to a storage zone. The classification algorithm discussed next provides the classes C_k ($k = 1, \ldots, m$), the number of storage locations required for the class I_k, and the storage medium (i.e., row depth) D_k corresponding to class k. In addition to what was previously introduced, the following notation is used in the classification algorithm:

$t_j =$ throughput for item $j = 1, \ldots, n$

$N_j =$ storage requirement for item $j = 1, \ldots, n$

$R_j =$ throughput to storage requirement ratio

$r_j =$ rank of item $j = 1, \ldots, n$

$L =$ length of storage zone in number of storage locations (i.e., capacity)

$k =$ index for class

$C_k =$ set of items in class $k = 1, \ldots, m$

$I_k =$ number of storage locations required for class $k = 1, \ldots, m$

$D_k =$ storage medium (i.e., row depth) for class $k = 1, \ldots, m$

Classification Algorithm

Input: n, t_j ($j = 1, \ldots, n$), N_j ($j = 1, \ldots, n$), d_j ($j = 1, \ldots, n$), L, d_{max}

Output: R_j, r_j, C_k ($k = 1, \ldots, m$), D_k ($k = 1, \ldots, m$), I_k ($k = 1, \ldots, m$)

Step 0. Let $k = 0$.

Step 1. For $j = 1, \ldots, n$, let $R_j = t_j/N_j$.

Step 2. For $j = 1, \ldots, n$ and rank R_j in descending order, let $r_j =$ rank of item j.

Step 3. For $l = 0, \ldots, d_{max}$, let $k = k + 1$, $C_k = \emptyset$, and $I_k = 0$. For $i = 1, \ldots, n$ and for all j such that $r_j = i$ and $d_j = l$:

(a) IF $I_k + N_j \leq L$, let $C_k = C_k \cup j$, $I_k = I_k + N_j$, and $D_k = d_j$.

(b) IF $I_k + N_j > L$, let $k = k + 1$, $C_k = \emptyset$, and $I_k = 0$. Go to (a).

Step 4. Stop.

The *classification algorithm* assigns items to classes such that the capacity of the corresponding storage zone is not exceeded. In fact, it is quite probable that each zone will contain a small number of unused storage locations. This is a result of not allowing items to be assigned to multiple classes and not attempting to solve the knapsack problem optimally. However, since a class-based storage policy is being implemented, it is not discouraging to have a small number of unused storage locations in each storage zone to accommodate fluctuations in inventory level. Recall that within each storage zone material is stored randomly; thus, a certain amount of "overflow" space is desirable.

The final step in the class-based layout procedure is to prioritize the classes for floor space allocation. Once again, items are ranked according to the throughput to storage requirement ratio R_j. The average class rank R_k^* is determined as follows:

1. Calculate R_j for each item $j = 1, \ldots, n$.
2. Rank the items in descending order, where r_j is the rank of item j.

3. Calculate the class rank for each class k:

$$R_k^* = \frac{\displaystyle\sum_{\forall j \in C_k} r_j}{|C_k|}$$

where $|C_k|$ is the cardinality of the set of items in class k.

The layout of the warehouse is constructed by placing storage zones corresponding to the classes with the two lowest average class ranks adjacent to the primary aisle, which was determined in phase 1. Additional storage zones and primary aisles are assigned to the layout according to the average class rank, the storage medium for each class, and the dimensions of the facility (see the case study in Section 14.4).

The steps of phase 3 are summarized next:

1. Apply the classification algorithm to determine the following:
    (a) The class assignments for items (i.e., storage zone assignments) C_k
    (b) The storage medium for each class (i.e., row depth) D_k
    (c) The number of storage locations required for each class I_k
2. Prioritize classes according to average class rank R_k^*.
3. Construct the layout by adding storage zones corresponding to classes and additional primary aisles.

14.4 PALLET STORAGE AND RETRIEVAL SYSTEM CASE STUDY

The procedure presented in Section 14.3 was motivated by several warehouse systems observed in a variety of industries. In this section, an industrial case study is presented to illustrate the class-based storage procedure for warehouse layout.

14.4.1 Case Study Background

The warehouse studied is a stand-alone facility that is adjacent to a manufacturing site. Initially, the building was used for consignment warehouse operations, in which whole pallets were stored and shipped to internal and external customers. The proposed use of the facility was for customer warehouse operations, in which case picking and custom packaging would be performed for external customers. To convert the facility from a consignment warehouse to a customer warehouse, two options were being considered. For the first option, the current zones and aisles could be used, which would minimize conversion costs. However, there was a strong feeling within the company that the current layout was not efficient for customer warehouse operations. The second option was to redesign the warehouse to make it more suitable for customer warehouse operations. A redesign of the current warehouse had to address the following questions:

- How much rack storage is needed?

- What is a suitable row depth for floor stacking?
- How many zones (and/or aisles) are needed?
- In what direction should aisles run?
- What storage policy should be used?
- What products should be stored on racks?

Storage was required for $n = 739$ products, referred to as brand-codes. Product was stored in cases, which were stored on pallets. The size of a case varied among brand-codes and, thus, the number of cases per pallet varied among brand-codes. The predetermined stack height also varied among brand-codes and ranged from two to five pallets high. The following information was collected for the case study:

- Stack height (in pallets) for each brand-code h_j
- An average inventory level for each brand-code s_j
- Throughput (i.e., number of cases stored and retrieved during a 14-week period) of each brand-code t_j
- Pallet, rack, and aisle dimensions
- Current layout of the warehouse

A "snapshot" of the current inventory was used to approximate average inventory levels. Although the data were assumed to be representative of storage requirements, it is recognized that the static model used does not capture the dynamic behavior of inventory levels. However, the class-based computational approach to warehouse layout provides the flexibility needed to accommodate the dynamic behavior of inventory levels, given a good static approximation of the inventory level.

To determine the desired measure for throughput, the flow of the customer warehouse process was considered. Once product is received, lift trucks transport it to storage one pallet at a time. Product remains in storage until it is needed in a case-pick station. At that time, an entire pallet of product is transported to case-pick, which was out of the scope of the study. Therefore, whole pallets are stored and retrieved one by one. Thus, throughput was measured in the number of pallets per week. This value was obtained by knowing the demand in case-pick for each brand-code (which came from

TABLE 14.2. Important Dimensions for the Customer Warehouse Layout

Object	Dimensions
Aisle	14 ft wide
Pallet	4 by 4 ft
Rack	4 by 8 ft (3 tier)
Column	4 ft wide, 4–28 ft deep
Slot	4 by 4 ft

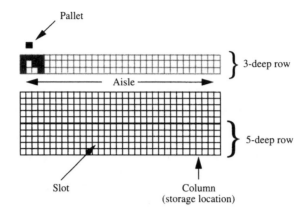

Figure 14.4. Top view of aisles, pallets, rows, columns, and slots.

the number of cases shipped per week) and assuming steady-state levels of inventory existed for each brand-code.

Two types of storage were considered in the warehouse, racks and floor stacking. Two racks (or two zones of floor storage) could be placed back to back, without being separated by an aisle. Racks had three tiers and a capacity of eight pallets per rack; the bottom tier could hold pallets stacked two-high and the second and third tiers could not hold stacked pallets. Several important dimensions are listed in Table 14.2.

Figure 14.4 illustrates aisles, pallets, zones, columns, and slots. Aisle lengths and building dimensions were predetermined. Two rows of any depth could be placed back to back without an aisle. A column could contain from one to seven slots (determined by the row depth) and was perpendicular to the adjacent aisle. Only one brand-code could be stored in each column and, thus, each column was considered a storage location. A slot is the smallest unit of floor space and can contain from one to five pallets depending on the stack height of the brand-code in the slot.

14.4.2 Application of the Computational Procedure

In phase 1 of the class-based layout procedure, the aisle layout and dimensions were to be determined. Based on the flow of material through the system, the initial primary aisle was determined by inspection. Three rectangular regions were identified for available storage, and the length of proposed storage zones was determined to be $L = 74$ slots (i.e., storage locations). The most desirable locations for storage were adjacent to the initial primary aisle. The next most desirable storage locations were in regions 1, 2, and 3, respectively. The initial primary aisle and proposed storage zones are shown in Figure 14.5.

In Phase 2 of the computational procedure, the storage medium was determined for each brand-code. The company wanted to provide two types of pallet storage, three-tier rack storage and floor stacking. Furthermore, floor stacking could range from three-deep rows to seven-deep rows. Row depth was a major concern of the company. The existing layout provided seven-deep floor stacking throughout the

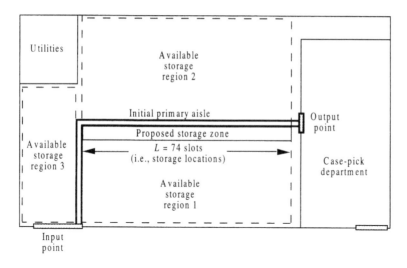

Figure 14.5. Initial primary aisle and proposed storage zone dimensions.

facility. However, many domain experts, ranging from lift truck operators to managers, felt that seven-deep rows throughout the warehouse would be inefficient for the customer warehouse operations. The minimum row depth for floor stacking was determined to be $d_{min} = 3$. Therefore, brand-codes with a calculated row depth less than 3 were assigned to rack storage. All brand-codes had stack heights ranging from two to six pallets and no minimum stack height h_{min} was established for floor stacking.

The data were downloaded to a spreadsheet program for implementation of the row depth algorithm. For each of the 739 brand-codes, the algorithm calculated the minimum row depth required to store the average inventory (plus $\alpha = 25\%$ allowance for honeycombing) in the minimum number of storage locations. Following is one iteration of the *row depth algorithm* for brand-code $j = 372$:

Input: $n = 739$, $h_{372} = 3$, $s_{372} = 8$, $\alpha = 0.25$, $h_{min} = 2$, $d_{max} = 7$, $d_{min} = 3$

Output: N_{372}, d_{372}

Step 1. Since $h_{372} > h_{min}$, go to Step 2.

Step 2. Let $N^* = \left\lceil \dfrac{(1+\alpha)s_j}{d_{max}h_j} \right\rceil = \left\lceil \dfrac{(1+0.25)(8)}{(7)(3)} \right\rceil = 1$

$$d_{372} = \left\lceil \frac{(1+\alpha)s_j}{N^* h_j} \right\rceil = \left\lceil \frac{(1+0.25)(8)}{(1)(3)} \right\rceil = 4$$

Step 3. Since $d_{372} > d_{min}$, go to Step 4.

Step 4. Let $N_{372} = 1$. Go to Step 5.

Step 5. Since $372 < 739$, let $j = j + 1 = 372 + 1 = 373$. Go to Step 1.

Phase 3 of the computational procedure allocated floor space to classes based on storage medium (i.e., row depth), required number of storage locations, and throughput. The ratio R_j was calculated for each brand-code and the brand-codes were ranked in descending order for each storage medium. Classes were formed for each storage medium by adding brand-codes until the number of storage locations required for the class I_k exceeded the storage zone capacity $L = 74$. A rack consumed two slots of floor space (equivalent to two storage locations of floor stacking) and provided six storage locations, each of which could contain one pallet. Thus, classes of rack storage were formed by adjusting L for the brand-codes assigned to rack storage (i.e., row depth $d_j < 3$). The classification algorithm generated 13 classes of brand-codes, based on storage medium and the ratio R_j. The average class rank R_k^* was determined for each of the 13 classes. Table 14.3 summarizes the classes for each storage medium.

14.4.3 Computational Results

The resulting layout required 13 storage zones for the 13 classes of brand-codes. The priority and dimensions of each storage zone are listed in Table 14.4. A letter (i.e., A, B, C, etc.) was added to the name of each storage zone to distinguish between multiple zones (and classes) of the same storage medium. Zone dimensions are given in floor slots, where a floor slot is 4 by 4 ft.

The storage zones are allocated floor space according to the priority listed in Table 14.4. The most desirable locations are adjacent to the initial primary aisle (see Figure 14.5). Following allocation of floor space to zones Rack A and 3-deep A, zones are allocated floor space in regions 1 and 2, respectively. This method of floor space allocation reduces travel distance for single-command storage or retrieval operations to higher priority classes. The proposed layout is shown in Figure 14.6.

14.4.4 Comparison of Existing and Proposed Designs

To justify the proposed warehouse layout, it was compared to the existing layout of the facility. As stated earlier, the existing layout consisted exclusively of seven-deep floor stacking, and primary aisles were arranged perpendicular to the proposed layout.

TABLE 14.3. Summary of Classes for Each Storage Medium

Storage Medium	Number of Classes	Number of Brand-codes	Percentage of Brand-codes	Number of Pallets	Percentage of Pallets
Three-tier racks	2	417	57	1,190	11
Three-deep floor	2	98	13	803	8
Four-deep floor	1	59	8	887	9
Five-deep floor	2	72	10	1,506	14
Six-deep floor	2	39	5	1,513	14
Seven-deep floor	4	54	7	4,705	44
Total	13	739	100	10,604	100

TABLE 14.4. Storage Zones for Warehouse Layout

Priority	Zone Name	Zone Length (L)	Zone Depth (D_k)
1	Rack A	74 slots (296 ft)	1 slot (4 ft)
2	3-deepA	74 slots	3 slots
3	5-deepA	74 slots	5 slots
4	7-deepA	74 slots	7 slots
5	6-deepA	74 slots	6 slots
6	4-deepA	74 slots	4 slots
7	7-deepB	74 slots	7 slots
8	7-deepC	74 slots	7 slots
9	3-deepB	74 slots	3 slots
10	5-deepB	74 slots	5 slots
11	6-deepB	74 slots	6 slots
12	7-deepD	74 slots	7 slots
13	Rack B	74 slots	1 slot

Limited rack storage was available for 800 pallets. Furthermore, a random storage policy was suggested for the existing layout.

The layouts were compared based on floor space utilization and material handling effort. Table 14.5 shows the floor space required for storing all 739 brand-codes in each warehouse. The current layout provided less rack storage than the proposed layout. However, the proposed layout drastically decreased the effects of honeycombing. As a result, the overall floor space requirement (including aisles) was much less for the proposed warehouse.

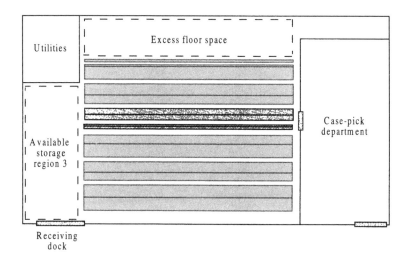

Figure 14.6. Proposed layout of the warehouse.

TABLE 14.5. Floor Space Requirements for Alternative Layouts

Layout	Rack Storage Space Required (ft^2)	Floor Stack Storage Space Required (ft^2)	Aisle Space Required (ft^2)	Total Space Required (ft^2)
Current	3200	90,720	32,816	126,736
Proposed	4736	71,040	29,008	104,784

To compare material handling effort, a random mix of 20 brand-codes was selected. The average weekly throughput (i.e., the number of pallets stored and retrieved) was obtained for each brand-code. For the current layout, the storage location for the brand-code was randomly generated and the distance traveled to store and retrieve a pallet was measured. The distance traveled to store and retrieve a pallet in the proposed system was determined by the location of the storage zone corresponding to the brand-code class. Table 14.6 summarizes the average weekly distance traveled to store and retrieve the 20 randomly selected brand-codes.

TABLE 14.6. Average Weekly Distance Traveled to Store and Retrieve 20 Randomly Selected Brand-codes

Brand-code	Average Throughput (pallets)	Current Storage and Retrieval Distance (ft)	Current Average Weekly Distance (ft)	Proposed Storage and Retrieval Distance (ft)	Proposed Average Weekly Distance (ft)
1	1.67	746	1,246	422	705
2	6.56	854	5,602	364	2,388
3	4.36	433	1,888	309	1,347
4	50.67	739	37,445	249	12,617
5	2.48	791	1,962	389	965
6	1.64	841	1,379	415	681
7	4.33	427	1,849	359	1,554
8	1.51	787	1,188	543	820
9	13.84	795	11,003	321	4,443
10	35.68	549	19,588	399	14,236
11	5.29	456	2,412	340	1,799
12	1.27	112	142	214	272
13	4.00	430	1,720	278	1,112
14	1.82	765	1,392	521	948
15	22.97	467	10,727	329	7,557
16	18.93	481	9,105	355	6,720
17	4.13	441	1,821	299	1,235
18	11.59	427	4,949	293	3,396
19	1.34	577	773	509	682
20	3.00	915	2,745	537	1,611
Total	197.08	12,033	118,936	7,445	65,088

14.4.5 Results

The proposed warehouse layout drastically improved on the current layout. The total floor space required to store all 739 brand-codes was decreased from 126,736 to 104,784 ft^2, or 17%. The average weekly distance traveled by lift truck to store and retrieve pallets for the random mix of 20 brand-codes was decreased from 118,936 to 65,088 ft^2, or 45%.

The floor space reduction was attributed to a decrease in honeycombing. Rather than floors stacking all material in seven-deep rows, a variety of row depths were provided in the proposed layout. The proposed layout also provided additional rack storage to hold the 57% of brand-codes that would be inefficiently stored in floor stacking.

The material handling reduction was attributed to the use of the class-based storage policy. Brand-codes were assigned to classes according to throughput. Therefore, fast-moving brand-codes were assigned to higher priority classes, which in turn were assigned to the most desirable storage zones (i.e., the storage zones that minimized the travel distance from input point to output point).

The proposed layout was presented to the company. Although the savings in floor space and material handling were attractive to management, a concern was raised about the extra effort needed to manage a class-based system. Since a relatively small number of storage zones were identified, and within each zone material is stored randomly, class-based storage would be relatively easy to implement. Each lift truck was already equipped with a bar-code scanner and database link. Thus, one alternative was to scan the lot and have the database return the class and storage zone of the pallet. Another alternative was to replace an unused character in the current brand-code with a character that identifies the storage zone for the pallet. However, with the projected savings in floor space and material handling, the company was in a position to justify facility and system conversion costs.

14.5 SUMMARY

This chapter presented a class-based storage procedure for warehouse layout. The method is targeted at practitioners of single-command lift truck pallet storage and retrieval. Three phases of the procedure were outlined: phase 1, determination of aisle layout and storage zone dimensions; phase 2, assignment of material to a storage medium; and phase 3, allocation of floor space. Two algorithms were provided for determining floor stack row depth and assigning items to classes. An industrial case study illustrating the procedure for a warehouse required storing 739 different products totaling more than 10,000 pallets of material. A new layout was developed and compared to the existing alternative. The proposed layout offered savings of more than 20,000 ft^2 of floor space and approximately a 45% reduction in material handling distance for a random mix of products.

Several assumptions were stated for bounding the problem addressed in this chapter. Although the method is easily generalized to accommodate most practical problems, the domain of application is somewhat limited. Future research should seek to enhance the procedure by considering issues such as multiple input/output points, alternative material handling devices, varying pallet sizes, stochastic inventory levels and throughput, and periodic reclassification of items.

REFERENCES

Ballou, R. H. (1967), Improving the physical layout of merchandise in warehouses, *Journal of Marketing*, Vol. 31, No. 3, pp. 60–64.

Eynan, A., and M. J. Rosenblatt (1994), Establishing zones in single-command class-based rectangular AS/RS, *IIE Transactions*, Vol. 26, No. 1, pp. 38–46.

Francis, R. L., L. F. McGinnis, Jr., and J. A. White (1992), *Facility Layout and Location: An Analytical Approach*, Prentice-Hall, Englewood Cliffs, NJ.

Harmatuck, D. J. (1976), A comparison for two approaches to stock location, *Logistics and Transportation Review*, Vol. 12, No. 4, pp. 282–284.

Lacksonen, T. A. (1994), Static and dynamic layout problems with varying areas, *Journal of the Operational Research Society*, Vol. 45, No. 1, pp. 59–69.

Lanker, K. E. (1988), Lift trucks, in J. A. Tompkins and J. D. Smith (Eds.), *The Warehouse Management Handbook*, McGraw-Hill, New York, pp. 321–371.

Larson, T. N., and A. Kusiak (1995), Work-in-process space allocation: A model and an industrial application, *IIE Transactions: Design and Manufacturing*, Vol. 27, No. 4, pp. 497–506.

Larson, T. N., March, H., and A. Kusiak (1997), A heuristic approach to warehouse layout with class-based storage, *IIE Transactions: Design and Manufacturing*, Vol. 29, No. 4, pp. 337–348.

Ratliff, H. D., and A. S. Rosenthal (1983), Order picking in a rectangular warehouse: A solvable case of the traveling salesman problem, *Operations Research*, Vol. 31, No. 3, pp. 507–521.

Roll, Y., and M. J. Rosenblatt (1983), Random vs. grouped storage policies and their effect on warehouse capacity, *Material Flow*, Vol. 1, pp. 191–199.

Rosenwein, M. B. (1994), An application of cluster analysis to the problem of locating items within a warehouse, *IIE Transactions*, Vol. 26, No. 1, pp. 101–103.

Ross, P. (1993a), The basics of pallet storage systems: Part 1, *Material Handling Engineering*, April, pp. 68–69.

Ross, P. (1993b), The basics of pallet storage systems: Part 2, *Material Handling Engineering*, May, pp. 61–63.

Smith, J. D., and J. E. Peters (1988), Warehouse space and layout planning, in J. A. Tompkins and J. D. Smith (Eds.), *The Warehouse Management Handbook*, McGraw-Hill, New York, pp. 91–114.

Tompkins, J. A. (1988), The challenge of warehousing, in J. A. Tompkins and J. D. Smith (Eds.), *The Warehouse Management Handbook*, McGraw-Hill, New York, pp. 1–14.

Tompkins, J. A., and J. A. White (1984), *Facilities Planning*, John Wiley, New York.

Wilson, H. G. (1977), Order quantity, product popularity and the location of stock in warehouses, *AIIE Transactions*, Vol. 19, No. 3, pp. 230–237.

QUESTIONS

14.1. What are the frequently used objectives of the warehouse layout?

14.2. What is a dedicated storage policy?

14.3. What is a class-based storage?

14.4. What is a storage zone?

14.5. What problems were addressed in the case study?

PROBLEMS

14.1. Model the problem addressed in the case study with one or more mathematical programming models.

14.2. Develop heuristic algorithms for solving the models developed in Problem 14.1.

CHAPTER 15

DESIGN FOR AGILITY

15.1 INTRODUCTION

Agile manufacturing is an industrial concept in industry that aims at achieving manufacturing flexibility, speed, and responsiveness. An agile manufacturing has at least the following four characteristics (Sheridan, 1993):

- Greater product customization—manufacturing to order but at a relatively low unit cost
- Rapid introduction of new or modified products
- Upgradable products—designed for disassembly, recyclability, and reconfigurability
- Dynamic reconfiguration of production processes—to accommodate swift changes in product designs or entire new product lines

Although the definition of agility has not been universally agreed on, some of its elements have been accepted throughout industry. The basic concept of agility is reflected in the description given in Brooke (1993): "In an 'agile' enterprise, products will be built quickly and cheaply for a customer based on detailed data received at the point of sale."

The above definition of agility has been accepted in this chapter. To be agile, products and manufacturing systems must be designed to be flexible and simple enough to allow easy reconfiguration and quick response to the changes in product designs. The agility concept has an impact on design of products and manufacturing systems.

The complexity of scheduling in manufacturing systems is an obstacle for achieving agility. Effective manufacturing scheduling reduces the manufacturing cost

448

and accelerates the response to the changing market. From a computational point of view, many manufacturing scheduling problems are NP hard. The time required to solve an NP-hard problem of reasonable size can be unrealistically long. For example, consider a problem that in the worst case requires 2^n microseconds to solve it, where n is the size of the problem. The problem of size of 10 will require no more than 0.001 second to solve it. A problem of size 40 could take as long as 12.7 days and of size 60, 366 centuries. In contrast to NP-hard problems, a problem of a polynomial time complexity can be solved within a reasonable amount of time.

Usually, manufacturing scheduling problems are solved with the constraints imposed by the existing designs of products and systems. Products and manufacturing systems are typically designed without considering the scheduling constraints.

Kusiak and He (1994) considered the schedulability issue in design of products and parts. They proposed five design rules aimed at improving schedulablilty of products and parts and studied the impact of each design rule on the quality of schedules in automated manufacturing.

The agility of manufacturing systems can be improved if enough consideration is given to concurrent design of products and systems with scheduling constraints. From the concurrent engineering point of view, a product should be designed for a manufacturing system, as much as the manufacturing system should be designed for the product to improve its performance and simplify the problem-solving process.

To date, no research on design of products and manufacturing systems for agility from a scheduling perspective has been reported in the literature. The success or failure of agile manufacturing is largely determined by the design of products and the system that manufactures them. This chapter attempts to establish design rules for agility using some results from scheduling theory. Four design agility rules are proposed in this chapter based on He and Kusiak (1996, 1997) and Kusiak and He (1997, 1998). The design rules are intended to support design of products and manufacturing systems to meet the requirements of agile manufacturing. Examples are provided to demonstrate the potential of using these rules. Procedures for implementing these rules are also discussed.

15.2 DESIGN RULES

15.2.1 Rule 1 (Modular Design): Decomposing a Complex System into Several Independent Units

The modular design rule is illustrated with a manufacturing scheduling problem modeled as a traveling salesman problem (TSP) [see model (15.1)–(15.6)]. Due to the large variety of products, different types of auxiliary equipment such as tools, grippers, feeders, and fixtures are required. The equipment needs to be set up when a changeover of a manufacturing system takes place. Frequently, the system changeover cost is sequence dependent and is a function of dissimilarity of product designs. It is desired that products be scheduled to minimize the overall changeover cost. It is known from scheduling theory (see, e.g., Baker, 1974, p. 94) that the single-machine (line) scheduling problem with sequence-dependent changeover cost is equivalent to the TSP:

$$\min \sum_{i=1}^{N} \sum_{j=1}^{N} c_{ij} x_{ij} \tag{15.1}$$

such that

$$\sum_{i=1}^{N} x_{ij} = 1 \qquad j = 1, \ldots, N \tag{15.2}$$

$$\sum_{j=1}^{N} x_{ij} = 1 \qquad i = 1, \ldots, N \tag{15.3}$$

$$u_i - u_j + N x_{ij} \leq N - 1 \quad i = 2, \ldots, N, \quad j = 2, \ldots, N; \, i \neq j \tag{15.4}$$

$$x_{ij} = 0, 1 \qquad i, j = 1, \ldots, N \tag{15.5}$$

$$u_i \geq 0 \qquad i = 1, \ldots, N \tag{15.6}$$

where

$N =$ number of products
$c_{ij} =$ system changeover cost when product i immediately precedes product j
$u_i =$ nonnegative variable

$$x_{ij} = \begin{cases} 1 & \text{if product } i \text{ immediately precedes product } j \\ 0 & \text{otherwise} \end{cases}$$

The objective function (15.1) minimizes the total changeover cost. Constraint (15.2) ensures that in a given schedule only one product immediately precedes product j. Constraint (15.3) imposes that product i is immediately followed by exactly one product. Constraint (15.4) is a subtour elimination constraint. Integrality is imposed by constraint (5). Constraint (15.6) ensures nonnegativity.

One way to simplify the manufacturing scheduling problem is to modify the scheduling constraints imposed by the existing system through design. The TSP is NP hard; however, it transforms into an assignment problem (AP) if the subtour elimination constraint (15.4) is removed. It is known that the AP can be solved in a polynomial time. Constraint (15.4) can be considered as a design constraint because it imposes that one system (line) is used for scheduling products. Removing constraint (15.4) means that the products can be separated into several groups based on the criterion of minimizing the total setup cost. Since the product groups are independent, the system can be decomposed into several independent units. The concept of Rule 1 is illustrated in Figure 15.1.

Therefore, the scheduling problem of minimizing the total changeover cost reduces to a problem with a polynomial time complexity, improving the responsiveness of the manufacturing system. In addition, at least three other benefits can be observed:

1. Better system balance due to fewer products to be produced in each system
2. Higher system availability—independence
3. Reduced total changeover cost

A procedure for implementing Rule 1 is proposed next:

Step 1. Solve the assignment problem.

Step 2. Decompose the system based on the product schedules obtained in Step 1.

Step 3. Perform trade-off analysis.

The implementation of Rule 1 is illustrated with two examples. First, consider a flexible manufacturing system that includes two milling machines, two lathes, a drilling machine, and two boring machines. Assume that four parts are produced in the system. The machine–part incidence matrix is as follows:

	P1	P2	P3	P4
Milling machine 1	1			1
Milling machine 2		1	1	
Lathe 1	1			1
Lathe 2		1	1	
Drilling machine		1	1	
Boring machine 1	1			
Boring machine 2		1	1	1

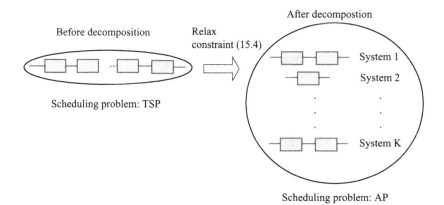

Figure 15.1. Concept of design rule 1.

Solving the AP, two production schedules {P1, P4} and {P3, P2} with two corresponding machining subsystems are formed:

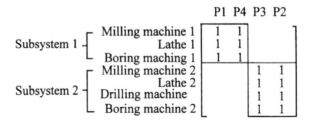

Note that to form two independent machining subsystems, the process plan of part P4 has been modified. The decision of whether the system should be decomposed should be made based on the trade-off analysis between the savings in total changeover cost and the cost due to the modification of the process plan of part P4.

The second example is concerned with chemical products. High cost and sequence-dependent changeovers characterize processing of chemical products. Another characteristic of the chemical process is that all products share the same process. There are two ways to produce M products. In Figure 15.2, all products are produced in a system with large reactors. The manufacturing scheduling problem in the system is the TSP.

Alternatively, the AP can be solved. Suppose that N schedules are generated. In this case, N subsystems with small reactors are used (see Figure 15.3). Since the production quantity is the same, the size of all reactors in Figure 15.2 is the total size of the reactors in Figure 15.3. For example, if $N = 3$, then, instead of purchasing a reactor with a capacity of 1200 kg, three reactors each of capacity 400 kg are purchased.

The trade-off analysis between the savings in total changeover cost and overhead cost increase due to the increase in the number of reactors leads to the selection of a proper system design.

15.2.2 Rule 2: Designing a Product with Robust Scheduling Characteristics

A product design is robust with respect to scheduling when the impact of any disruption on a production schedule due to changes in the product mix or production demand is low.

Concurrent design of products and manufacturing systems has an impact on efficient scheduling and hence the agility of a manufacturing system. Of key

Figure 15.2. Chemical process with large reactors.

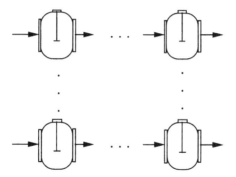

Figure 15.3. The *N* parallel subsystems with small reactors.

importance is that designs of products must be robust enough to support efficient control and scheduling of manufacturing systems. Assembly systems are developed to ensure their long-term flexibility. In many cases, manufacturers are driven to use more automation due to technological constraints or a competitive pressure. The manufacturing equipment is often expensive. Assembly systems have to support a wide range of products over a long period of time. Therefore, the new product designs are expected to be robust enough to allow the system to accommodate changes in product mix and production demand and be reconfigurable.

Next, product designs with linear assembly structures to be produced in a just-in-time (JIT) assembly line are used to illustrate the robust product design rule. A product has a linear assembly structure if the assembly operations are arranged in a linear order. Figure 15.4 presents an example of linear and nonlinear assembly structures.

A family of products with a linear assembly structure is a collection of products where each product has a linear assembly structure with identical basic operations and one or more optional operations. Figure 15.5 shows the example of a product family with basic operations o_1, o_2, and o_3 and optional operations o_4 and o_5.

Note that a product cannot have a downstream optional operation without having an optional operation that directly precedes it, for example, in Figure 15.5, product P1 cannot have operation o_5 without having operation o_4. This assumption is common in some industries; for example, in the automotive industry, cars of a product family

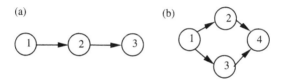

Figure 15.4. Assembly structure of products: (*a*) linear assembly structure; (*b*) nonlinear asssembly structure.

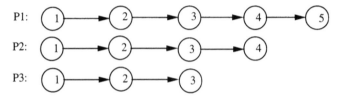

Figure 15.5. Example of a product family.

share the same basic body. In order to have a cruise control option the car must have an automatic transmission option installed.

The stacked product structure discussed in Andreasen et al. (1988) is a special case of the product design in Figure 15.5. In the stacked product structure most of the components are laid in stacks and are finally secured by the internal cohesion of the components (Andreasen et al., 1988). Examples of stacked products are an appliance plug and a suction pump.

To show the robust property of product designs with linear assembly structure, the design and scheduling approach for a JIT assembly line is developed next.

15.2.2.1 *Designing and Scheduling an Assembly Line.* Traditionally, design of assembly lines has been accomplished by line balancing methods. The fundamental assembly line balancing problem is to assign operations to an ordered sequence of stations so that the precedence relations are satisfied and some measures of performance are optimized (e.g., minimize the cycle time or minimize the number of stations; Ghosh and Gagnon, 1989). Most solution approaches developed for the assembly line balancing problem are heuristics due to the NP hardness of the problem. The assembly line scheduling problem is to determine the sequence of products that minimizes some performance measures (e.g., throughput time). Pinedo (1992) discussed several objectives for scheduling flexible assembly systems, including the minimization of completion time.

For a single product, scheduling is not necessary in a balanced assembly line. In a multiple-product assembly line, the line balancing and scheduling are of major concern. In the latter case, products are scheduled to achieve manufacturing efficiency. The assembly line balancing and scheduling problems have been treated as two separate problems (Yano and Rachamadugu, 1991; Dar-El and Cucuy, 1977; Dar-El and Cother, 1975; Thomopoulos, 1967), each difficult to solve. It can be shown that giving a special consideration to a product design, line design and scheduling can be simplified.

In this section, a JIT assembly line without buffers between stations is considered. The JIT assembly line is not balanced with respect to a cycle time. Rather, the products follow a no-delay schedule. In fact, for a single product, the production efficiency (measured by the total idle time) achieved by line balancing with a fixed cycle time is identical with the no-delay schedule of a JIT assembly line. For example, consider a

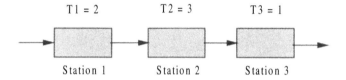

Figure 15.6. Layout of the balanced assembly line.

single product line. Suppose that the line is balanced with a cycle time 3, as shown in Figure 15.6.

In Figure 15.6, there are three stations and the corresponding processing times for each station are 2, 3, and 1 time units, respectively. Since each product spends 3 time units at each station, the total idle time for assembly of one unit product is 3 time units $[(3 - 2) + (3 - 3) + (3 - 1) = 3]$. Assume that instead of balancing the cycle time, the products are dispatched to the line in fixed time intervals and each product is sent to the next station immediately after processing is finished at the current station. This schedule is called a no-delay schedule. Figure 15.7 shows the no-delay schedule for the assembly line in Figure 15.6.

One can see from Figure 15.7 that the idle time between processing of any two units of the product is 3 time units. In the no-delay schedule in Figure 15.7, the idle time on station 1 can be interpreted as a dispatching time interval.

The previous example shows that scheduling products in a JIT assembly line is equivalent to finding a no-delay schedule that is an NP-hard problem (Van Der Veen and Van Dal, 1991). However, by considering the special structure of the assembly line, solving the no-delay scheduling problem can be simplified. An instance of the no-delay scheduling problem is represented by an $N \times M$ nonnegative matrix $T = [ts_{ik}]$, where ts_{ik} denotes the processing time of product i on station k. When processing times are semiordered, that is, $ts_{ik} \leq ts_{i+1,k}$ for $i = 1, \ldots, N - 1, k = 1, \ldots, M$, the mean completion time objective function can be minimized by the shortest processing time (SPT) rule (Van Der Veen and Van Dal, 1991).

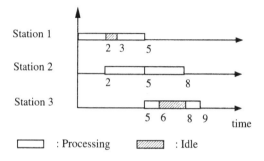

Figure 15.7. No-delay schedule.

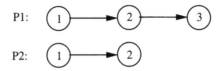

Figure 15.8. Assembly structure of two products.

For example, consider two products P1 and P2 with linear assembly structure shown in Figure 15.8.

The assembly line design for products P1 and P2 is shown in Figure 15.9.

The processing times are semiordered, that is, $ts_{ik} \geq ts_{i+1,k}$, $i = 1, 2$, $k = 1, 2$; therefore, the processing order is (P2, P1).

15.2.2.2 Algorithm for Designing and Scheduling Assembly Lines. In this section, an algorithm for designing and scheduling JIT assembly lines is developed. Assume that the line is designed and scheduled for a family of products with linear assembly structure shown in Figure 15.5. Before the design algorithm is presented, the following assumptions are made:

1. All stations are equally effective in performing the operations assigned.
2. No buffer capacity is allowed between stations.
3. The number of stations is set with a consideration given to the budget, floor space, and other operational limitations.
4. Products are assembled in batch sizes of 1.

For the purpose of balancing, a roughly equal load of operations need to be assigned to each station, where the load is the total processing time of the operations assigned to the station. A dynamic programming algorithm for minimizing the sum of the squares of the deviations of the station load from a target load is presented.

Notation

$i = $ product index

$j, e = $ operation index

$k, s = $ station index

$N = $ number of products

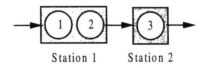

Figure 15.9. Assembly line for products P1 and P2.

$M =$ number of stations

$J =$ number of operations

$t_j =$ processing time of operation j

$I =$ set of integers

$V_i = \{v_{i1}, v_{i2}, \ldots, v_{ij}, \ldots, v_{iJ}\} =$ operation vector of product i

where

$$v_{ij} = \begin{cases} 1 & \text{if operation } j \text{ is required by product } i \\ 0 & \text{otherwise} \end{cases}$$

Let $t = (\sum_{j=1}^{J} t_j)/M$ be the target station load. The dynamic programming algorithm is presented next (Johri, 1992).

Dynamic Programming Algorithm 15.1. Let $z_0 = 0$ and $z_M = J$. The problem of assigning operations to M stations can be thought of as finding a strictly increasing sequence of $M - 1$ integers $z_1, z_2, \ldots, z_{M-1}$ between 1 and J such that station k consists of operations $z_{k-1} + 1$ to z_k, $k = 1, 2, \ldots, M$, or more formally,

$$Z = \{z_1, z_2, \ldots, z_{M-1} \mid 0 < z_1 < z_2 < \cdots < z_{M-1} < J; z_k \in I\}.$$

Let $C(k, j)$ denote the minimal sum of the squares of the deviations from assigning the first j operations to k stations. Then, $C(M, J)$ is the minimal sum of squares of the deviations in the final solution to the problem. Problem $C(k, j)$ can be solved recursively for $k = 1, 2, \ldots, n$ as follows:

$$C(k, j) = \min \left[\left(t - \sum_{e=z_{k-1}+1} t_e \right)^2 + C(k - 1, z_{k-1}) \right] \quad z_{k-1} = k - 1, k, \ldots, j - 1 \quad (15.7)$$

Note that for $j = k, k + 1, \ldots, J - (M - k)$, the following boundary conditions exist:

$$C(1, j) = \left(t - \sum_{e=1}^{j} t_e \right)^2 \quad j = 1, 2, \ldots, J - M + 1 \quad (15.8)$$

Let S be the sequence of products in the decreasing order of $|V_i|$, $S = \{1, 2, \ldots, N: |V_1| \geq |V_2| \geq \cdots \geq |V_N|\}$. Given the linear assembly structure of products, it is not difficult to see that S always satisfies the semiordered property, irrespective of the structure of the assembly line. Hence, S is the optimal schedule that minimizes the mean completion time.

The following example is used to illustrate the assembly line design and scheduling algorithm.

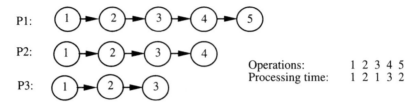

P1:   ① → ② → ③ → ④ → ⑤

P2:   ① → ② → ③ → ④

| Operations: | 1 2 3 4 5 |
| Processing time: | 1 2 1 3 2 |

P3:   ① → ② → ③

Figure 15.10. Structure of products and processing time data for three products.

Example 15.1. Design an assembly line and determine a production schedule for the products shown in Figure 15.10. Assume that the required number of stations $M = 2$. From Figure 15.10, $N = 3$ and $J = 5$.

The target load $t = (\Sigma_{j=1}^{5} t_j)/M = (1 + 2 + 1 + 3 + 2)/2 = \frac{9}{2} = 4.5$. For $k = 1$,

$$C(1, 1) = (t - t_1)^2 = (4.5 - 1)^2 = (3.5)^2 = 12.25$$

$$C(1, 2) = (t - t_1 - t_2)^2 = (4.5 - 1 - 2)^2 = (1.5)^2 = 2.25$$

$$C(1, 3) = (t - t_1 - t_2 - t_3)^2 = (4.5 - 1 - 2 - 1)^2 = (0.5)^2 = 0.25$$

$$C(1, 4) = (t - t_1 - t_2 - t_3 - t_4)^2 = (4.5 - 1 - 2 - 1 - 3)^2 = (-2.5)^2 = 6.25$$

For $k = 2$,

$$C(2, 5) = \min\{(t - t_2 - t_3 - t_4 - t_5)^2 + C(1, 1), (t - t_3 - t_4 - t_5)^2 + C(1, 2),$$

$$(t - t_4 - t_5)^2 + C(1, 3), (t - t_5)^2 + C(1, 4)\}$$

$$= \min\{(4.5 - 2 - 1 - 3 - 2)^2 + 12.5, (4.5 - 1 - 3 - 2)^2 + 2.25,$$

$$(4.5 - 3 - 2)^2 + 0.25, (4.5 - 2)^2 + 6.25\}$$

$$= \min\{24.5, 4.5, 0.5, 12.5\}$$

$$= 0.5$$

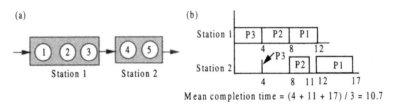

(a)

→ ① ② ③ → ④ ⑤ →

Station 1   Station 2

(b)

| Station 1 | P3 | P2 | P1 |

4   8   12

P3

| Station 2 | | P2 | | P1 |

4   8   11 12   17

Mean completion time = $(4 + 11 + 17) / 3 = 10.7$

Figure 15.11. Optimal line design and schedule for the data in Example 15.1: (a) assembly line; (b) schedule.

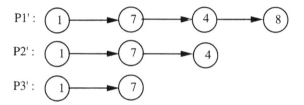

Figure 15.12. New designs of products.

Therefore, the optimal assignments of operations are for station 1, o_1, o_2, o_3, and station 2, o_4, o_5. The optimal schedule of products $S = \{P3, P2, P1\}$. The optimal layout of the assembly line and the optimal schedule of products are shown in Figure 15.11.

15.2.2.3 Robust Characteristics

1. Easy Scheduling and Rescheduling. The linear structure of products simplifies scheduling irrespective of the changes in the product mix or demand. For example, suppose that the production of product 2 is canceled. Since the set {P3, P2, P1} is semiordered, the minimal mean completion time schedule is {P3, P1}. The schedule generated accommodates easily change in the production demand. Since the sequential relationship among products is fixed, that is, it has the semiordered processing time property, the units of each product need not to be equally divided among all the cyclic schedules. For example, assume that 50 units of product P1, 20 units of product P2, and 10 units of product P3 are produced. Before the 10th unit of product 3 is produced, the schedule is {P3, P2, P1}. After the 10th unit of product 3 is produced, the schedule is {P2, P1}.

2. Easy Reconfiguration. The linear assembly structure of products improves the system reconfigurability. For example, consider the product designs with the linear assembly structure in Figure 15.10 and the corresponding assembly system in Figure 15.11a. Assume that the design of products is modified as shown in Figure 15.12. Assume that the operations of new product are assigned to the existing stations, as shown in Figure 15.13.

Irrespective of the assignment of operations to the stations, the semiordered processing time property is always satisfied for the linear assembly structure. Therefore, the optimal schedule can be easily determined as {P3′, P2′, P1′}.

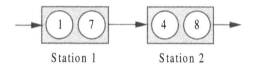

Figure 15.13. Assignment of operations to two stations.

15.2.3 Rule 3: Streamlining the Flow of Products in an Assembly Line

15.2.3.1 Scheduling a Streamlined Assembly Line. The design of an assembly line determines the traffic and direction of the product flow. The complexity of scheduling, sequencing, and controlling of the traffic increases with the number of routings.

Consider the example of three products shown in Figure 15.14. Suppose that the assembly line is balanced for the three products (see Figure 15.15). The routings for each product are as follows:

Product 1: route 1 = (station 1, station 2, station 3)

Product 2: route 1 = (station 1, station 2, station 3)

route 2 = (station 1, station 2, station 1, station 3)

Product 3: route 1 = (station 1, station 2, station 3)

route 2 = (station 2, station 1, station 3)

Since the routings for each product are not identical and products may flow in different directions, the line in Figure 15.15 can be viewed as a job shop. In comparison to a flow shop, the flow of products in a job shop is more complex. The complexity of scheduling the job shop can be reduced if the flow of products can be streamlined. The redesigned assembly line is shown in Figure 15.16.

The flow of products in the assembly system in Figure 15.16 is simpler than that in the assembly system in Figure 15.15. The streamlined assembly line design can be implemented by the procedures presented next.

15.2.3.2 Designing a Streamlined Assembly Line. In traditional assembly systems, dedicated assembly equipment was used for mass production, and line balancing methods were important in maximizing equipment utilization and reducing production cost. Modern assembly systems employ reprogrammable hardware such as robots, automatic guided vehicles, reconfigurable fixtures, and so on, and a agile delivery of products is the guiding design principle. The traditional line balancing

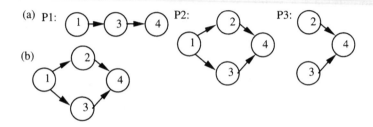

Figure 15.14. Product assembly structures: (*a*) three assembly structures; (*b*) superimposed product structure.

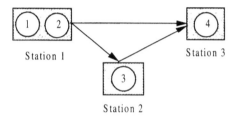

Figure 15.15. Nonstreamlined assembly line.

methods are no longer suitable for design of agile assembly systems due to the following reasons:

1. High computational complexity.
2. Perfect balancing cannot be achieved due to the variation of products.
3. Setup is ignored due to assembly of different products.

Another limitation of traditional assembly line balancing methods is that parallel stations are not considered. In practice, parallel stations are used to improve the line balancing efficiency and reliability of the system.

Setup reduction has not been given adequate consideration in the design of assembly system. This is because the assembly line has been traditionally designed for a few products in large volume.

In an assembly system, a robot might require special tools, that is, grippers. Some operations may share the same tool. Thus, if these operations are assigned to the same station, only one tool need be used; however, if these operations are assigned to different stations, one tool must be used for each station. Furthermore, if successive operations that require the same tool are assigned to the same station, tool changes between these operations can be avoided. Otherwise, the station will always incur a tool change between successive operations for each product. The setup time is significant and must be considered in the design of an assembly line.

A two-step procedure for the design of a streamlined assembly line is presented next. The first step is to convert the nonlinear superimposed assembly structure into a linear structure using a forward procedure. The second step is to assign operations to stations to minimize the total dissimilarity cost and station cost using a dynamic programming algorithm.

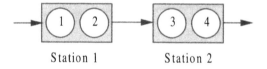

Figure 15.16. Streamlined assembly line.

The following terminology is used throughout this section. An *assembly line* includes a number of serially arranged *production stages*, each stage consisting of one station or a number of stations in parallel. At each *station* a *robot* performs a portion of the total assembly work on the product by adding components to the existing subassembly (the terms station and robot are used interchangeably).

Define:

$G =$ precedence graph of superimposed assembly structure
$S =$ sequence of operations
$O =$ set of operations that have no preceding operations in G
$e =$ last operation in S
$d_{ij} =$ dissimilarity cost between operations i and j

The computation of dissimilarity cost d_{ij} follows the definition in Kusiak (1990). The requirement for auxiliary equipment by each product i is represented by the column vector

$$E_i = [e_{1i}, e_{2i}, \ldots, e_{qi}, \ldots]^T$$

where

$$e_{qi} = \begin{cases} 1 & \text{if auxiliary equipment } q \text{ is required by product } P_i \\ 0 & \text{otherwise} \end{cases}$$

The dissimilarity cost between any two distinct products P_i and P_j can be expressed as

$$d_{ij} = \sum_{q=1}^{B} w_q \delta(e_{qi}, e_{qj})$$

where

$$\delta(e_{qi}, e_{qj}) = \begin{cases} 1 & \text{if } e_{qi} \neq e_{qj} \\ 0 & \text{otherwise} \end{cases}$$

$w_q =$ weight of auxiliary equipment q
$B =$ total number of products

The value of weight w_q can be assigned to an auxiliary equipment q depending on its importance in the assembly process. A way to determine the value of weight w_q for all q is to set it proportional to the estimated setup time caused by the auxiliary equipment q.

Forward Procedure

   Step 1. Initialize O and select e arbitrary from O.

Step 2. Compute $j^* = \arg \min\{d_{ej}: j \in O\}$ and set $e = j^*$. Update O.

Step 3. If $O = \varnothing$, stop; else go to step 2.

The forward procedure is illustrated with the following example.

Example 15.2.   The assembly structure of three products and the superimposed assembly structure are shown in Figure 15.17. The equipment–operation incidence matrix and dissimilarity cost matrix are shown in (15.9) and (15.10), respectively:

$$
e_{qi} = \begin{array}{c} \\ e_1 \\ e_2 \\ e_3 \\ e_4 \\ e_5 \\ e_6 \end{array}
\begin{array}{cccc} 1 & 2 & 3 & 4 \\ \left[\begin{array}{cccc} 1 & & 1 & \\ & 1 & 1 & 1 \\ 1 & 1 & & 1 \\ 1 & & 1 & \\ & 1 & & 1 \\ 1 & & 1 & \end{array}\right] \end{array}
\tag{15.9}
$$

$$
d_{ij} = \begin{array}{c} \\ 1 \\ 2 \\ 3 \\ 4 \end{array}
\begin{array}{cccc} 1 & 2 & 3 & 4 \\ \left[\begin{array}{cccc} 0 & 5 & 2 & 5 \\ & 0 & 5 & 0 \\ & & 0 & 5 \\ & & & 0 \end{array}\right] \end{array}
\tag{15.10}
$$

Step 1. Initialize $O = \{1\}$; select $e = 1$ and $S = \{1\}$.

Step 2. Compute $j^* = 1$; set $e = 1$, $S = \{1\}$. Update $O = \{2, 3\}$.

Step 3. Since $O \neq \varnothing$, go to step 2.

Step 2. Compute $j^* = \min\{d_{12}, d_{13}\} = \min\{5, 2\} = 3$; set $e = 3$, $S = \{1, 3\}$. Update $O = \{2\}$.

Step 3. Since $O \neq \varnothing$, go to step 2.

Step 2. Compute $j^* = \min\{d_{32}\} = 2$; set $e = 2$, $S = \{1, 3, 2\}$. Update $O = \{4\}$.

Step 3. Since $O \neq \varnothing$, go to step 2.

(a)   (b)

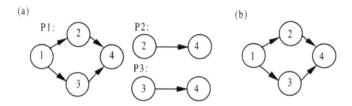

Figure 15.17. Assembly structures: (a) individual products; (b) superimposed assembly structure.

Step 2. Compute $j^* = \min\{d_{24}\} = 4$; set $e = 4$, $S = \{1, 3, 2, 4\}$. Update $O = \varnothing$.
Step 3. Since $O = \varnothing$, stop.

The nonlinear superimposed assembly structure in Figure 15.17b is then converted into the linear assembly structure in Figure 15.18. A dynamic programming algorithm is developed to assign operations to stages so that the total of the total dissimilarity cost and station cost is minimized.

Notation

$N =$ number of products
$i =$ product index
$j, v, w =$ operation indices
$k =$ stage index
$D_i =$ production demand of product i, number of products/day
$c =$ processing capacity of a station (robot), time unit/day
$m =$ number of stages required
$t_j =$ processing time of operation j

$$t_{ij} = \begin{cases} t_j & \text{if operation } j \text{ is required by product } i \\ 0 & \text{otherwise} \end{cases}$$

$t(v, w) = \sum_{i=1}^{N} \sum_{j=v}^{w} D_i t_{ij}$, the processing requirement for operations $v, v + 1, \ldots, w$
$n(v, w) = \cup \, t(v, w)/c'$, the number of stations required for operations $v, v + 1, \ldots, w$

Dynamic Programming Algorithm 15.2.   Let $C(k, j)$ be the minimal cost of partitioning the first j operations into k stages ($k \leq j$). Let $C(n, m)$ be the minimum cost solution of the problem. Let $D(G, s) = \sum_{i \in G} \sum_{j > i: j \in G} d_{ij}$ be the dissimilarity cost if the set of operations G are assigned to the stage s. Then $C(k, j)$ can be computed for $k = 2, 3, \ldots, n$ by the following recursive procedure $C(k, j) = \min\{C(k - 1, z_{i-1}) + D(z_{i-1} + 1: j, k) + n(z_{i-1} + 1, j)\}$ for $j = k, k + 1, \ldots, m - (n - k)$ with the boundary conditions $C(1, j) = D(1: j, 1) + n(1, j), j = 1, 2, \ldots, m - n + 1$.

Example 15.3.   Consider the assembly structure in Figure 15.18. For the convenience of demonstration, index the operations in the linear superimposed assembly structure as follows: $1 \rightarrow 2 \rightarrow 3 \rightarrow 4$.
The processing time matrix and cost matrix are (15.11) and (15.12), respectively:

Figure 15.18. Linear superimposed assembly structure.

$$t(v, w) = \begin{matrix} 1 \\ 2 \\ 3 \\ 4 \end{matrix} \begin{matrix} 1 & 2 & 3 & 4 \\ \begin{bmatrix} 100 & 190 & 340 & 480 \\ & 90 & 240 & 380 \\ & & 150 & 290 \\ & & & 140 \end{bmatrix} \end{matrix} \quad (15.11)$$

$$n(v, w) = \begin{matrix} 1 \\ 2 \\ 3 \\ 4 \end{matrix} \begin{matrix} \begin{bmatrix} 3 & 5 & 9 & 12 \\ & 3 & 6 & 10 \\ & & 4 & 8 \\ & & & 4 \end{bmatrix} \end{matrix} \quad (15.12)$$

The remaining data required are as follows:

$$D_1 = 30 \qquad D_2 = 20 \qquad D_3 = 40$$

$$c = 40 \qquad m = 3$$

$$t_1 = 2 \qquad t_2 = 1 \qquad t_3 = 3 \qquad t_4 = 2$$

For $k = 1$,

$$C(1, 1) = D(1: 1, 1) + n(1, 1) = 0 + 3 = 3$$

$$C(1, 2) = D(1: 2, 1) + n(1, 2) = 5 + 5 = 10$$

For $k = 2$,

$$C(2, 2) = C(1, 1) + D(2: 2, 2) + n(2, 2) = 3 + 0 + 3 = 6$$

$$C(2, 3) = \min\{[C(1, 1) + D(2: 3, 2) + n(2, 3)], [C(1, 2) + D(3: 3, 2) + n(3, 3)]\}$$

$$= \min\{[3 + 5 + 6], [10 + 0 + 4]\}$$

$$= \min\{14, 14\}$$

$$= 14$$

For $k = 3$,

$$C(3, 4) = \min\{[C(2, 2) + D(3: 4, 3) + n(3, 4)], [C(2, 3) + D(4: 4, 3) + n(4, 4)]\}$$

$$= \min\{[6 + 5 + 8], [14 + 0 + 4]\}$$

$$= \min\{19, 18\}$$

$$= 18$$

By backtracking, the following solution is found:

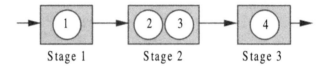

Figure 15.19.   Layout of the assembly line.

Stage 1: operation 1, 3 robots

Stage 2: operations 2 and 3, 6 robots

Stage 3: operation 4, 4 robots

The structure of the assembly line is shown in Figure 15.19.

15.2.4   Rule 4: Reduce the Number of Stations in an Assembly Line

It is known from flow shop scheduling theory that when the number of machines increases to 3, the scheduling problem becomes NP hard (Garey and Johnson, 1979). Instead of designing long assembly lines, one should attempt to build shorter ones. By reducing the length of a line, the complexity of the scheduling problem can be reduced. In assembly systems, different operations can be assigned to one station and multiple robots can be used. Figure 15.20 shows two different designs of an assembly system.

The configuration of the assembly system impacts on the complexity of the scheduling problem. Certain configurations of assembly systems have a potential for being undecidable, in the sense that no algorithm exist to compute their behavior. In addition to high computational complexity, long assembly lines are undesirable in terms of the line balancing efficiency and are more likely to have behavioral problems (Chase, 1975). Buxey et al. (1973) pointed out that other things being equal, line efficiency is greater for shorter lines and the number of stations should be minimized whenever possible.

By shortening the line, the scheduling problem might be reduced to a simpler one-, two-, or three-machine flow shop problem. For a one-station line, the scheduling problem becomes a single-machine scheduling problem and can be easily solved, for example, using the shortest processing time rule. For a two-station line, the scheduling problem is equivalent to a two-machine flow shop problem and can be solved, for example, using Johnson's algorithm. For a three-station line, the problem is to assign operations to the stations so that Johnson's algorithm can be extended. Design procedures for one-, two-, or three-station lines are developed next.

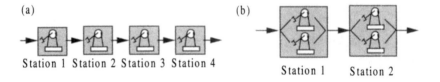

Figure 15.20.   Two different designs of an assemblu system: (*a*) long line; (*b*) short line.

15.2.4.1 *Development of Design Approaches for Short Lines*

Notation

$N =$ number of products

$M =$ number of stations

$J =$ number of operations

$i =$ product index

$j =$ operation index

$k =$ station index

$t_j =$ processing time of operation j

$$t_{ij} = \begin{cases} t_j & \text{if operation } j \text{ is required by product } i \\ 0 & \text{otherwise} \end{cases}$$

$ts_{ik} =$ processing time of product i at station k

1. One-Station Line Design. In an one-station assembly line, all assembly operations are performed on one station. Therefore, the scheduling problem becomes a single-machine scheduling problem and can be easily solved, for example, using the shortest processing time rule.

2. Two-Station Line Design. The scheduling problem that minimizes the makespan for a two-machine flow shop can be solved optimally by Johnson's algorithm (Johnson, 1954). In this section, the design procedure based on Johnson's algorithm for a two-station assembly line is developed.

Consider a product family with a linear assembly structure. Let P be the set of feasible partitions of operations among the two stations. A feasible partition $p = \{p(\text{I}), p(\text{II})\}$, where $p \in P$, and $p(\text{I})$ and $p(\text{II})$ are sets of operations assigned to station 1 and station 2, respectively. A partition is feasible if each station contains at least one operation and precedence relations among operations are not violated. The objective of the design procedure is to find a $p^* = \{p^*(\text{I}), p^*(\text{II})\}$ so that the makespan is minimal. Then the size of P is $J - 1$. Therefore, a two-station line procedure can be implemented as an exhaustive search method.

Example 15.4. Consider the design of a two-station line for the family of three products in Figure 15.21.

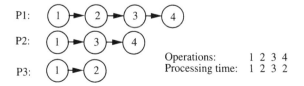

Figure 15.21. Assembly structure and the processing times.

Since $J = 4$, there are three possible designs (see Figure 15.22). For each line design, the makespan is computed. The optimal design with the minimum makespan of 13 time units is shown in Figure 15.22b.

3. Three-Station Line Design. Johnson's algorithm for the two-machine flow shop problem can be extended to a special case of the three-machine flow shop problem (French, 1982). If:

Either $\quad \min\{ts_{i1} \mid i = 1, \ldots, N\} \geq \max\{ts_{i2} \mid i = 1, \ldots, N\}$

$$(15.13)$$

Or $\quad \min\{ts_{i3} \mid i = 1, \ldots, N\} \geq \max\{ts_{i2} \mid i = 1, \ldots, N\}$

that is, the maximum processing time on the second machine is no greater than the minimum processing time on either the first or the third one, an optimal schedule can be found by letting

$$a_i = ts_{i1} + ts_{i2} \qquad b_i = ts_{i2} + ts_{i3}$$

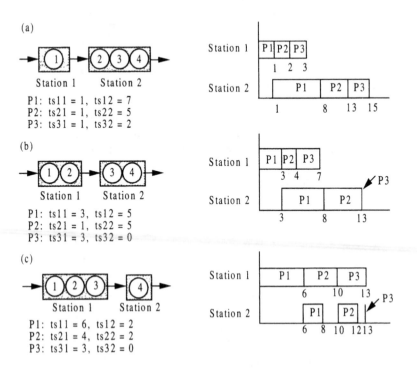

Figure 15.22. Line designs for Example 15.4: (a) line design 1; (b) line design 2; (c) line design 3.

and scheduling the products as if they were to be processed on two machines only but with the processing time of each product being a_i and b_i on the first and second machines, respectively.

If one views each station of a serial assembly line as a machine, then a three-station assembly line can be designed so that the processing times of the products satisfy condition (15.13). In this case, the scheduling problem can be easily solved.

Suppose that the condition $\min\{ts_{i3}: i = 1, \ldots, N\} \geq \max\{ts_{i2}: i = 1, \ldots, N\}$ is used to design the three-station assembly line. The following heuristic procedure is developed for products with linear assembly structures.

Three-Station Line Design Procedure

Step 0.   *Initialization.* Let I be the set of products to be produced. Set

$$a = i^* = \arg \max \left\{ \sum_{j=1}^{J} t_{ij}: i = 1, \ldots, N \right\}$$

Let $t_{\max}$ be the maximum processing time on station 2 and $t_{\min}$ be the minimum processing time on station 3. Set $t_{\max} = 0$ and $t_{\min} = 0$.

Step 1.   *Assignment of operations of product a.* Compute $t_{ave} = (\sum_{j=1}^{J} t_{aj})/3$, where t_{ave} is the average processing time of product a on a station. Let ts_{ik} be the processing time of product i at station k. Assign sequentially operations of product a to stations 1 and 2 such that ts_{a1} and $ts_{a2} \leq t_{ave}$. Assign the remaining operations to station 3. Let Q_k, $k = 1, 2, 3$, be the set of operations assigned to station k. Set $t_{\max} = ts_{a2}$ and $t_{\min} = ts_{a3}$.

Step 2.   For $i \in I \setminus \{a\}$, do:

If $\sum_{j \in Q_2} t_{ij} > t_{\max}$, then set $t_{\max} = \sum_{j \in Q_2} t_{ij}$.
If $\sum_{j \in Q_3} t_{ij} < t_{\min}$, then set $t_{\min} = \sum_{j \in Q_3} t_{ij}$.
If $t_{\min} < t_{\max}$, then remove operations in Q_2 from left to right to Q_3
until $t_{\min} \geq t_{\max}$.

Step 3.   Compute $a_i = ts_{i1} + ts_{i2}$, $b_i = ts_{i2} + ts_{i3}$.

Example 15.5.   Consider the design of a three-station line for the products shown in Figure 15.23.

Step 0.   *Initialization.* Set

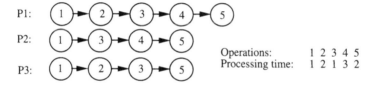

P1:   (1) ► (2) ► (3) ► (4) ► (5)

P2:   (1) ► (3) ► (4) ► (5)

P3:   (1) ► (2) ► (3) ► (5)

Operations: 1 2 3 4 5
Processing time: 1 2 1 3 2

Figure 15.23. Assembly structure of the products and their processing times.

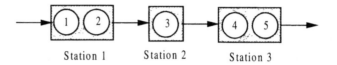

Figure 15.24. Three-station line design.

$$I = \{1, 2, 3\} \qquad a = i^* = \max\{9, 7, 6\} = 1 \qquad t_{\max} = t_{\min} = 0$$

Step 1. $t_{\text{ave}} = \frac{9}{3} = 3$. Assign operations of product 1 to stations 1, 2, and 3:

$$Q_1 = \{1, 2\} \quad ts_{11} = 3$$

$$Q_2 = \{3\} \qquad ts_{12} = 1$$

$$Q_3 = \{4, 5\} \quad ts_{13} = 5$$

Set $t_{\max} = ts_{12} = 1$, $t_{\min} = ts_{13} = 5$.
Step 2.   For $i = 2$ (product 2),

$$ts_{21} = 1 \qquad ts_{22} = 1 \qquad ts_{23} = 5$$

For $i = 3$ (product 3),

$$ts_{31} = 3 \qquad ts_{32} = 1 \qquad t_{\min} = ts_{33} = 2$$

since $ts_{33} < t_{\min}$. Since $t_{\min} > t_{\max}$, the three-station line design is completed.
Step 3. $a_1 = 3 + 1 = 4$, $b_1 = 1 + 5 = 6$
$\qquad a_2 = 1 + 1 = 2$, $b_2 = 1 + 5 = 6$
$\qquad a_3 = 3 + 1$, $b_3 = 1 + 2 = 3$

The three-station line design is shown in Figure 15.24.

15.3   PRODUCT DIFFERENTIATION

15.3.1   Delayed Product Differentiation

Delayed product differentiation is a design concept aimed at increasing product variety and manufacturing efficiency. It is based on delaying the time when a product assumes its identify, that is, a particular product model at a particular stage of a particular manufacturing process. Although the general concepts of delayed product differentiation have been published in the literature (Matherm, 1987; Whitney, 1993; Lee, 1994), its implementation has not been discussed. The concept of delayed

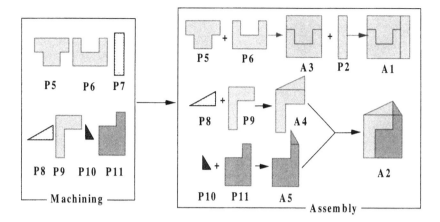

Figure 15.25. Delayed product differentiation.

product differentiation was discussed in Lee (1994) as a valuable approach to improving performance of a supply chain.

Normally, a manufacturing process involves multiple stages, each requiring different parts or subassemblies. Increasing the level of part commonality at an early stage of manufacturing process may delay the differentiation of products. Commonality here is defined as the use of a component by several different products. When used properly, part commonality may decrease the inventory cost, manufacturing cost, material handling cost, and so on. Otherwise, it may adversely impact the performance of a manufacturing system. In this chapter, an implementation approach of the product differentiation concept is developed.

The delayed product differentiation concept can be implemented by accomplishing the product differentiation at the assembly stage (see Figure 15.25).

In delayed product differentiation, common and simple parts are machined and then delivered to the assembly system to form product variants. The delayed product differentiation concept is cited as an assembly-driven strategy in Whitney (1993). Some design strategies (e.g., modular product design) allow for delayed product differentiation by a number common parts serving numerous product models.

15.3.2 Early Product Differentiation

The opposite of delayed product differentiation is early product differentiation (see Figure 15.26). In early product differentiation, unique parts are often used (He and Kusiak, 1996).

Designing parts according to the delayed product differentiation concept is referred to as differential design and the design of parts related to early product differentiation is referred to as integral design.

Although the number of parts in differential design is larger than that in integral design, the total number of different parts can be reduced if common parts are shared by differential designs. Differential design implies breakdown of a unique part into

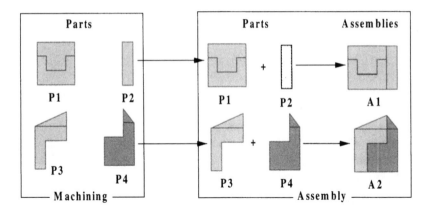

Figure 15.26. Early product differentiation.

several common parts. Table 15.1 summarizes the advantages and disadvantages of differential designs (Redford and Chal, 1994).

Most products are designed by following differential and integral design concepts. While from the viewpoint of assembly differential product structures are preferred, good judgment is needed to ensure that the requirements for assembly do not override valid requirements of other aspects of manufacture.

Redford and Chal (1994) provided qualitative guidelines for the rationalization of product structures. Their guidelines consider one product design at a time. No collective impact of multiple-product designs is considered. In this chapter, we attempt to quantify the impact of the delayed product differentiation design on the performance of manufacturing systems and provide a quantitative basis for the rationalization of product structures.

A good example of combining the two strategies in retail industry is the production of T-shirts. Some producers used the concept of late differentiation by limiting the number of sizes of T-shirts to a few (size standardization). The early differentiation concept was followed in supplying neutrally colored T-shirts close to the points of sale and imprinting there the final colors as demanded by sales. The late differentiation principle reduced the level of in-process inventory per T-shirt size, while early differentiation reduced the market response time.

TABLE 15.1. Advantages and Disadvantages of Differential Designs

Advantages
    Use of favorably priced semifinished materials and standard parts
    Simpler subassemblies and parts
    Reduced time and cost of maintaining the products
Disadvantages
    Need for tighter quality control
    More interfaces between parts
    Higher potential for decreased reliability

15.3.3   Manufacturing Performance and the Design of Products

One difficulty in implementing the delayed product differentiation strategy is that the management and design teams could be reluctant to proceed with a new design without the evidence of benefits in terms of improved manufacturing efficiency, faster response time to the market, reduced manufacturing cycle time, and so on. Thus, scheduling methodologies can contribute here.

Design for manufacturing requires design engineers to take a broader perspective than the product functionality and performance. It also requires generalization of the definition of cost used for evaluation of alternative designs, which often includes only the material cost of a product and direct labor in assembly.

Youssef (1992) pointed out that timeliness in creating goods and services is essential in the competitive environment. Stephen and Tatikonda (1992) showed that product introduction time affects its competitiveness. Therefore, the time required to manufacture products should be incorporated into the evaluation of alternative designs. The impact of design decisions on manufacturing performance should be considered at the early product design stage.

Andreasen et al. (1988) showed that from an assembly point of view the optimal design of a part could only be achieved by considering various design alternatives, thus providing some degree of design freedom. Considering various form divisions can create the design alternatives.

In this chapter, it is assumed that a fixed number of standard parts are available. Designers intend to replace the unique (integral) designs of parts with differential designs that contain some standard parts. They face the decision of selecting appropriate differential designs that improve manufacturing performance. For example, consider the integral design of part P in Figure 15.27.

The design in Figure 15.27 is replaced with the design in Figure 15.28.

The design in Figure 15.28 contains three parts. If these parts are standard and available in the system or can be delivered from suppliers, then the design in Figure 15.28 can reduce the design and manufacturing time. However, if the assembly time required by the design in Figure 15.27 is relatively long, then using the design in Figure 15.28 may not be appropriate. Given the machining and assembly time, designers would like to know which design to select. Here, the trade-off was

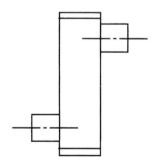

Figure 15.27. Integral design of part P.

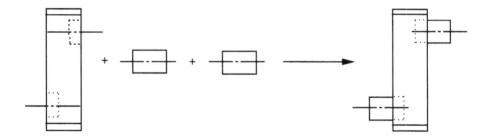

Figure 15.28. Differential design of part P in Figure 15.27.

considered for a single design. If the number of designs involved in the decision process increases, the problem becomes complex.

In addition to the manufacturing cycle time, the efficiency of the assembly process is also affected by differential designs. Differential designs impact the efficiency of assembly process by affecting the number of stations in the assembly system and balancing the assembly system. A differential design may increase the number of assembly operations to be performed and the assembly time and hence requires additional stations in the system or affects balancing the system. The degree of this impact depends on the structure of the assembly system.

15.4 SUMMARY

Agility is the ability of a company to produce a variety of products quickly and at a low cost. This demands that products and manufacturing systems be simple, robust, and flexible enough to quickly respond to the changing market requirements.

The complexity of manufacturing scheduling is an obstacle to achieving agility. This chapter has generated rules that would allow to design products and systems with agility from a scheduling perspective. The aim is to design products beyond the manufacturing process, as illustrated in Figure 15.29. Many companies have been quite successful in integrating design of products with manufacturing process design. However, when it comes to broader system concerns, the "design over the wall" paradigm needs to be addressed. The agility aspect is one of the system factors that ought to be considered in design.

Four design agility rules were proposed in this chapter. The first rule deals with decomposition of a manufacturing system. This rule simplifies the scheduling problem and reduces the total changeover cost. The second rule is concerned with design of products with robust scheduling characteristics. Product designs with robust scheduling characteristics can improve the responsiveness of a manufacturing system to the changes in the product demand and product mix and reconfigurability of the system. The third rule results in a streamlined assembly line that has the type of product flow that simplifies scheduling. The fourth rule emphasizes the reduction of

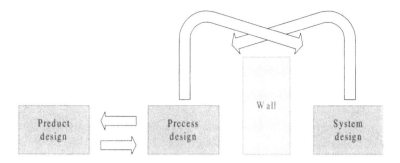

Figure 15.29. Design for manufacturing environment.

the number of stations in an assembly line. Examples were provided to demonstrate the potential from applying these rules. Procedures for implementing these rules were discussed. In the final part of the chapter two product differentiation strategies were discussed. The delayed product differentiation relies on making the product unique at the late manufacturing stage. It is often accomplished by using standard components for building products. The early differentiation strategy aims at making the product distinct early in the production process.

REFERENCES

Andreasen, M. M., S. Kahler, and T. Lund (1988), *Design for Assembly*, IFS Publications, Bredford, United Kingdom.

Baker, K. R. (1974), *Introduction to Sequencing and Scheduling*, John Wiley, New York.

Brooke, L. (1993), "Is agile the answer?" *Clinton's Automotive Industries*, Vol. 173, No. 8, pp. 26–28.

Buxey, G. M., N. D. Slack, and R. Wild (1973), Production flow line system design—A review, *AIIE Transactions*, Vol. 5, No. 1, pp. 37–48.

Chase, R. B. (1975), Strategic considerations in assembly line selection, *California Management Review*, Vol. 18, No. 1, pp. 17–23.

Dar-El, E. M., and R. F. Cother (1975), Assembly line sequencing for model-mix, *International Journal of Production Research*, Vol. 13, No. 5, pp. 463–477.

Dar-El, E. M., and S. Cucuy (1977), Assembly line sequencing for balanced assembly lines, *OMEGA*, Vol. 5, No. 3, pp. 333–342.

French, S. (1982), *Sequencing and Scheduling: An Introduction to the Mathematics of the Job Shop*, John Wiley, New York.

Garey, M. R., and D. S. Johnson (1979),*Computers and Intractability*, W. H. Freeman, New York.

Ghosh, S., and R. J. Gagnon (1989), A comprehensive literature review and analysis of the design, balancing and scheduling of assembly systems, *International Journal of Production Research*, Vol. 27, No. 4, pp. 637–670.

He, D., and A. Kusiak (1996), Performance analysis of modular products, *International Journal of Production Research*, Vol. 34, No. 1, pp. 253–272.

He, D., and A. Kusiak (1997), Design of assembly systems for modular products, *IEEE Transactions on Robotics and Automation*, Vol. 13, No. 5, 1997, pp. 646–655.

He, D., and A. Kusiak (1998), Designing an assembly system for modular products, *Computers and Industrial Engineering*, Vol. 31, No. 1, pp. 37–52.

Johnson, S. M. (1954), Optimal two-and three-stage production schedules with set-up times included, *Naval Research Logistics Quarterly*, Vol. 1, No. 1, pp. 61–68.

Johri, P. K. (1992), Optimal partitions for shop floor control in semiconductor wafer fabrication, *European Journal of Operational Research*, Vol. 59, No. 2, pp. 294–297.

Kusiak, A. (1990), *Intelligent Manufacturing Systems*, Prentice-Hall, Englewood Cliffs, NJ.

Kusiak, A., and W. He (1994), Design of components for schedulability, *European Journal of Operational Research,* Vol. 76, No. 1, pp. 49–59.

Kusiak, A., and D. He (1997), Design for agile assembly: An operational perspective, *International Journal of Production Research*, Vol. 35, No. 1, pp. 157–178.

Kusiak, A., and D. He (1998), Design for agility: A scheduling perspective, *Robotics and Computer-Integrated Manufacturing Systems*, Vol. 14, No. 4, pp. 415–427.

Lee, H. L. (1994), Design for supply chain management: Concepts and examples, in *The Northwestern University Manufacturing Management Symposium Series '94*, Manufacturing Logistic, Evanston, IL.

Mather, H. (1987), Logistics in manufacturing: A way to beat the competition, *Assembly Automation,* Vol. 7, No. 4, pp. 175–178.

Pinedo, M. (1992), Scheduling of flexible assembly systems, in A. Kusiak (Ed.), *Intelligent Design and Manufacturing*, John Wiley, New York, pp. 449–468.

Redford, A., and J. Chal (1994), *Design for Assembly: Principles and Practice*, McGraw-Hill, London.

Sheridan, J. H. (1993), Agile manufacturing: Beyond lean production, *Industrial Week*, Vol. 242, No. 8, pp. 30–46.

Stephen, R. R., and M. V. Tatikonda (1992), Time management in new product development: Case study findings, *Journal of Manufacturing Systems*, Vol. 11, No. 5, pp. 359–368.

Thomopoulos, N. T. (1967), Line balancing sequencing for mixed-model assembly, *Management Science*, Vol. 14, No. 2, pp. 59–75.

Van Der Veen, J. A. A., and R. Van Dal (1991), Solvable cases of the no-wait flow-shop scheduling problem, *Journal of Operational Research Society*, Vol. 42, No. 11, pp. 971–980.

Whitney, D. E. (1993), Nippondenso Co. Ltd: A case study of strategic product design, *Research in Engineering Design*, Vol. 5, No. 1, pp. 1–20.

Yano, C. A., and R. Rachamadugu (1991), Sequencing to minimize work overload in assembly lines with product options, *Management Science*, Vol. 37, No. 5, pp. 572–586.

Youssef, M. A. (1992), Agile manufacturing: A necessary condition for competing in global markets, *Industrial Engineering*, Vol. 24, No. 12, pp. 18–20.

QUESTIONS

15.1. What are the basic characteristics of an agile manufacturing system?

15.2. What is accomplished by the modular design rule (Rule 1)?

15.3. What is accomplished by the robust product design rule (Rule 2)?

15.4. What is accomplished by the streamlining the flow of products rule (Rule 3)?

15.5. What is early product differentiation?

15.6. What is late product differentiation?

15.7. What is integral design?

PROBLEMS

15.1. An analyst in the AA Corporation is designing an assembly system for six products, given the matrix (15.14) of sequence-dependent changeover times:

$$
\begin{array}{c}
\text{Product number} \\
\text{Product number}
\begin{array}{c}
1 \\ 2 \\ 3 \\ 4 \\ 5 \\ 6
\end{array}
\begin{bmatrix}
\infty & 7 & 3 & 15 & 5 & 9 \\
4 & \infty & 2 & 10 & 9 & 3 \\
6 & 9 & \infty & 11 & 1 & 7 \\
7 & 3 & 1 & \infty & 8 & 8 \\
2 & 14 & 2 & 7 & \infty & 6 \\
4 & 11 & 7 & 9 & 3 & \infty
\end{bmatrix}
\end{array}
\qquad (15.14)
$$

(a) Develop two system layout alternatives: serial and modular.

(b) Compare the total changeover time for the two alternatives.

(c) Provide the reasons for the difference in the total changeover time in (b).

15.2. The BB Corporation is developing a machine cell to produce a family of five different parts that are being designed. Four machines are considered as the candidates for the manufacturing cell. Specify operational requirements for the five parts so that the resultant scheduling problem could be a special case of the three-machine job shop problem that can be solved with Johnson's scheduling algorithm of Chapter 7.

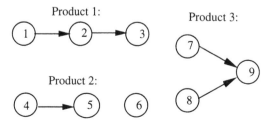

Figure 15.30. Precedence graphs representing three products.

**TABLE 15.2. Parts and Their Processing
Times**

Part Number	Processing Time
1	4
2	4
3	2
4	4
5	3
6	1
7	3
8	3
9	2

15.3. Design two variants of the same product, one following the early differentiation
principle and the other according to the late differentiation principle. Develop
two measures for quantification of the benefits from the two design principles.

15.4. The three products in Figure 15.30, each with three assembly operations, have
been assigned to two assembly stations by solving a capacity balancing model.
The station–operation assignments are:

Station 1: operations 1, 6, 4, 7, 9

Station 2: operations 2, 3, 5, 8

  (a) Is this station–operation assignment a valid schedule?

  (b) For the designs in Figure 15.30, draw a Gantt chart of a feasible
schedule and determine its makespan.

  (c) What product design changes, if any, would be necessary to
generate an optimal schedule with the shortest maximum flow?

Hint: Use the data in Table 15.2 for (b) and (c).

CHAPTER 16

SUPPLIER EVALUATION

16.1 INTRODUCTION

Supply chain management is a systems approach to managing the information, materials, and services from raw-materials suppliers through factories and warehouses to the end customer (Chase et al., 1998). The search for new tools to manage the supply chain is under way. This chapter deals with one of the key topics in the array of supply chain issues—the supplier evaluation problem. Supplier evaluation emphasizes attributes that encourage partnerships and continuous improvement throughout the supply chain. Studies show that quality and service considerations tend to dominate price and delivery criteria (Wilson, 1994). The traditional "three quotes" policy of buying focused on the purchase price of the product. However, current trends in purchasing and supplier evaluation focus on the total cost of the product, which includes the initial price as well as various direct and indirect costs associated with quality and service. Although customers' perceptions of what is important have changed, many firms have neglected to update their procedures for supplier evaluation.

Supply source selection decisions must be based on various supplier information, including (1) customer ratings of supplier quality performance, (2) customer ratings of supplier delivery performance, (3) total lead time to company by commodity, (4) past price paid, and (5) engineering support/design service capabilities (Monczka et al., 1992). Efforts to reengineer supplier evaluation should focus on this information as well as information that is deemed important to different commodities and functional perspectives. The resulting supplier evaluation system must meet the needs of all the purchasing categories (i.e., commodities) within the domain of the firm.

Much of the research in supplier evaluation has focused on studies of changes in evaluation criteria (such as those cited above) and methods for weighting those criteria

(Cook and Johnston, 1992; Thompson, 1991; Nydick and Hill, 1992). More progressive supplier evaluation systems have been introduced for total quality and just-in-time environments (Giunipero and Brewer, 1993) and for total cost supplier selection (Smytka and Clemens, 1993). However, methods for evaluating supplier–customer relationships and building supply chain alliances are scarce.

This chapter summarizes the results of an industrial case study performed by a cross-functional team to redesign and implement a supplier evaluation system at an industrial corporation. The team membership was constructed to adequately represent the functional areas pertinent to this case study. The objective of the supplier performance team was to identify the key characteristics in a supplier–customer relationship and exploit these characteristics in a system that fosters strong supply chain alliances. To accommodate the commodity team concept, the system should be flexible in its approach to supplier evaluation. Based on the results of this study, a larger study involving multiple suppliers was planned.

16.2   KEY CHARACTERISTICS OF THE SUPPLIER–CUSTOMER RELATIONSHIP

To provide a foundation for a flexible supplier evaluation system, the supplier performance team needs to follow an information collection procedure. This section summarizes the process, including a discussion of the results.

16.2.1   Information Collection Process

Rather than simply replicate the existing "one-sided" view of supplier evaluation, a methodology for development and expansion of supplier alliances is discussed based on an industrial case study. Each member of the cross-functional team assembled was asked to respond to the following question: *What is important to you in a relationship with a supplier?*

Each member provided several responses, and the team arranged the key characteristics of supplier–customer relationships into groups that shared common properties, such as issues relating to quality. Initially, the team identified approximately 85 characteristics. Following discussion and refinement of the list, 61 key characteristics resulted (see the Appendix). Each characteristic was then classified as quantitative, qualitative, or subjective based on the following definitions:

> *Qualitative:* There are observable criteria for evaluating the characteristic; however, no meaningful numeric measure of the characteristic exists.
>
> *Example:* "Supports or has capability for electronic commerce." Observable criteria might be fax capability, an Internet connection, and so on, however, a numeric measure is not appropriate.

> *Quantitative:* A numeric measure to describe the characteristic exists.

Example: "Responds in a timely manner to problems." Response time can be measured, for example, in days.

Subjective: Criteria are not necessarily observable and evaluation may rely on individual opinions based on prior experiences.

Example: "Sales personnel have knowledge of customer processes and products." This is likely to depend on the salesperson and generally is judged based on prior experience.

16.2.2 Key Characteristics

A survey was developed to gain a broader perspective of how individuals view the list of characteristics that was developed by the supplier performance team. Approximately 150 surveys were distributed to individuals in all functional areas that were identified by the team. After approximately six weeks, 66 completed surveys were collected. Although the total number of completed surveys was satisfactory, the distribution of responses across functional areas was not uniform (i.e., 28 of the responses came from Purchasing). Table 16.1 summarizes the survey responses by function.

The survey gathered information on (1) the perceived importance of each characteristic and (2) the potential to evaluate a supplier on each characteristic. For each characteristic (in each section), the individual assigned a score of "na," 1, 3, or 9. For importance, the scores corresponded to "not applicable" (na), "slightly important" (1), "moderately important" (3), and "extremely important" (9). For evaluation potential, the scores corresponded to "not applicable" (na), "slight potential for evaluation" (1), "moderate potential for evaluation" (3), and "extreme potential for evaluation" (9). The reason for using the scale 1, 3, and 9 was threefold: (a) it follows the ranking used in the quality function deployment (QFD) methodology, (b) the QFD software was used to collect the data, and (c) the employees were familiar with this scale. The data from the surveys were stored in a database and a series of reports were obtained.

TABLE 16.1. Survey Responses by Function

Functional Area	Number of Survey Responses
Advanced Operations Engineering	7
Applications Engineering	7
Design-to-Cost	1
Electrical Design	7
Manufacturing/Production	5
Material Verification	3
Mechanical Design	3
Procurement Quality Assurance	4
Purchasing	28
Warehousing	1

16.2.3 Importance of Characteristics

A report was generated to list the characteristics with "extreme" importance to each functional group. If more than half of the individuals in a given functional area (i.e., Applications Engineering) rated the characteristic as "extremely important" (i.e., 9), the characteristic was listed as extremely important for the function. This information was gathered in an attempt to detect patterns of perceived importance across functional areas. If subgroups of characteristics could be established for different functions, it was believed that a focus could be defined for supplier evaluation.

The decomposition algorithm (Kusiak and Wang, 1993), similar to Algorithm 10.3 of Chapter 10, was applied in an attempt to identify groups of functions and characteristics. However, due to the high density of the function–characteristic importance matrix, subgroups of characteristics could not be identified. Therefore, the characteristics were ranked based on overall importance. To obtain the ranking, a count of the number of respondents (out of 66) that rated the characteristic "extremely important" was obtained. The characteristics were then sorted in descending order by the count and ranked 1–61, with the highest count being ranked first. The rank of each characteristic is included in the Appendix. It was noted that the overall importance inferred from this ranking is skewed by uneven representation of different functional areas. For example, any characteristic that is generally viewed "extremely important" to Purchasing may appear to have higher overall importance.

16.2.4 Potential for Evaluating Characteristics

Although it is interesting to study the information from Section 16.2.3, individual perceptions of the potential to evaluate suppliers on each of the characteristics are much more useful. The data from the second part of the survey was used to generate a list of characteristics with high potential for evaluation for each functional area. Analogous to Section 16.2.3, if more than half of the individuals in a given functional area rated the characteristic 9, it was considered to have high potential for evaluation in that functional area.

Once again, a decomposition algorithm was used to identify groups of functions and characteristics. As expected, the evaluation potential matrix was much sparser and the algorithm successfully identified three subgroups of key characteristics corresponding to the functions listed in Table 16.2. The group number corresponding to each characteristic is listed in the Appendix. A letter O denotes an overlapping characteristic and a letter N corresponds to characteristics that did not have high potential for evaluation in any of the functional areas.

16.2.5 Discussion

Several interesting conclusions were drawn from the analysis of the surveys. The results clearly illustrated the difference between perceived importance and potential for evaluation. Although 57 of the key characteristics were considered "extremely important" in at least one functional area, almost half of those (25) did not appear to have "high potential for evaluation." In other words, many characteristics were

TABLE 16.2. Job Functions Corresponding to Key Characteristic Subgroups

Subgroup of Key Characteristics	Corresponding Function(s)
Group 1	Purchasing
Group 2	Material Verification
	Procurement Quality Assurance
	Warehousing
Group 3	Advanced Operations Engineering
	Applications Engineering
	Electrical Design
	Manufacturing/Production
	Mechanical Design
No group	Design-to-Cost

considered extremely important but cannot be easily evaluated. Many of these characteristics, such as "displays willingness to do business with the customer," were considered subjective by the team.

The survey data also suggested that focus groups can be formed and effectively used for supplier evaluation. The decomposition algorithm defined three subgroups of key characteristics based on job function. When evaluating suppliers, the commodity team may allocate specific key characteristics to individual team members based on these subgroups, rather than proceed with the current practice of requiring every team member to evaluate suppliers on all criteria. This would allow for more detailed analysis of key characteristics and, when appropriate, supplier–customer partnerships that address continuous improvement issues.

Based on the information obtained during this phase of the initiative, the team began to build a comprehensive supplier evaluation model. A flexible method of evaluating suppliers was needed to support the commodity team concept. The model also needed to consider the key characteristics of the supplier–customer relationship as well as the perceived importance and evaluation potential of these characteristics.

16.3 BUILDING A COMPREHENSIVE MODEL

The proposed supplier rating system achieves a balance between flexibility and standardization. Although it is designed for supplier selection, improvement, and certification, it can be adapted and extended to meet the needs of independent commodity teams. The system is designed around a set of core supplier capabilities, three supplier rating matrices, and a multicriteria-based weighting scheme.

16.3.1 Supplier Capabilities

Six core supplier capabilities were identified for emphasis in the proposed system (see Heinritz, 1971): (1) past performance, (2) engineering capabilities, (3) manufacturing capabilities, (4) management capabilities, (5) price, and (6) environmental protection. The key characteristics of the supplier–customer relationship were then mapped to the six core capabilities as described below.

16.3.1.1 Past Performance. The prior actions of the supplier are considered in past performance. The key characteristics of past performance are summarized into three categories: (1) schedule, (2) quality, and (3) subjective. Most key characteristics of past performance are currently monitored using existing supplier evaluation measures for delivery accuracy and quality. Unlike the other core capabilities, information is available for rating a supplier on past performance. Table 16.3 lists the key characteristics associated with each category of past performance. Note that the numbers correspond to the characteristics listed in the Appendix.

16.3.1.2 Engineering Capabilities. The supplier's success in developing new technology is dependent on its engineering capabilities. Due to the varying complexity of products (e.g., screws versus printed circuit boards), engineering capabilities may vary considerably between commodity teams. Key characteristics can provide general guidelines for the three categories: (1) design support, (2) processes, and (3) tools. In practice, however, engineering capabilities are expected to be very "team specific." Table 16.4 lists key characteristics corresponding to design support. Commodity teams should determine the specific processes and tools needed by suppliers to provide adequate engineering capabilities.

16.3.1.3 Manufacturing Capabilities. The supplier's success in bringing products to market is dependent on its manufacturing capabilities. As with engineering capabilities, manufacturing capabilities may vary considerably between commodity teams. The categories of key characteristics corresponding to manufacturing capabilities are (1) support, (2) processes, and (3) tools. Once again, manufacturing capabilities are expected to be very team specific. Table 16.5 lists possible key characteristics of manufacturing capabilities. In practice, the categories include key processes and tools determined by the commodity team.

16.3.1.4 Management Capabilities. Greater detail can be included in describing management capabilities, which are more common across commodity teams than engineering and manufacturing capabilities. The key characteristics of

TABLE 16.3. Key Characteristics of Past Performance

Category	Key Characteristics
Schedule	48, 50
Quality	27, 28, 29, 30, 31, 32, 33, 34, 35
Subjective	1, 9, 13, 18, 46

TABLE 16.4. Key Characteristics of Engineering Capabilities

Category	Key Characteristics
Design Support	52, 53, 54, 55, 56, 57, 58, 59, 60, 61
Processes	Determined by commodity team
Tools	Determined by commodity team

management capabilities are summarized in seven categories: (1) quality systems, (2) program management, (3) financial stability, (4) information management, (5) communications capability, (6) cultural enhancement, and (7) total quality management (TQM) approach.

Quality systems must emphasize statistical process control data, ISO 9000 certification, and customer satisfaction. Program management must achieve total confidence in the system to satisfy customer needs with regard to costs, technical expertise and delivery performance and quality through the life cycle of the product. Financial stability must be illustrated through cost accounting system receivables, payables, inventories, estimating and tracking of costs (i.e., labor, overhead, and capital), operating budgets, and forecasts. An agile information management system must be in place to enable employees to maintain a world class status through continuous improvement and quick response. The supplier must have a system installed for electronic communication throughout the supply chain. Through cultural enhancement, teams must be directly responsible for all aspects of problem solving and empowered to implement solutions that foster employee ownership. Finally, a TQM approach should emphasize value-added concepts, verify elimination of non-value-added activities, benchmark critical processes, and utilize best practices. Table 16.6 lists the key characteristics corresponding to each of these categories.

16.3.1.5 Price. The supplier's ability to provide a competitive price is the fifth core capability. The commodity team may consider price relative to a supplier's competitors and simply rate the supplier low, medium, or high. Alternatively, the actual price of the product can be used to determine a standardized score. In either case, price should be explicitly rated without considering key characteristics of other core capabilities. Table 16.7 lists key characteristics related to price.

TABLE 16.5. Key Characteristics of Manufacturing Capabilities

Category	Key Characteristics
Support	36, 37, 38, 39, 49
Processes	Determined by commodity team
Tools	Determined by commodity team

TABLE 16.6. Key Characteristics of Management Capabilities

Category	Key Characteristics
Quality systems	19, 25, 26
Program management	3, 4, 5, 6, 7, 8, 10, 11, 47
Financial stability	40, 44
Information management	14, 20, 21, 22, 23, 24
Communications capability	2
Cultural enhancement	17
TQM approach	15, 16

16.3.1.6 Environmental Awareness. The supplier must be aware of the customer's environmental responsibilities. To be successful at the sixth core capability, environmental protection, the supplier should be willing to disclose ingredients of trade secrets, dispose of containers from the customer, and dispose of outdated and/or failed product. The key characteristics of environmental protection are also listed in Table 16.7. In addition, the commodity team may include characteristics that are relevant to specific products or regulations.

16.3.2 Supplier Rating Matrices

The proposed system is based on the premise that different technologies, commodities, and supplier relationships will emphasize different core capabilities for supplier evaluation. Therefore, a weighting scheme must be employed to prioritize supplier capabilities and the system must consider multiple dimensions (i.e., the technology life cycle and the relationship life cycle). Figure 16.1 illustrates the three dimensions of the proposed system.

16.3.2.1 Technology Life-Cycle/Supplier Capability Matrix. The tech-nology life-cycle/supplier capability matrix (see Figure 16.2) captures the relative importance of each supplier capability for each phase of the technology life cycle. The team assigns weights to each capability (e.g., past performance, engineering capabilities) to establish priorities for evaluating the overall performance of a supplier. The model assumes that the relative importance of a given capability, such as price, varies across the technology life cycle. For example, a product in the introduction phase is likely to be an improvement over existing technology. Therefore, in

TABLE 16.7. Key Characteristics of Price and Environmental Protection

Core Capability	Key Characteristics
Price	41, 42, 43, 45, 51
Environmental awareness	12

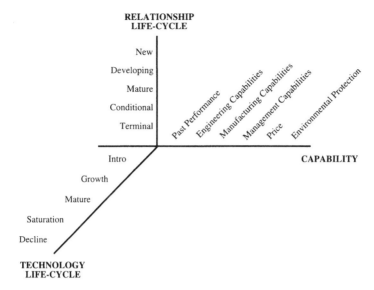

Figure 16.1. Dimensions of the proposed system.

evaluating the supplier of the product, price may carry a low weight and engineering capabilities may carry a high weight. In this case, the customer is willing to pay more for the product but expects superior engineering support to improve and develop the product into later phases of the technology life cycle. Similarly, the supplier of a product in the saturation or decline phase of the technology life cycle may focus on price and past performance, with very little weight on engineering, manufacturing, and management capabilities.

TECHNOLOGY LIFE-CYCLE/CAPABILITY MATRIX

	PAST PERFORMANCE	ENGINEERING CAPABILITIES	MANUFACTURING CAPABILITIES	MANAGEMENT CAPABILITIES	PRICE	ENVIRONMENT PROTECTION	
TECHNOLOGY LIFE CYCLE PHASE					LOW MEDIUM HIGH	COMMITED NEUTRAL HARMFUL	
							TOTAL
INTRO							0
GROWTH							0
MATURE							0
SATURATION							0
DECLINE							0

Figure 16.2. Technology life-cycle/capability matrix.

The commodity team assigns a weight to each of the supplier capabilities in each phase of the technology life cycle. A total of 100 points was divided among the six capabilities in each row of the matrix. Key characteristics of each supplier capability (see Section 16.3.1) are considered in determining the overall weight. The commodity team can also integrate its own key characteristics, particularly under engineering and manufacturing capabilities.

16.3.2.2 Relationship Life-Cycle/Supplier Capability Matrix.

The relationship life-cycle/supplier capability matrix (see Figure 16.3) captures the relative importance of each supplier capability for each phase of the supplier–customer relationship. This perspective reveals the supplier capabilities that are most important to forming alliances. Once again, the commodity team assigns a weight to each capability in each phase of the relationship life cycle. The model assumes that the relative importance of a given capability varies as the relationship develops and eventually terminates. For example, developing relationships may emphasize engineering and management capabilities and terminal relationships may emphasize price. If a supplier–customer relationship is classified in one of the five phases (new, developing, mature, conditional, and terminal), a strategy for building an alliance can be determined by focusing on the capabilities that carry the highest priority in the desired phase of the alliance. For example, if a relationship is classified as conditional and the commodity team wants to establish a mature relationship, the supplier should focus on the capabilities that carry the highest weights in the "mature" row of the matrix—perhaps management capabilities and past performance. This method of propagating a supplier's status through the phases of the matrix facilitates continuous improvement.

As with the technology life-cycle/supplier capability matrix, the commodity team assigns weights to each of the supplier capabilities in each phase of the relationship

Figure 16.3. Relationship life-cycle/capability matrix.

life cycle. A total of 100 points is divided among the six capabilities in each row of the matrix. The key characteristics outlined in Section 16.3.1, as well as those established by the commodity team, are used to determine the overall weight.

16.3.2.3 *Supplier Rating Matrix.* The supplier rating matrix (see Figure 16.4) is used to summarize the commodity team's ratings of a supplier on each of the six core capabilities. The scoring system is determined by each commodity team and should include the key characteristics cited in Section 16.3.1, as well as additional characteristics developed by the team. For consistency, a 10-point maximum score is recommended for each capability. The commodity team can then determine the total score for a supplier (1000 points possible) in each of the phases of the technology life cycle and each of the phases of the relationship life cycle by multiplying the respective matrices. (Note that to calculate the total score, the transpose of the supplier rating matrix is multiplied by the technology life-cycle and relationship life-cycle matrices.) Section 16.4 illustrates how the proposed supplier rating system exploits a supplier's ability to provide products in different phases of the technology life cycle and provides a mechanism for building alliances through continuous improvement.

Supplier Name:_____ Supplier Code:_____ Commodity Code:_____

SUPPLIER RATING MATRIX

	PAST PERFORMANCE			ENGINEERING CAPABILITIES			MANUFACTURING CAPABILITIES			MANAGEMENT CAPABILITIES						PRICE	ENVIRONMENT PROTECTION
	SCHEDULE	QUALITY	SUBJECTIVE	DESIGN SUPPORT	PROCESSES	TOOLS	SUPPORT	PROCESSES	TOOLS	QUALITY SYSTEMS	PROGRAM MANAGEMENT	FINANCIAL STABILITY	INFORMATION MANAGEMENT	COMMUNICATIONS CAPABILITY	CULTURAL ENHANCEMENT	TQM APPROACH	LOW MEDIUM HIGH	COMMITTED NEUTRAL HARMFUL
Rating																		

TECHNOLOGY LIFE-CYCLE SUPPLIER RATING	
INTRO	/ 1000
GROWTH	/ 1000
MATURE	/ 1000
SATURATION	/ 1000
DECLINE	/ 1000

RELATIONSHIP LIFE-CYCLE SUPPLIER RATING	
NEW	/ 1000
DEVELOPING	/ 1000
MATURE	/ 1000
CONDITIONAL	/ 1000
TERMINAL	/ 1000

Figure 16.4. Supplier rating matrix.

16.4 SYSTEM IMPLEMENTATION

The strategy for implementing the proposed supplier rating system focuses on three activities: (1) developing a tool for commodity teams, (2) disseminating commodity team information through a bulletin board system (BBS), and (3) developing an intelligent supplier evaluation system.

16.4.1 Tool for Commodity Teams

The flexibility of the tool allows one to adopt it as a standard while providing commodity teams the latitude to incorporate their existing practices. The key issue in implementation is obtaining user acceptance. This section presents a case study of the proposed system conducted by the ABC firm's commodity team. The activities of the team are summarized in this section to illustrate the use of the supplier rating system.

16.4.1.1 Supplier Evaluation. The first activity of the ABC commodity team was to assign weights to each capability in each row of the technology life-cycle and relationship life-cycle matrices. This was accomplished through group discussion until consensus was reached. The weighted matrices are shown in Figures 16.5 and 16.6.

The team then mapped its own list of 41 key characteristics (not necessarily the same as the key characteristics in the Appendix) into the six core capabilities. For each key characteristic, the team evaluated three suppliers: Supplier X, Supplier Y, and Supplier Z. Team members assigned each supplier a rating from 1 to 4 (4 being best) on the 41 characteristics and the average was calculated for each characteristic and then for the supplier. The overall average for the supplier constituted the team's

TECHNOLOGY LIFE-CYCLE/CAPABILITY MATRIX

TECHNOLOGY LIFE-CYCLE PHASE	PAST PERFORMANCE (SCHEDULE / QUALITY / SUBJECTIVE)	ENGINEERING CAPABILITIES (DESIGN SUPPORT / PROCESSES / TOOLS)	MANUFACTURING CAPABILITIES (SUPPORT / PROCESSES / TOOLS)	MANAGEMENT CAPABILITIES (QUALITY SYSTEMS / PROGRAM MANAGEMENT / FINANCIAL STABILITY / INFORMATION MANAGEMENT / COMMUNICATIONS CAPABILITY / CULTURAL ENHANCEMENT / TQM APPROACH)	PRICE (LOW MEDIUM HIGH)	ENVIRONMENT PROTECTION (COMMITED NEUTRAL HARMFUL)	TOTAL
INTRO	5	35	20	25	10	5	100
GROWTH	15	20	20	20	20	5	100
MATURE	30	10	15	15	25	5	100
SATURATION	30	10	10	10	35	5	100
DECLINE	20	5	10	5	55	5	100

Figure 16.5. Technology life-cycle/capability matrix for the ABC Commodity Team.

RELATIONSHIP LIFE-CYCLE/CAPABILITY MATRIX

RELATIONSHIP LIFE-CYCLE PHAS	PAST PERFORMANCE	ENGINEERING CAPABILITIES	MANUFACTURING CAPABILITIES	MANAGEMENT CAPABILITIES	PRICE LOW MEDIUM HIGH	ENVIRONMENT PROTECTION COMMITED NEUTRAL HARMFUL	TOTAL
NEW	5	35	15	30	10	5	100
DEVELOPING	15	25	20	25	10	5	100
MATURE	30	10	20	15	20	5	100
CONDITIONAL	25	10	10	15	35	5	100
TERMINAL	20	5	10	5	55	5	100

Figure 16.6. Relationship life-cycle/capability matrix for the ABC Commodity Team.

current approach to supplier evaluation. Thus, each supplier received a score from 1 to 4 and the supplier with the highest score was typically selected.

The supplier's score for each capability was then calculated using the ratings on the 41 key characteristics and the team's mapping of these characteristics to the six core capabilities. For example, the scores for the key characteristics corresponding to manufacturing capabilities were averaged and multiplied by 2.5 to obtain the score out of 10 possible points (as suggested in Section 16.3.2.3). Figure 16.7 shows the supplier rating matrix for Supplier X. The supplier receives a score corresponding to each phase of the technology life cycle and each phase of the relationship life cycle. Based on the scores, Supplier X is most effective, supplying declining technology in a terminal relationship. Table 16.8 summarizes ratings for each supplier using both the team's previous method and the proposed supplier evaluation system.

16.4.1.2 Source Selection.
The proposed system provides much more information for performing source selection than the previous method. Obviously, Supplier Y is preferred regardless of technology and the stage of the relationship. However, if the commodity team is considering purchasing declining technology, Supplier X is clearly preferred over Supplier Z. Using the previous method of supplier evaluation, the opposite is true. This is due to the team's perspective of the relative importance of price (for declining technology, price was assigned 55 of the possible 100 weighting points) and the suppliers' respective scores on that core capability. Similarly, Supplier X is superior in a terminal relationship. Thus, if both suppliers are providing a declining technology in what is considered a terminal relationship, the commodity team should select Supplier X as a sole source for the product.

The proposed system can be used to draw similar conclusions for products in all phases of the technology life cycle and suppliers in all phases of the relationship life cycle. In general, the proposed system provides greater flexibility and allows the team

Supplier Name: Supplier X _____   Supplier Code: _____   Commodity Code: _____

SUPPLIER RATING MATRIX

	PAST PERFORMANCE			ENGINEERING CAPABILITIES			MANUFACTURING CAPABILITIES			MANAGEMENT CAPABILITIES							PRICE	ENVIRONMENT PROTECTION
	SCHEDULE	QUALITY	SUBJECTIVE	DESIGN SUPPORT	PROCESSES	TOOLS	SUPPORT	PROCESSES	TOOLS	QUALITY SYSTEMS	PROGRAM MANAGEMENT	FINANCIAL STABILITY	INFORMATION MANAGEMENT	COMMUNICATIONS CAPABILITY	CULTURAL ENHANCEMENT	TQM APPROACH	LOW MEDIUM HIGH	COMMITED NEUTRAL HARMFUL
Rating	4.4			6			5.1			6							8.33	6.61

TECHNOLOGY LIFE-CYCLE SUPPLIER RATING	
INTRO	608 / 1000
GROWTH	614 / 1000
MATURE	604 / 1000
SATURATION	630 / 1000
DECLINE	691 / 1000

RELATIONSHIP LIFE-CYCLE SUPPLIER RATING	
NEW	615 / 1000
DEVELOPING	592 / 1000
MATURE	587 / 1000
CONDITIONAL	640 / 1000
TERMINAL	691 / 1000

Figure 16.7. Supplier rating matrix for supplier X.

to exploit the strengths of suppliers that were not apparent in the previous supplier evaluation procedure.

16.4.1.3 Monitoring the Progress of a Supplier–Customer Alliance.

Perhaps the greatest benefit of the proposed system is that it facilitates continuous improvement and enables commodity teams to identify opportunities to build supply chain alliances. Once again, consider the supplier ratings in Table 16.8. Supplier Y is strong in all core capabilities except, perhaps, price. The strength of Supplier Y is also apparent in the consistency of scores across all phases of the technology and relationship life cycles. The opportunity for an effective alliance is evident by the "good fit" between the commodity team's objectives (as portrayed by the weighting matrices) and the supplier's strengths.

The proposed system can also be used to target continuous improvement initiatives. Consider Supplier X, which is relatively strong in a terminal relationship. Suppose that the commodity team would like to develop a mature relationship with the supplier. Based on the objectives of the commodity team (i.e., the weights in the relationship life-cycle/capability matrix), the supplier must focus on past performance (current score of 4.4) and manufacturing capabilities (current score of 5.1) to raise its score for the mature phase of the relationship life cycle. This strategy can be used to

TABLE 16.8. Summary of Supplier Ratings

Rating System	Supplier X	Supplier Y	Supplier Z	Possible Score
Existing system	2.38	3.22	2.63	4
Technology life cycle				
Introduction	608	793	642	1000
Growth	614	790	602	1000
Mature	604	789	559	1000
Saturation	630	783	542	1000
Decline	691	773	531	1000
Relationship life cycle				
New	615	795	643	1000
Developing	592	795	619	1000
Mature	587	791	567	1000
Conditional	640	784	554	1000
Terminal	691	773	531	1000

propagate suppliers through the phases of the relationship life cycle and, thus, build supplier–customer alliances.

16.4.2 Bulletin Board

After gaining acceptance, the supplier rating matrices was included as components of a supplier alliance BBS. This medium will disseminate information throughout the supply chain. The system will allow suppliers, design engineers, and managers, among others, to view the perspectives, goals, and priorities of individual commodity teams. Design engineers can use the information to justify part selection; suppliers can use the information to focus on areas for continuous improvement; and managers can use the information for establishing business goals and benchmarking.

16.4.3 Intelligent Supplier Evaluation System

The discussed supplier rating system is the foundation for an intelligent supplier evaluation system. The system evaluates the pool of available suppliers in the context of the key characteristics, independent supplier ratings, and the relationship and technology life cycles. As a result, preferred alliances are identified and strategies for advancing the alliance will be outlined. Computational intelligence is employed to evaluate commodity team rules and assist in weighting supplier capabilities. As with the BBS, suppliers will have access to information generated by the system to aid continuous improvement. Future additions to the system include a supplier perspective that addresses the key characteristics of the customer's role in the alliance. Figure 16.8 illustrates the architecture of the intelligent supplier evaluation system.

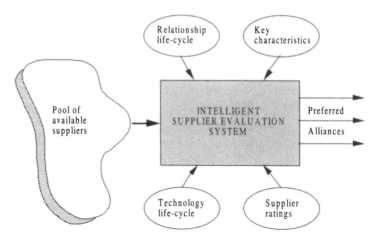

Figure 16.8. Architecture of an intelligent supplier evaluation system.

16.5 SUMMARY

This chapter presented a flexible supplier evaluation system developed in the context of a case study at an industrial company. The activities of the supplier performance team were motivated by the need to reengineer the supplier evaluation process to consider criteria that focus on continuous improvement and supply chain alliances. Key characteristics of the supplier–customer relationship were identified and studied through the team's analysis and an internal survey. The key characteristics became the foundation for six core supplier capabilities. Two perspectives, the technology life cycle and the relationship life cycle, add flexibility to evaluation of the core capabilities. The proposed system facilitates more comprehensive evaluation than previous methods. Three suppliers were rated to illustrate supplier evaluation, source selection, and strategies for building supply chain alliances.

Assigning weights to the technology life-cycle and relationship life-cycle matrices is an important step in the process. In the case study, weights were determined through team discussion until consensus was reached. It was assumed that all members of the team had equal power in determining the weight. However, this assumption did not consider group dynamics or the relevance of different job functions to the objective. When feasible, a procedure based on other established methods, such as the analytical hierarchy process (Saaty, 1980), could be developed to help commodity teams determine weights.

Supplier–customer relationships are evolving from "three quotes" with minimal interaction to supply chain alliances. To realize the benefit of lower total product cost, it is essential that suppliers and customers have tools for identifying and building strong partnerships. This case study confronted the issue of supplier evaluation in a manner that focuses on the supplier–customer relationship and methods for building supply chain alliances through continuous improvement.

APPENDIX: KEY CHARACTERISTICS OF THE SUPPLIER–CUSTOMER RELATIONSHIP

Number	Key Characteristic	Rank	Group
1	Displays a willingness to do business with the customer	8	N
2	Has capability for electronic commerce	50	N
3	Responds in a timely manner to problems	1	3
4	Responds in a timely manner to resolutions for rejected parts	9	2
5	Responds in a timely manner to prices and bids	26	1
6	Responds in a timely manner to requests for quotations	27	1
7	Responds in a timely manner to corrective action requests	18	2
8	Responds in a timely manner to technical information	16	O
9	Sales personnel have knowledge of the customer's processes and products	53	N
10	Notifies the customer of "heads-up" issues	12	2
11	Provides order confirmation/acknowledgment	42	N
12	Complies with the Environmental Protection Agency	39	N
13	Displays ethical behavior	15	3
14	Exhibits procedures for business functions	56	2
15	Practices TQM, CPI (Continuous Production Improvement) program emphasizing lead time reduction	39	N
16	Practices TQM, CPI program emphasizing cost/product improvement	32	N
17	Embraces high-performance work systems (HPWSs)	60	N
18	Commits to the customer's objectives	32	N
19	Practices SPC (Statistical Process Control)	32	2
20	Utilizes production planning and control methods for shop floor control	45	N
21	Utilizes production planning and control methods for order management	45	N
22	Utilizes production planning and control methods for master scheduling	50	N
23	Utilizes production planning and control methods for inventory accuracy	42	N
24	Utilizes production planning and control methods for bill of material accuracy	45	N
25	Achieves process capability index of 1.3	56	2
26	Provides traceability of parts, material, and processes	28	2
27	Maintains certified supplier status	32	2
28	Meets dimensional requirements	3	O
29	Meets specifications	1	O
30	Complies with purchase order requirements	4	O
31	Provides required documentation	10	2
32	Provides quality packaging	25	2
33	Provides markings on component	22	O
34	Meets finish requirements	6	O
35	Provides accurate count/quantity	14	2
36	Adheres to the customer's routing instructions	36	2
37	Includes standardized (necessary) packing slip data	31	2
38	Provides packaging flexibility	53	2
39	Uses bar codes	61	N

(*continued*)

Number	Key Characteristic	Rank	Group
40	Shows financial stability	28	N
41	Provides competitive pricing	18	1
42	Shares cost data	42	N
43	Lot/piece is sensitive to delivery and pricing	41	N
44	Performs financial resource planning	59	N
45	Absorbs rework cost	45	N
46	Is flexible to the customer's schedules	18	1
47	Provides accurate and detailed information	5	N
48	Provides on-time delivery	6	1
49	Is willing to inventory the customer's stock	53	N
50	Is willing and able to expedite delivery	12	1
51	Maintains minimum buy requirements	56	N
52	Provides samples upon request	36	3
53	Shares technology with the customer	24	3
54	Participates in integrated product and process development (IPPD)	50	N
55	Participates during design stage	36	3
56	Maintains technical expertise	10	O
57	Provides technical support	16	O
58	Adheres to specifications	22	2
59	Is a technology leader	28	3
60	Makes tools available	49	N
61	Accepts the customer's specifications	18	2

REFERENCES

Chase, R. B., N. J. Aquilano, and F. R. Jacobs (1998), *Production and Operations Management: Manufacturing and Services*, Irwin/McGraw-Hill, Boston, MA.

Cook, W. D., and D. A. Johnston (1992), Evaluating suppliers of complex systems: A multiple criteria approach, *Journal of the Operational Research Society*, Vol. 43, No. 11, pp. 1055–1061.

Giunipero, L. C., and D. J. Brewer (1993), Performance based evaluation systems under total quality management, *International Journal of Purchasing and Materials Management*, Vol. 29, pp. 35–41.

Heinritz, S. F. (1971), *Purchasing: Principles and Applications*, Prentice-Hall, Englewood Cliffs, NJ.

Kusiak, A., and J. Wang (1993), Decomposition of the design process, *ASME Transactions: Journal of Mechanical Design*, Vol. 115, No. 4, pp. 687–695.

Larson, T. N., A. Kusiak, D. Clymer, and G. Thompson (1996), Analyzing the supplier/customer relationship: An industrial case study, *Proceedings of the 5th IE Research Conference*, Minneapolis, MN, May, pp. 175–181.

Monczka, R. M., E. L. Nichols, Jr., and T. J. Callahan (1992), Value of supplier information in the decision process, *International Journal of Purchasing and Materials Management*, Vol. 28, pp. 20–30.

Nydick, R. L., and R. P. Hill (1992), Using the analytic hierarchy process to structure the supplier selection procedure, *International Journal of Purchasing and Materials Management*, Vol. 28, pp. 31–36.

Saaty, T. (1980), *The Analytic Hierarchy Process*, McGraw-Hill, New York.

Smytka, D. L. and M. W. Clemens (1993), Total cost supplier selection model: A case study, *International Journal of Purchasing and Materials Management*, Vol. 29, pp. 42–49.

Thompson, K. N. (1991), Scaling evaluative criteria and supplier performance estimates in weighted point prepurchase decision models, *International Journal of Purchasing and Materials Management*, Vol. 27, pp. 27–36.

Wilson, E. J. (1994), The relative importance of supplier selection criteria: A review and update, *International Journal of Purchasing and Materials Management*, Vol. 30, pp. 35–41.

QUESTIONS

16.1. What information about suppliers is of interest to a potential customer company?

16.2. What is a good way of defining supplier characteristics?

16.3. What methods would you use for the evaluation of supplier characteristics?

16.4. How are metrics associated with supplier characteristics?

PROBLEMS

16.1. A multidisciplinary team has been formed to design a system for measuring relationships between suppliers and the AA Corporation. The team has recommended that an association between supplier characteristics and components purchased be developed.

(a) Make a list of a few components.

(b) Develop a list of supplier characteristics.

(c) Create a matrix, with rows being supplier characteristics and columns being components.

(d) Fill in the matrix in (c) using the following sale: 9, strongly applicable; 3, moderately applicable; 1, weakly applicable; blank, does not apply.

(e) Analyze the data in the matrix using known analysis tools, such as computing the total in each row and column, applying one of the decomposition algorithms of Chapter 10, and so on. Interpret the results.

16.2. Define an application scenario and fill in the matrices in Figures 16.2–16.7.

CHAPTER 17

DATA MINING

17.1 INTRODUCTION

Researchers and practitioners currently deal with the world of assumptions, axioms, formulas, models, algorithms, and other constructs defined in abstract spaces. The world of tomorrow is likely to be highly data driven. To a large degree, the data (symbolic and numeric) will define spaces for processing by the new and existing tools. Data mining is emerging as an area of computational intelligence that offers new theories, techniques, and tools for the data-based spaces. It has gained considerable attention among practitioners and researchers. The growing volume of data available in digital form spurs this accelerated interest. As a new discipline, data mining relies to a large degree on techniques and tools borrowed from other areas (e.g., decision analysis, cluster analysis, genetic algorithms, and neural networks).

One of the few theories developed specifically for data mining is the rough set theory. Rough set theory is a formal approach to data analysis and data mining (Pawlak, 1991). It has found many applications in industry, service organizations, and health care systems for:

- Reduction of data sets
- Finding hidden data patterns
- Generation of decision rules from data sets

Rough set theory has found many applications in industry, service organizations, and health care (Kowalczyk and Slisser, 1997), software engineering (Ruhe, 1996), edge detection (Wojcik, 1993a), data filtration (Skowron, 1994), and clinical decision making (Slowinski and Stefanowski, 1993; Tsumoto, 1997).

498

Medical diagnosis has seen numerous applications of rough set theory used to predict diagnosis of different diseases. Some of the most widely cited references on medical diagnosis include acute appendicitis (Carlin et al., 1998), progressive encephalopathy (Paszek et al., 1996), selective vagotomy (Pawlak et al., 1998), diabetes (Stepaniuk et al., 1998), and meningoencephalitis (Tsumoto and Ziarko, 1996).

A comprehensive comparative analysis of prediction methods included in Kononenko et al. (1998) indicates that automatically generated diagnostic rules outperform the diagnostic accuracy of physicians. The authors' claim is supported by a comprehensive review of the literature on four diagnostic topics: localization of primary tumor, prediction of reoccurrence of breast cancer, thyroid diagnosis, and rheumatoid prediction. Berry and Linoff (1997) and Groth (1998) surveyed many engineering and business applications of data mining.

In this chapter, the application of rough set theory in decision making is discussed. The name rough set theory may sound similar to fuzzy set theory; however, both handle uncertainty of data differently (Pawlak, 1997). The relationship of rough set theory to Boolean reasoning, evidence theory, and discriminant analysis is well established. Rough set theory has found a well-deserved place in decision theory and is becoming fundamental in data mining.

17.2 BACKGROUND AND DEFINITIONS

Rough set theory is based on the assumption that data and information are associated with every object of the universe of discourse (Pawlak, 1997). *Objects* described by the same properly selected information (referred in this chapter as *features*) are indiscernible. The adjective *properly* is of key importance, as the algorithms presented in this chapter will select a subset of all features necessary to characterize a category of objects.

The basic construct in rough set theory is called a *reduct*. It is defined as a minimal sufficient subset of features RED $\subseteq A$ such that (Shan et al., 1995):

(a) Relation $R(\text{RED}) = R(A)$; that is, RED produces the same categorization of objects as the collection A of all features.

(b) For any $g \in \text{RED}$, $R(\text{RED} - \{g\}) \neq R(A)$; that is, a reduct is a minimal subset of features with respect to the property (a).

By definition, a "reduct" represents an alternative and simplified way of representing the same features. It is easy to see that a reduct has the same properties as a key defined in relational database theory (with respect to a specific instance of a relation only). In this context a reduct can be called an empirical key.

The term reduct was initially defined for sets rather than objects with input and output features or decision tables with decision features (attributes) and outcomes. Reducts of the objects in a decision table have to be computed with the consideration given to the value of the output feature. The original definition of reduct considers features only. In this research, each reduct is viewed from four perspectives:

- Feature
- Feature value
- Object
- Rule perspective

The reduct definition and the meaning of the perspectives are illustrated next. Consider the data in Table 17.1 for eight objects with four input features $F_1, \ldots, F_4$ and output feature O.

The sets of *o-reducts* for the first two objects in Table 17.1, the *reduct generation algorithm* discussed in the next section, are shown in Figure 17.1.

The entry "x" in each reduct implies that the corresponding feature is not considered in determining the feature output of an object. To obtain an *o*-reduct, one input feature at a time is considered, and it is determined whether this feature uniquely identifies all objects. A single conflict between the output features of any two objects disqualifies that input feature from producing a single-feature *o*-reduct.

Consider the following *o*-reduct represented from set 1 in Figure 17.1:

$$1 \text{ x x x } 2 \tag{17.1}$$

The *o*-reduct in (17.1) can be also expressed as the decision rule

IF the value of input feature $F_1 = 1$

THEN the value of output feature $O = 2$ \qquad (17.2)

and therefore is called an *r-reduct* (rule-reduct).

The data set in Table 17.1 was relatively small so all possible reducts could be computed even by inspection. In many data mining applications data sets are large and the need for a systematic procedure of computing reducts is apparent due to the high computational complexity. The reduct generation algorithm presented next was compiled based on the developments in Pawlak (1991).

TABLE 17.1. Eight Objects

Object No.	F_1	F_2	F_3	F_4	O
1	1	0	0	2	2
2	0	1	1	0	1
3	1	1	0	2	2
4	1	2	0	2	2
5	0	2	1	0	1
6	0	0	1	0	1
7	1	3	0	2	2
8	0	3	1	0	1

```
          1  x  x  x  2
Set 1     x  x  0  x  2
          x  x  x  2  2


          0  x  x  x  1
Set 2     x  x  1  x  1
          x  x  x  0  1
```

Figure 17.1. Two sets of *o*-reducts for the data in Table 17.1.

17.3   REDUCT GENERATION ALGORITHM

The *reduct generation (RG) algorithm* produces reducts for objects with input and output features.

The RG Algorithm (based on Pawlak, 1991)

   Step 0. Initialize object number $i = 1$.

   Step 1. Select object i and find a set of *o*-reducts with one feature only.

   If found, go to Step 3; otherwise go to Step 2.

   Step 2. For object i, find an *o*-reduct with $m - 1$ features, where m is the number of input features. This step is accomplished by deleting one feature at a time.

   Step 3. Set $i = i + 1$. If all objects have been considered, stop; otherwise go to Step 1.

The version of the reduct generation algorithm enumerates *o*-reducts with one or $m - 1$ features only. It could be easily modified to produce all possible reducts; however, the computing cost would be excessive. In fact, reducts with more that one feature are needed only for objects that do not produce single-feature reducts. In this case, considering any k out of m features would result in $m!/k!(m - k)!$ reducts. In the case that a relatively large number of objects would require multifeature *o*-reducts, one of the following actions could be taken:

- All *o*-reducts with $1 < k < m$ features could be computed.
- Other methods (e.g., heuristic algorithms) could be used to process the data.
- Other properties of the data could be identified.

The reduct generation algorithm is illustrated in Example 17.1.

Example 17.1.   Determine *o*-reducts for the data in Table 17.2.

   Step 0. Set object number $i = 1$.

   Step 1. No *o*-reduct with one feature is found for object 1. For example, $F_1 = 0$ appears under $O = 0$ and $O = 1$.

   Step 2. By dropping one of the four features of object 1 at a time, the following three *o*-reducts are identified 0 x 0 2 0, 0 1 x 2 0, and 0 1 0 x 0. Each of the following sets

TABLE 17.2. Five-Object Data Set

Object No.	F_1	F_2	F_3	F_4	O
1	0	1	0	2	0
2	1	1	0	2	2
3	0	0	0	1	0
4	0	1	1	0	1
5	0	0	1	3	0

of the features $(F_1, F_3, F_4) = (0, 0, 2)$, $(F_1, F_2, F_4) = (0, 1, 2)$, and $(F_1, F_2, F_3) = (0, 1, 0)$ uniquely identifies object 1. Note that deleting feature F_1 did not produce the o-reduct x 1 0 2 0 due to the fact that $(F_2, F_3, F_4) = (1, 0, 2)$ appears for $O = 0$ and $O = 2$.

Step 3. Set $i = 2$ and go to Step 1.

Step 1. For object 2, o-reduct 1 x x x 2 is identified. Note that only one object with $O = 2$ exists.

Step 3. Set $i = 3$, and go to Step 1.

Step 1. For object 3, o-reducts x 0 x x 0 and x x x 1 0 are identified.

Step 3. Set $i = 4$, and go to Step 1.

Step 1. For object 4, o-reduct x x x 0 1 is identified.

Step 3. Set $i = 5$, and go to Step 1.

Step 1. For object 5, o-reducts x 0 x x 0 and x x x 3 0 are identified.

Step 3. Since all objects have been considered, stop.

The o-reducts generated for the objects in Table 17.2 are shown in Table 17.3.

Table 17.3 includes a subset of all o-reducts. For all o-reducts generated for the data in Table 17.22 see Table 17.23 in Appendix 17.1.

The nine o-reducts in Table 17.3 could be used in a number of different ways. One has to remember that each of them represents a corresponding object, as indicated by the data in Table 17.2. Any combination of o-reducts, one for each object, could be chosen to represent the five objects. Assume that we have selected the o-reducts 1, 2, 4, 6, and 8 from Table 17.3, as shown in Table 17.4

TABLE 17.3. o-Reducts of Objects in Table 17.2

Object No.	o-Reduct No.	F_1	F_2	F_3	F_4	O
2	1	1	x	x	x	2
4	2	x	x	x	0	1
	3	0	x	0	2	0
1	4	0	1	x	2	0
	5	0	1	0	x	0
5	6	x	0	x	x	0
	7	x	x	x	3	0
3	8	x	0	x	x	0
	9	x	x	x	1	0

TABLE 17.4. Sets of *o*-Reducts for Each Object in Table 17.3

Object No.	*o*-Reduct No.	F_1	F_2	F_3	F_4	O
2	1	1	x	x	x	2
4	2	x	x	x	0	1
1	4	0	1	x	2	0
5	6	x	0	x	x	0
3	8	x	0	x	x	0

It can be noted in Table 17.4 that only three input features F_1, F_2, and F_4 out of four are needed to unambiguously define the objects with output feature O. For example, the output feature $O = 2$ is determined for $F_1 = 1$, $O = 1$ for $F_4 = 0$. In addition to the reduced number of features, the solution in Table 17.4 indicates that *o*-reducts 6 and 8 are identical and can be merged. The latter illustrates the *data reduction* aspect of data mining, specifically feature reduction as well as rule (reduct) reduction. The power of data reduction becomes apparent for data sets with a large number of features. The concept of a reduct incorporated in a data reduction scheme to be discussed later allows for a meaningful reduction in the number of features.

In order to analyze the results of *feature extraction*, the feature values for the objects in Table 17.4 have been replaced with the values of Table 17.2 and shown in Table 17.5.

Of course, the data in Table 17.5 can be expressed as *decision rules*, for example, the rule (*r*-reduct) corresponding to object 2 is

$$\text{IF input feature } F_1 = 1$$

$$\text{THEN output feature } O = 2 \tag{17.3}$$

In order to express robustness of representing objects with the selected features, a measure called an *object redundancy factor (ORF)* is introduced. The ORF is the number of times an object can be independently represented with the reduced number of features minus 1. For an object with single-feature reducts ORF = $k - 1$, where k is the number of features included in the reduced object. This measure will also reflect the user's confidence in predictions generated in the *decision-making phase*.

Two other definitions used in the feature extraction algorithm presented in the next section are provided:

TABLE 17.5. Original Feature Values from Table 17.2

Object No.	F_1	F_2	F_3	F_4	O
2	1	1	0	2	2
4	0	1	1	0	1
1	0	1	0	2	0
5	0	0	1	3	0
3	0	0	0	1	0

Group is a set of objects with identical outcome features in an object–feature matrix.

Value conflict arises when identical feature values of any different objects correspond to different outcomes.

17.4 FEATURE EXTRACTION ALGORITHM

In this section, a heuristic *feature extraction (FE) algorithm* is presented for systematic selection of a desired set of features. A user defines the term *desired*, often, as the minimal set of features.

Notation

i = row index of the object–feature matrix

j = feature index

F_j = feature j

n = number of input features

q = number of object groups

g_p = number of features in object–feature matrix $p, p = 1, \ldots, q - 1$

Without loss of generality this algorithm assumes that single-feature o-reducts can be generated for some objects. If such reducts did not exist, a slight modification of the feature extraction algorithm would be needed. Rather than generating single-feature o-reducts, o-reducts with more features would have to be created.

FE Algorithm

Step 0. Group rows of the object–feature matrix according to the value of the output feature.

Step 1. Set object number $i = 1$.

Step 2. Select object i.

Step 3. Set feature number $j = 1$.

Step 4. Select feature F_j from an object–feature matrix.

Step 5. Compare the value of feature F_j with all feature values in the matrix. If any such feature matches the feature F_j, replace the value of all matching features with x; otherwise maintain the value of feature F_j and the corresponding features.

Step 6. If $j \leq m$, then set the iteration number $j = j + 1$ and go to Step 4; otherwise output a partial solution matrix generated for object i and go to Step 7.

Step 7. If $i < g_p$, then set $i = i + 1$ and go to Step 2; otherwise go to Step 8.

Step 8. Generate the final solution.

In Step 8 *o*-reducts with the minimum possible number of features are generated for the objects with all x elements. If needed, new columns (features) with all x entries are added one at a time. If all single columns are exhausted and no *o*-reducts are generated, combinations of column pairs, triplets, and so on, are considered. At the same time the number of features in the reducts sought will increase. This process will lead to determining all required reducts, if they exist.

The feature extraction algorithm is illustrated with two cases:

- Integer data
- Real numbers with tolerances

17.4.1 Integer Number Case

Example 17.2.   Consider the data in Table 17.6. The steps of the feature extraction algorithm are illustrated next.

Step 0. The rows of the matrix in Table 17.6 are arranged according to the output values shown in Table 17.7. Note that the objects in Table 17.7 have two numbers (e.g., 1[2]). The first number denotes the object number after grouping, while the second number, in square brackets, denotes the original object number.

Step 1. Object number i is set to 1.

Step 2. Object $i = 1$ is selected.

Step 3. Feature number is set to $j = 1$.

Step 4. Feature F_1 is selected from the object–feature matrix in Table 17.7.

Step 5. The value of feature F_1 is compared with all features corresponding to the decision outcome $O = 2$. Since the values of feature F_1 differ for the two outputs $O = 1$ and $O = 2$, the value of feature $F_1 = 0$ for object 1 and the corresponding features are retained.

TABLE 17.6. Test Data

Object No.	F_1	F_2	F_3	F_4	O
[1]	1	0	0	2	2
[2]	0	1	1	0	1
[3]	1	1	0	2	2
[4]	1	2	0	2	2
[5]	0	2	1	0	1
[6]	0	0	1	0	1
[7]	1	3	0	2	2
[8]	0	3	1	0	1

TABLE 17.7. Grouped Matrix

Object No.	F_1	F_2	F_3	F_4	O
1[2]	0	1	1	0	1
2[5]	0	2	1	0	1
3[6]	0	0	1	0	1
4[8]	0	3	1	0	1
5[1]	1	0	0	2	2
6[3]	1	1	0	2	2
7[4]	1	2	0	2	2
8[7]	1	3	0	2	2

Steps 3–5 are repeated for the remaining features of the object 1. If any feature value in the object–feature matrix matches that of feature F_j, the value of feature F_j and the corresponding features are replaced with x; the iteration number j is set to $j + 1 = 1 + 1 = 2$.

Step 6. The partial solution matrix generated after all features of object $i = 1$ have been considered is shown in Table 17.8.

Steps 1–6 for object $i = 2$ have resulted in the partial solution shown in Table 17.9.

TABLE 17.8. Partial Solution 1

Object No.	F_1	F_2	F_3	F_4	O
1[2]	0	x	1	0	1
2[5]	0	2	1	0	1
3[6]	0	0	1	0	1
4[8]	0	3	1	0	1
5[1]	1	0	0	2	2
6[3]	1	x	0	2	2
7[4]	1	2	0	2	2
8[7]	1	3	0	2	2

TABLE 17.9. Partial Solution 2

Object No.	F_1	F_2	F_3	F_4	O
1[2]	0	x	1	0	1
2[5]	0	x	1	0	1
3[6]	0	0	1	0	1
4[8]	0	3	1	0	1
5[1]	1	0	0	2	2
6[3]	1	x	0	2	2
7[4]	1	x	0	2	2
8[7]	1	3	0	2	2

TABLE 17.10. Final Solution

Object No.	F_1	F_2	F_3	F_4	O
1[2]	0	x	1	0	1
2[5]	0	x	1	0	1
3[6]	0	x	1	0	1
4[8]	0	x	1	0	1
5[1]	1	x	0	2	2
6[3]	1	x	0	2	2
7[4]	1	x	0	2	2
8[7]	1	x	0	2	2

Repeating Steps 1–6 for the remaining two objects has resulted in the solution represented in Table 17.10.

Step 8. It is clearly visible in the matrix in Table 17.10 that column F_2 can be deleted, as it does not add value to the proper classification of objects. In fact, the value of the output could be uniquely identified based on each the three features F_1, F_3, and F_4 of Table 17.10. Since we have single-feature reducts only, the number of input features in a row that are different than x indicate the number of times an object can be uniquely identified. In other words, the redundancy factor of each object in Table 17.10 is ORF = 3 – 1 = 2.

17.4.2 Real Number Case

The case with real numbers could be treated in the same way as the integer number case illustrated in Example 17.2. However, we will complicate it by incorporating a tolerance (error) of each feature value. For example, the tolerance of 5% implies that the values 2.02 and 1.98 fall into the same category and are considered as equivalent.

Example 17.3.   Consider the data in Table 17.11 that carries a 5% tolerance. The steps of the feature extraction algorithm are illustrated next.

Step 0. The rows of the matrix in Table 17.11 are grouped according to the output value as shown in Table 17.12.

TABLE 17.11. Continuous Data with Tolerances

Object No.	F_1	F_2	F_3	F_4	O
[1]	1.02	0.05	2.98	2.03	2
[2]	2.03	3.04	1.04	2.03	1
[3]	0.99	0.95	3.04	1.97	2
[4]	2.03	2.05	3.11	3.01	2
[5]	0.03	1.97	0.96	2.02	1
[6]	0.04	0.05	1.04	1.04	1
[7]	0.99	3.04	1.04	1.04	2
[8]	1.02	0.97	0.94	3.01	1

TABLE 17.12. Grouped Data

Object No.	F_1	F_2	F_3	F_4	O
1[2]	2.03	3.04	1.04	2.03	1
2[5]	0.03	1.97	0.96	2.02	1
3[6]	0.04	0.05	1.04	1.04	1
4[8]	1.02	0.97	0.94	3.01	1
5[1]	1.02	0.05	2.98	2.03	2
6[3]	0.99	0.95	3.04	1.97	2
7[4]	2.03	2.05	3.11	3.01	2
8[7]	0.99	3.04	1.04	1.04	2

Step 1. Object number i is set to 1.

Step 2. Object $i = 1$ is selected.

Step 3. Feature number is set to $j = 1$.

Step 4. Feature F_1 is selected from the matrix in Table 17.12.

Step 5. The value of feature F_1 is compared with all features in the matrix in Table 17.12. Since the value of feature $F_1 = 2.03$ for $O = 1$ (object 1) is identical to the feature of object 7 in the $O = 2$ group, the value of each of these two features is changed to x.

Steps 3–5 are repeated for the three remaining features of object 1.

Step 6. The partial solution resulting from the consideration of object 1 is shown in Table 17.13.

Repeating Steps 1–6 for objects 2 and 3 has resulted in the partial solution represented in Table 17.14.

In Steps 3–5 it has been determined that all features of object 4 match the values of features corresponding to the objects with output $O = 2$. The resultant solution is presented in Table 17.15. Note that the shadowing in Table 17.15 is the reverse of the shadowing in the previous tables of this example. It is visible in Table 17.15 that no o-reduct has been generated for any of the three objects 1, 4, and 8 (the shadowed object numbers in Table 17.15). Each of the five remaining objects is represented with

TABLE 17.13. Partial Solution 1

Object No.	F_1	F_2	F_3	F_4	O
1[2]	x	x	x	x	1
2[5]	0.03	1.97	x	x	1
3[6]	0.04	0.05	x	1.04	1
4[8]	1.02	0.97	x	3.01	1
5[1]	1.02	0.05	2.98	x	2
6[3]	0.99	0.95	3.04	x	2
7[4]	x	2.05	3.11	3.01	2
8[7]	0.99	x	x	1.04	2

TABLE 17.14. Partial Solution 2

Object No.	F_1	F_2	F_3	F_4	O
1[2]	x	x	x	x	1
2[5]	0.03	x	x	x	1
3[6]	0.04	0.05	x	1.04	1
4[8]	1.02	0.97	x	3.01	1
5[1]	1.02	0.05	2.98	x	2
6[3]	0.99	0.95	3.04	x	2
7[4]	x	x	3.11	3.01	2
8[7]	0.99	x	x	1.04	2

one of the two features F_1 and F_2. In order to generate o-reducts for the objects 1, 4, and 8, an additional feature has to be considered. We arbitrarily choose feature F_4.

For the three objects 1, 4, and 8 numerous two-feature o-reducts spanning over the three features can be generated, thus resulting in alternative solutions. One of those alternative final solutions is presented in Table 17.16. The redundancy factor for each object in Table 17.16 is ORF = 0. In this case one out of four features was removed.

Another way of handling continuous data is through conversion into a discrete form. Assigning integer values to continuous intervals transforms continuous data in a discrete form. To illustrate this transformation, consider the data in Table 17.17.

TABLE 17.15. Intermediate Solution

Object No.	F_1	F_2	F_3	F_4	O
1[2]	x	x	x	x	1
2[5]	0.03	x	x	x	1
3[6]	0.04	x	x	x	1
4[8]	x	x	x	x	1
5[1]	x	x	2.98	x	2
6[3]	x	x	3.04	x	2
7[4]	x	x	3.11	x	2
8[7]	x	x	x	x	2

TABLE 17.16. Final Solution

Object No.	F_1	F_2	F_3	F_4	O
1[2]	2.03	3.04	1.04	2.03	1
2[5]	0.03	1.97	0.96	2.02	1
3[6]	0.04	0.05	1.04	1.04	1
4[8]	1.02	0.97	0.94	3.01	1
5[1]	1.02	0.05	2.98	2.03	2
6[3]	0.99	0.95	3.04	1.97	2
7[4]	2.03	2.05	3.11	3.01	2
8[7]	0.99	3.04	1.04	1.04	2

TABLE 17.17. Continuous Input Data

Object No.	F_1	F_2	F_3	F_4	O
1	1.02	0.05	1.98	2.04	2
2	0.02	0.97	0.94	1.01	1
3	0.99	0.95	0.04	1.97	2
4	1.04	2.05	2.11	2.03	2
5	0.03	1.97	0.96	0.34	1
6	0.04	0.05	1.04	3.05	1
7	1.03	3.04	1.64	2.04	2
8	0.05	2.96	1.03	0.03	1

The assignment of integer values to continuous intervals is defined in Table 17.18. The data in Table 17.17 can be converted into an integer form using the transformation defined in Table 17.18. Applying the feature extraction algorithm to the transformed data results in the matrix shown in Table 17.19.

Rather than using a heuristic feature extraction algorithm, a feature extraction model is presented next.

TABLE 17.18. Data Intervals

Continuous Interval	Discrete Value
0.00–0.34	0
0.35–1.05	1
1.06–2.11	2
2.12–3.05	3

TABLE 17.19. Output Matrix

Object No.	F_1	F_2	F_3	F_4	O
2	0	1	1	1	1
5	0	2	1	0	1
6	0	0	1	3	1
8	0	3	1	0	1
1	1	0	2	2	2
3	1	1	0	2	2
4	1	2	2	2	2
7	1	3	2	2	2

17.5 FEATURE EXTRACTION MODEL

In this section, an integer programming model for the extraction of features from the data set is presented. The features are implicitly extracted by selecting reducts.

17.5.1 Integer Programming Formulation

The model considers a distance measure d_{ij} between reducts i and j. For an o-reduct–feature incidence matrix with n features in each o-reduct, the *distance measure* is defined as

$$d_{ij} = \sum_{k=1}^{n} \delta(a_{ik}, a_{jk}) \tag{17.4}$$

where $\delta(a_{ik}, a_{jk}) = v_k$, $k = 1, \ldots, r$; r = value count. Note that the distance measure d_{ij} considers input features only. The value of v_k, $k = 1, \ldots, 4$, could be defined in a number of ways, for example:

$$v_1 = 0 \quad \text{for } a_{ik} = a_{jk} = \text{x}$$

$$v_2 = 0 \quad \text{for } 2|a_{ik} - a_{jk}| \, / \, | \, a_{ik} + a_{jk} | \le T_k |a_{ik} + a_{jk}| / 2$$

$$v_3 = 1 \quad \text{for } 2|a_{ik} - a_{jk}| \, / \, | \, a_{ik} + a_{jk} | > T_k |a_{ik} + a_{jk}| / 2$$

$$v_4 = M \quad \text{for all other cases}$$

where

T_k = relative tolerance of feature k,
M = arbitrary large number

For $v_1 = 0$, $v_2 = 0$, $v_3 = 1$, $v_4 = 10$, the distance measure d_{ij} defined in (17.4) becomes the *Hamming distance*. The value of the Hamming distance measure for the two o-reducts 1 and 3,

$$1 \; [\; \text{x} \; \text{x} \; 0 \; \text{x} \;] \qquad 3 \; [\; \text{x} \; \text{x} \; 0 \; \text{x} \;]$$

is $d_{13} = 0 + 0 + 1 + 0 = 1$.

In order to formulate an integer programming model for feature extraction, the following notation is used:

$n =$ number of features
$m =$ number of o-reducts
$l =$ number of objects

R_q = set of reducts for object number q, $q = 1, \ldots, l$, where $|\bigcup\limits_{q=1}^{l} R_q| = m$

$$x_i = \begin{cases} 1 & \text{if reduct } i \text{ is selected} \\ 0 & \text{otherwise} \end{cases}$$

$$y_{ij} = \begin{cases} 1 & \text{if } o\text{-reduct } i \text{ and reduct } j \text{ have been selected} \\ 0 & \text{otherwise} \end{cases}$$

The objective function (17.5) maximizes the total distance between any two o-reducts i and j (Kusiak and Tseng 1999):

$$\max \sum_{i=1}^{m} \sum_{j=1}^{m} d_{ij} \, y_{ij} \tag{17.5}$$

subject to

$$\sum_{i \in R_q} \sum_{j=1}^{m} y_{ij} = 1 \quad \text{for all } q = 1, \ldots, l; \, i \neq j \tag{17.6}$$

$$\sum_{j=1}^{m} y_{ij} \leq x_i \quad \text{for all } i = 1, \ldots, m; \, i \neq j \tag{17.7}$$

$$x_i = 0, 1 \quad \text{for all } i = 1, \ldots, m \tag{17.8}$$

$$y_{ij} = 0, 1 \quad \text{for all } i = 1, \ldots, m, j = 1, \ldots, m \tag{17.9}$$

Constraint (17.6) ensures that for each object (a set of o-reducts) exactly one o-reduct is selected. The consistency between variable x_i and y_{ij} is imposed by constraint (17.7). Constraint (17.8) and (17.9) ensure integrality.

Model (17.5)–(17.9) will be applied to the following two cases of the o-reduct–feature matrix:

- Equal-weight case
- Unequal-weight case

17.5.2 Equal-Weight Case

Example 17.4. Extract a minimum number of features from the one-attribute o-reducts from the reducts in Table 17.20. The content of Table 17.20 was obtained from Table 17.23 of Appendix 17.1.

TABLE 17.20. One-Attribute o-Reducts

Object No.	o-Reduct No.	F_1	F_2	F_3	F_4	O
1	1	x	x	0	x	2
	2	x	x	x	2	2
5	3	x	x	0	x	2
2	4	x	x	x	3	0
	5	1	x	x	x	1
4	6	x	2	x	x	1
	7	x	x	2	x	1
	8	x	x	x	0	1

The objects in Table 17.20 have been grouped (sorting is not necessary) according to the value of the output feature, which pays off during analysis of the solution matrix. The distance measures d_{ij} for the eight reducts are shown in the matrix.

$$
\begin{array}{c}
\text{\textit{o}-Reduct} \\
\begin{array}{cccccccc}
1 & 2 & 3 & 4 & 5 & 6 & 7 & 8
\end{array} \\
\text{\textit{o}-Reduct}
\begin{array}{c}
1\\2\\3\\4\\5\\6\\7\\8
\end{array}
\left[
\begin{array}{cccccccc}
0 & \infty & 1 & 2 & 2 & 2 & 1 & 2 \\
\infty & 0 & 2 & 1 & 2 & 2 & 2 & 1 \\
1 & 2 & 0 & 2 & 2 & 2 & 1 & 2 \\
2 & 1 & 2 & 0 & 2 & 2 & 1 & 1 \\
2 & 2 & 2 & 2 & 0 & \infty & \infty & \infty \\
2 & 2 & 2 & 2 & \infty & 0 & \infty & \infty \\
1 & 2 & 1 & 1 & \infty & \infty & 0 & \infty \\
2 & 1 & 2 & 1 & \infty & \infty & \infty & 0
\end{array}
\right]
\end{array} \qquad (17.10)
$$

All ∞ entries in matrix (17.10) correspond to reducts belonging to the same object and are excluded from consideration as only one reduct for each object is selected. Solving model (17.5)–(17.9) for the above data with LINDO software produces the following solution:

$$y_{13} = 1 \qquad y_{31} = 1 \qquad y_{42} = 1 \qquad y_{71} = 1$$

and

$$x_1 = 1 \qquad x_3 = 1 \qquad x_4 = 1 \qquad x_7 = 1$$

The best way to interpret this solution is to concentrate on the subscripts of the decision variable x. In this case the subscripts 1, 3, 4, and 7 indicate the o-reducts selected that are highlighted in the matrix

$$
\begin{array}{ccc}
\text{Object} & \text{o-Reduct} & F_1 F_2 F_3 F_4 \quad O \\
\begin{array}{c} 1 \\ \\ 3 \\ 2 \\ 4 \end{array}
&
\begin{array}{c} 1 \\ 2 \\ 3 \\ 4 \\ 5 \\ 6 \\ 7 \\ 8 \end{array}
&
\begin{bmatrix}
x & x & 0 & x\,2 \\
x & x & x & 2\,2 \\
x & x & 0 & x\,2 \\
x & x & x & 3\,0 \\
1 & x & x & x\,1 \\
x & 2 & x & x\,1 \\
x & x & 2 & x\,1 \\
x & x & x & 0\,1
\end{bmatrix}
\end{array}
\qquad (17.11)
$$

The o-reducts 1 and 3 are identical and can be merged. The solution shows that the features F_3 and F_4 have been selected. These two features correspond to three o-reducts (rules): merged o-reduct (1, 3), o-reduct 4, and o-reduct 7, which in turn uniquely represent the eight single-feature o-reducts in matrix (17.10). The input file for this example that was used by LINDO is shown in Appendix 17.2.

Note that LINDO generates only one first optimal solution, which in fact does not have to contain the minimum number of features. To obtain a solution that guarantees the minimum number of features, the following constraint needs to be incorporated in model (17.5)–(17.9):

$$
\sum_{j \notin R_q} d_{ij} x_i - \min \left(\sum_{j \notin R_q} d_{kj} \right) x_k \le 0 \quad \text{for } k \in R_q;\ i \ne k, i \in R_q;
$$

$$
\text{all } i = 1, \ldots, m, \text{ all } j = 1, \ldots, m \qquad (17.12)
$$

In fact, the final three lines in the input file of Appendix 17.2 represent this constraint.

17.5.3 Unequal-Weight Case

In Example 17.4, an assumption was made that all features and objects (o-reducts) are of the same importance. Here, a more general scenario is considered when the features and objects carry weights. The value of the weight is assumed to be reverse proportional to the importance of a feature or an o-reduct. For example, one object may be assigned a lower weight than another one because of the higher frequency with which it appears. Also, some attributes might be more important than others and therefore be assigned a lower weight.

The weight assigned to each feature is denoted as w_k and each object (reduct) as f_i. Using two weights w_k and f_i, a *transformed matrix* $[e_{ik}]$ is generated from the original reduct–input feature matrix. The entries of the transformed matrix are defined as follows:

$$
e_{ik} = f_i \bullet (w_k \bullet v_k)
$$

where

$e_{ik} =$ entry of transformed o-reduct–feature matrix

$f_i =$ weight of o-reduct i

$w_k =$ weight of feature k

$$v_k = \begin{cases} 1 & \text{if feature } k \neq x \\ 0 & \text{otherwise} \end{cases}$$

Based on the entries e_{ik} of the transformed matrix, the *weighted distance measure* is computed as

$$D_{ij} = t_{ij} \bullet s_{ij}$$

where t_{ij} is the coefficient between o-reducts i and j, which is defined as

$$t_{ij} = \begin{cases} \displaystyle\sum_{k=1}^{n} (e_{ik} + e_{jk}) & \text{for } i \neq j \\ \\ 0 & \text{otherwise} \end{cases}$$

The unequal-weight case is illustrated in Example 17.5.

Example 17.5. Given the object– (reduct–) feature matrix (see Table 17.21) with the corresponding weights, select features with the maximum total weighted distance measure. Using the data in Table 17.21, the following transformed o-reduct–input feature matrix $[e_{ik}]$ is created:

TABLE 17.21. Expanded Data Set from Table 17.20

Object No.	F_1	F_2	F_3	F_4	O	Weight w_i
1	x	x	0	x	2	1
	x	x	x	2	2	1
2	x	x	x	3	0	1
	1	x	x	x	1	0.92
4	x	2	x	x	1	0.92
	x	x	2	x	1	0.92
	x	x	x	0	1	0.92
5	x	x	0	x	2	0.60
Weight f_i	0.60	1	1	0.80		

$$
\begin{array}{cc|cccc|}
\text{Object} & o\text{-Reduct} & F_1 & F_2 & F_3 & F_4 \\
1 & 1 & 0 & 0 & 1 & 0 \\
 & 2 & 0 & 0 & 0 & 0.8 \\
5 & 3 & 0 & 0 & 0.6 & 0 \\
2 & 4 & 0 & 0 & 0 & 0.8 \\
4 & 5 & 0.55 & 0 & 0 & 0 \\
 & 6 & 0 & 0.92 & 0 & 0 \\
 & 7 & 0 & 0 & 0.92 & 0 \\
 & 8 & 0 & 0 & 0 & 0.74
\end{array}
\tag{17.13}
$$

From (17.13), the matrix $[t_{ij}]$ of coefficients is created:

$$
o\text{-Reduct}\quad
\begin{array}{c|cccccccc|}
 & 1 & 2 & 3 & 4 & 5 & 6 & 7 & 8 \\
1 & 0 & 1.8 & 1.6 & 1.8 & 1.55 & 1.92 & 1.92 & 1.74 \\
2 & 1.8 & 0 & 1.4 & 1.6 & 1.35 & 1.72 & 1.72 & 1.54 \\
3 & 1.6 & 1.4 & 0 & 1.4 & 1.15 & 1.52 & 1.52 & 1.34 \\
4 & 1.8 & 1.6 & 1.4 & 0 & 1.35 & 1.72 & 1.72 & 1.54 \\
5 & 1.55 & 1.35 & 1.15 & 1.35 & 0 & 1.47 & 1.47 & 1.29 \\
6 & 1.92 & 1.72 & 1.52 & 1.72 & 1.47 & 0 & 1.84 & 1.66 \\
7 & 1.92 & 1.72 & 1.52 & 1.72 & 1.47 & 1.84 & 0 & 1.66 \\
8 & 1.74 & 1.54 & 1.34 & 1.54 & 1.29 & 1.66 & 1.66 & 0
\end{array}
\tag{17.14}
$$

(column heading: o-Reduct; row heading: o-Reduct)

The matrix $[D_{ij}]$ of the weighted distance measures is shown as

$$
o\text{-Reduct}\quad
\begin{array}{c|cccccccc|}
1 & 0 & \infty & 1.6 & 3.6 & 3.1 & 3.84 & 1.92 & 3.48 \\
2 & \infty & 0 & 2.8 & 1.6 & 2.7 & 3.44 & 3.44 & 1.54 \\
3 & 1.6 & 2.8 & 0 & 2.8 & 2.3 & 3.04 & 1.52 & 2.68 \\
4 & 3.6 & 1.6 & 2.8 & 0 & 2.7 & 3.44 & 1.72 & 1.54 \\
5 & 3.1 & 2.7 & 2.3 & 2.7 & 0 & \infty & \infty & \infty \\
6 & 3.84 & 3.44 & 3.04 & 3.44 & \infty & 0 & \infty & \infty \\
7 & 1.92 & 3.44 & 1.52 & 1.72 & \infty & \infty & 0 & \infty \\
8 & 3.48 & 1.54 & 2.68 & 1.54 & \infty & \infty & \infty & 0
\end{array}
\tag{17.15}
$$

(column heading: o-Reduct; row heading: o-Reduct)

Solving model (17.5)–(17.9) for the data in matrix (17.15) produces the following solution:

$$y_{28} = 1 \quad y_{37} = 1 \quad y_{48} = 1 \quad y_{73} = 1$$

and

$$x_2 = 1 \quad x_3 = 1 \quad x_4 = 1 \quad x_7 = 1$$

This solution translates into the o-reducts 2, 3, 4, and 7 being selected with the corresponding features F_3 and F_4. Due to the weights, the reducts selected differ from the solution represented in matrix (17.11), although the resultant features are the same.

The integer programming model (17.5)–(17.9) is useful; however, it has a limitation—it cannot be solved for large number reducts and features due to high computational complexity. To solve large-scale reduct selection problems, the previously discussed feature extraction algorithm can be used.

17.5.4 Discussion

Where Does the Power of Feature Extraction Come From? The two observations stated next may be useful in providing an answer to this question.

   Observation 1.   The feature extraction algorithm eliminates features that do not contribute to the differentiation between the objects represented with the features. Recall that in a reduct the identical feature values that belong to different objects and different outcomes are removed.

   Observation 2.   The feature extraction algorithm eliminates features that may cause unnecessary variability in the relationship between the feature values and outcomes. This is due to the deletion of features sending "conflicting signals" to the outcome feature accomplished by removal of features on the "equality basis" stated in Observation 1.

Is It Necessary to Generate All Reducts? The simple answer is no. It is known that generating single-feature o-reducts involves low computational effort. However, generating o-reducts with more than one feature involves more effort, especially when all possible reducts are generated. The integer programming approach presented in this chapter allows for careful control of the type and number of o-reducts that need to be generated. The feature reduction algorithm avoids direct generation of o-reducts, thus decreasing the computational effort. In addition, new reducts are generated as needed rather than all in advance.

How Can Data for the Integer Programming Model Be Generated? Incorporate single-feature o-reducts and generate two-feature o-reducts first for all objects. If the solution is not satisfactory, add multifeature o-reducts to all previously generated reducts.

Why Should Feature Extraction Be Chosen over Other Methods? Regression analysis, neural networks, and expert systems share some commonality with rough set theory. However, there are three fundamental differences between the first two approaches and the feature extraction approach. First, both neural networks and regression make decisions for all cases with an error, while our approach makes accurate decisions when it has sufficient "knowledge." Second, neural networks and regression models are "population based" while the feature extraction approach follows an "individual (data object) based paradigm." The two population-based tools determine features that are common to a population. The feature extraction concept of rough set theory identifies unique features of an object and sees whether these unique features are shared with other objects. It is obvious that the two paradigms differ, and in general the features derived by each of the two paradigms are not identical. Third, each of the two population-based methods uses a fixed set of features to arrive at a decision. In the feature extraction approach the same set of features may apply to a group of objects.

The population-based tools (i.e., neural networks and regression analysis) and the data mining models and algorithms discussed in this chapter all use a training mode and a decision-making mode. Decision rules generated by feature extraction algorithms provide a link to expert systems.

17.6 DECISION MAKING

The results produced by solving the feature extraction model or using the heuristic feature extraction algorithm explain why human decision makers cannot accurately predict outcomes. For example, consider the data in Table 17.6, where a human decision maker tends to consider all feature values at hand while making a prediction. Some of those features may introduce a "noise," for example, feature F_2 (Table 17.6) in predicting outcome $O = 1$ or $O = 2$ or feature F_3 in predicting outcome $O = 2$. Using the reduced matrix produced by solving the feature extraction model, a *decision-making algorithm* can be developed. The decision-making algorithm would use only a subset of all features in making a prediction (decision). Feature extraction partitions the original data set into categories, with each category being identifiable by a specific set of features.

The concept of feature extraction is summarized in Figure 17.2, where a large data set (object–feature matrix) is transformed into a reduced data set, called here a *decision table*. The number of features and objects in the decision table is much smaller than in the original data set. Each object (row) in the original data set is represented in the decision table in many different ways, corresponding to the alternative solutions in Figure 17.2.

Figure 17.2 illustrates the learning phase of data mining. In the decision-making phase a decision-making algorithm matches the feature values of a new case for which the outcome is not known with the solutions generated by the feature extraction algorithm (see Figure 17.3). If the match is satisfactory, the new object is assigned an outcome equal to the outcomes of the matching objects. Matching the features of the new case with more than the solution produced by the feature extraction algorithm increases the redundancy of the decision made. The value of the object redundancy factor depends on the criticality of a decision made. It is obvious that applications that involve human safety (e.g., medicine, avionics) call for high redundancy factors.

17.7 DATA FARMING

Data farming is a concept parallel to *data mining* (Hackathorn, 1999). In contrast to data mining, data farming assumes that the information necessary for decision making does not exist (e.g., on the web), and an effort is needed to create a proper environment for data collection. The analogy to farming activities, such as preparing the field, seeding crops, cultivating the soil, and harvesting the crop, is apparent. Cultivating a few seeds of data will eventually produce a harvest of information. The information being collected can be organized in time using various techniques and tools such as graphs, matrices, virtual reality models, mathematical models, and other means.

Some examples of business areas that can greatly benefit for data farming are (Hackathorn, 1999):

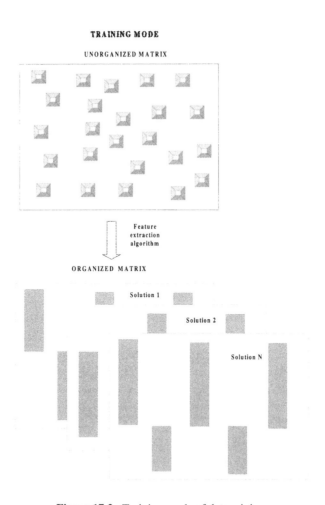

Figure 17.2. Training mode of data mining.

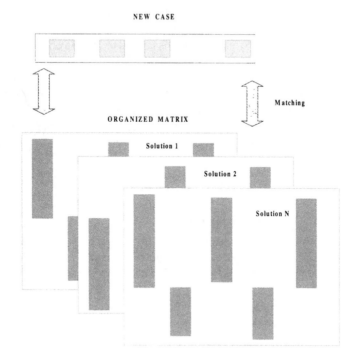

Figure 17.3. Decision-making mode of data mining.

- Customer relationships (see Chapter 16)—understanding the current needs of customers and anticipating their future needs
- Supply-chain management—managing value chain from suppliers through distributors to end customers
- Competitive analysis—understanding and monitoring established competitors within a company's market and detecting emerging competitors
- Technology trends—tracking developments in key technologies of a company's business and forecasting market impacts on these new technologies
- Deregulation—understanding current and future impacts of industry deregulation and exploiting opportunities emerging from the deregulation
- Global economics and politics—monitoring trends and technical events worldwide for threads to existing operations and for opportunities for future markets

The web content as a whole is not well organized, and therefore it has to be refined and transformed into a usable form. The process of refining web information involves the following four phases (see Figure 17.4):

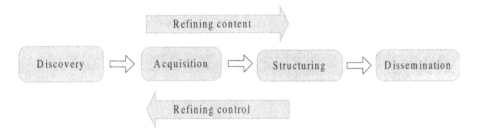

Figure 17.4. Process of refining information.

- Discovery—involves exploration of available web resources to find items that relate to specific topics. It involves work beyond searching generic directory services (e.g., Yahoo) or indexing services (e.g., Alta Vista). The goal is to locate individuals and organizations that create content important to the corporation. As the source information content (including the web) changes, the discovery is a continuous process.

- Acquisition—the collection and maintenance of information identified by its source. A key issue here is the efficiency of reviewing the information by a human.

- Structuring—the analysis and transformation of content into a useful form and format, for example, processed web pages, spreadsheets, and database tables.

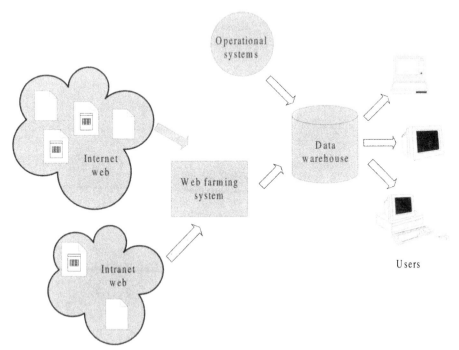

Figure 17.5 The data warehouse as the center of web farming process.

- Dissemination—the packaging and delivery of information to the users either directly or through a data warehouse.

Figure 17.5 shows the flow of information from various sources through the web farming system and data warehouse to the users. The primary source of the web farming system is the global web. Enterprise's Intranet can supplement this information source. Regardless of its source, most information acquired by the web farming system is not in a form immediately suitable for incorporation into the data warehouse.

17.8 SUMMARY

The initial intent of rough set theory was to define an approximation space based on the equivalence relation. In this chapter, the equivalence relation was generalized to features with real numbers and tolerances. The notion of alternative solutions was introduced to emphasize robustness of decision making. An integer programming model was developed to extract features with the properties imposed by a user. The approach presented in the chapter shows that formal methods offer many avenues for extraction of useful knowledge from large data sets.

APPENDIX 17.1: INPUT DATA AND CORRESPONDING O-REDUCTS

TABLE 17.22. Input Data

Object No.	F_1	F_2	F_3	F_4	O
1	0	1	0	2	2
2	0	0	1	3	0
3	0	1	1	1	1
4	1	2	2	0	1
5	0	0	0	1	2

TABLE 17.23. The List of All Possible o-Reducts for the Data in Table 17.22

Object No.	F_1	F_2	F_3	F_4	O	o-Reduct No.	F_1	F_2	F_3	F_4	O
1	0	1	0	2	2	1	x	x	0	x	2
						2	x	x	x	2	2
						3	0	x	0	x	2
						4	0	x	x	2	2
						5	x	1	0	x	2
						6	x	1	x	2	2
						7	x	x	0	2	2
						8	0	1	0	x	2
						9	0	1	x	2	2
						10	x	1	0	2	2

TABLE 17.23. Continued

Object No.	F_1	F_2	F_3	F_4	O	o-Reduct No.	F_1	F_2	F_3	F_4	O
2	0	0	1	3	0	11	x	x	x	3	0
						12	0	x	x	3	0
						13	x	0	1	x	0
						14	x	0	x	3	0
						15	x	x	1	3	0
						16	0	0	1	x	0
						17	0	0	x	3	0
						18	x	0	1	3	0
3	0	1	1	1	1						
						19	x	1	1	x	1
						20	x	1	x	1	1
						21	x	x	1	1	1
						22	0	1	1	x	1
						23	0	1	x	1	1
						24	0	x	1	1	1
						25	x	1	1	1	1
4	1	2	2	0	1	26	1	x	x	x	1
						27	x	2	x	x	1
						28	x	x	2	x	1
						29	x	x	x	0	1
						30	1	2	x	x	1
						31	1	x	2	x	1
						32	1	x	x	0	1
						33	x	2	2	x	1
						34	x	2	x	0	1
						35	x	x	2	0	1
						36	1	2	2	x	1
						37	1	2	x	0	1
						38	x	2	2	0	1
5	0	0	0	1	2	39	x	x	0	x	2
						40	0	x	0	x	2
						41	x	0	0	x	2
						42	x	0	x	1	2
						43	x	x	0	1	2
						44	0	0	0	x	2
						45	0	0	x	1	2
						46	x	0	0	1	2

APPENDIX 17.2: INPUT FILE OF EXAMPLE 17.4

```
MIN 999Y12+Y13+20Y14+20Y15+20Y16
+10Y17+20Y18+999Y21+20Y23+10Y24+20Y25
+20Y26+20Y27+10Y28+Y31+20Y32+20Y34
+20Y35+20Y36+10Y37+20Y38+20Y41+10Y42+20Y43
+20Y45+20Y46+10Y47+10Y48+20Y51
```

```
+20Y52+20Y53+20Y54+999Y56+999Y57
+999Y58+20Y61+20Y62+20Y63+20Y64+999Y65
+999Y67+999Y68+10Y71+20Y72+10Y73+10Y74
+999Y75+999Y76+999Y78+20Y81+10Y82+20Y83
+10Y84+999Y85+999Y86+999Y87
st
Y12+Y13+Y14+Y15+Y16+Y17+Y18
+Y21+Y23+Y24+Y25+Y26+Y27+Y28=1
Y31+Y32+Y34+Y35+Y36+Y37+Y38=1
Y41+Y42+Y43+Y45+Y46+Y47+Y48=1
Y51+Y52+Y53+Y54+Y56+Y57+Y58
+Y61+Y62+Y63+Y64+Y65+Y67+Y68
+Y71+Y72+Y73+Y74+Y75+Y76+Y78
+Y81+Y82+Y83+Y84+Y85+Y86+Y87=1
-X1+Y12+Y13+Y14+Y15+Y16+Y17+Y18<=0
-X2+Y21+Y23+Y24+Y25+Y26+Y27+Y28<=0
-X3+Y31+Y32+Y34+Y35+Y36+Y37+Y38<=0
-X4+Y41+Y42+Y43+Y45+Y46+Y47+Y48<=0
-X5+Y51+Y52+Y53+Y54+Y56+Y57+Y58<=0
-X6+Y61+Y62+Y63+Y64+Y65+Y67+Y68<=0
-X7+Y71+Y72+Y73+Y74+Y75+Y76+Y78<=0
-X8+Y81+Y82+Y83+Y84+Y85+Y86+Y87<=0

17X1-14X2<=0
13X3-11X4<=0
13X5-11X6<=0

END
INTEGER 64
```

Note: The last three constraints in the input file represent constraint (17.12).

REFERENCES

Berry, M. J. A., and G. Linoff (1997), *Data Mining Techniques: For Marketing, Sales, and Customer Support*, John Wiley, New York.

Carlin, U. S., J. Komorowski, and A. Ohrn (1998), Rough set analysis of patients with suspected of acute appendicitis, in *Proceedings of IPMU'98*, Paris, France, pp. 1528–1533.

Groth, R. (1998), *Data Mining: A Hands-On Approach for Business Professionals*, Prentice-Hall, Upper Saddle River, NJ.

Hackathorn, R. (1999), Web mining, *DB2 Magazine*, Vol. 4, No. 2, pp. 47–53.

Kowalczyk, W., and F. Slisser (1997), Modeling customer retention with rough data models, in *Proceedings of the First European Symposium on PKDD '97*, Trondheim, Norway, pp. 4–13.

Kusiak, A., and T. L. Tseng (1999), Modeling approach to data mining, in *Proceedings of the Industrial Engineering and Production Management Conference*, Glasgow, Scotland, July, pp. 1–13.

Paszek, P., and A. Wakulicz-Deja (1996), Optimization of diagnosis in progressive encephalopathy applying the rough set theory, in *Proceedings of the Fourth European Congress on Intelligent Techniques in Soft Computing*, Aachen, Germany, Vol. 1, pp. 192–196.

Pawlak, Z. (1991), *Rough Sets: Theoretical Aspects of Reasoning About Data*, Kluwer, Boston, 1991.

Pawlak, Z. (1997), Rough sets and data mining, in T. Chandra, S. R. Leclair, J. A. Meech, B. Varma, M. Smith, and B. Balachandran, (Eds.), *Proceedings of the Australiasia-Pacific Forum on Intelligent Processing and Manufacturing of Materials*, Vol. 1, Gold Coast, Australia, pp. 663–667.

Pawlak, Z., K. Slowinski, and R. Slowinski (1998), Rough classification of patients after highly selective vagotomy for duodenal ulcer, *International Journal of Man-Machine Studies*, Vol. 24, pp. 413–433.

Ruhe, G. (1996), Qualitative analysis of software engineering data using rough sets, in *Proceedings of the Fourth International Workshop on Rough Sets, Fuzzy Sets, and Machine Discovery*, Tokyo, Japan, pp. 292–299.

Shan, N., W. Ziarko, H. J. Hamilton, and N. Cercone (1995), Using rough sets as tools for knowledge discovery, in U. M. Fayyad and R. Uthurusamy, (Eds.), *Proceedings of the First International Conference on Knowledge Discovery and Data Mining*, AAAI Press, Menlo Park, CA, pp. 263–268.

Skowron, A. (1994), Data filtration: A rough set approach, in *Proceedings of the International Workshop on Rough Sets and Knowledge Discovery*, Banff, Alberta, Canada, pp. 108–118.

Slowinski, R., and J. Stefanowski (1993), Rough classification with valued closeness relation, in *New Approaches in Classification and Data Analysis*, Springer-Verlag, New York, pp. 482–489.

Stepaniuk, J., M. Urban, and E. Baszun-Stepaniuk (1998), The application of the rough set based data mining technique in the prognosis of the diabetic nephropathy prevalence, in *Proceedings of the Seventh International Workshop on Intelligent Information Systems*, Marlborg, Poland, pp. 388–391.

Szczuka, M., and D. Slezak (1997), Hyperplane-based neural networks for real-valued decision tables, in *Proceedings of the Rough Sets Society Conference (RSSC'97)*, Raleigh, NC, pp. 265–268.

Tsumoto, S. (1997), Extraction of experts' decision process from clinical databases using rough set model, in *Proceedings of the First European Symposium on PKDD '97*, Trondheim, Norway, pp. 58–67.

Tsumoto, S., and W. Ziarko (1996), The application of rough sets–data mining based technique to differential diagnosis of meningoencphaltis, in *Proceedings of the 9th International Symposium on Foundations on Intelligent Systems*, Zakopane, Poland, pp. 438–447.

Wojcik, Z. M. (1993a), Edge detector free of the detection/localization tradeoff using rough sets, in *Proceedings of the International Workshop on Rough Sets and Knowledge Discovery*, Banff, Alberta, Canada, pp. 421–438.

Wojcik, Z. M. (1993b), Rough sets for intelligent image filtering, *Proceedings of the International Workshop on Rough Sets and Knowledge Discovery*, Banff, Alberta, Canada, pp. 399–410.

QUESTIONS

17.1. What is data mining?

17.2. What theories and methodologies are used in data mining?

17.3. What is a reduct?

17.4. What is the relationship between a reduct, a decision rule, and features?

17.5. What algorithm is used to generate reducts?

17.6. What does the feature extraction algorithm accomplish?

17.7. What is the objective of the feature extraction model?

17.8. What are some other objective functions and constraints that could be used in the feature extraction model?

17.9. What is data farming?

17.10. What are the differences between data mining and data farming?

PROBLEMS

17.1. For the data in Table 17.24:

(a) Design a heuristic algorithm and determine a reduct with the minimum number of features.

(b) Solve the feature extraction model.

(c) Compare the reducts generated in (a) and (b).

17.2. For the data in Table 17.25:

(a) Design a heuristic algorithm and determine a reduct with the minimum number of features.

(b) Solve the feature extraction model.

(c) Compare the reducts generated in (a) and (b).

17.3. Consider the data in Table 17.20. Find a set of weights for objects and features that would result in a reduct (set of features) different than the one in matrix (17.11).

17.4. For the data in Table 17.20:

(a) Apply and solve the generalized *p*-median model of Chapter 10.

(b) Solve the feature extraction model.

(c) Compare the reducts and the features generated in (a) and (b).

TABLE 17.24. Two-Attribute o-Reducts

Object No.	F_1	F_2	F_3	F_4	O
	0	x	0	x	2
	0	x	x	2	2
1	x	1	0	x	2
	x	1	x	2	2
	x	x	0	2	2
	0	x	x	3	0
2	x	0	1	x	0
	x	0	x	3	0
	x	x	1	3	0
	x	1	1	x	1
3	x	1	x	1	1
	x	x	1	1	1
	1	2	x	x	1
	1	x	2	x	1
4	1	x	x	0	1
	x	2	2	x	1
	x	2	x	0	1
	x	x	2	0	1
	0	x	0	x	2
5	x	0	0	x	2
	x	0	x	1	2
	x	x	0	1	2

TABLE 17.25. Three-Attribute o-Reducts

Object No.	F_1	F_2	F_3	F_4	O
	0	1	0	x	2
1	0	1	x	2	2
	x	1	0	2	2
	0	0	1	x	0
2	0	0	x	3	0
	x	0	1	3	0
	0	1	1	x	1
3	0	1	x	1	1
	0	x	1	1	1
	x	1	1	1	1
	1	2	2	x	1
4	1	2	x	0	1
	x	2	2	0	1
	0	0	0	x	2
5	0	0	x	1	2
	x	0	0	1	2

INDEX